Theoretical studies of ecosystems

Theoretical studies of ecosystems

THE NETWORK PERSPECTIVE

Editors

M. HIGASHI
Faculty of Science and Technology, Ryukoku University, Kyoto, Japan

T. P. BURNS
Department of Zoology, Kyoto University, Kyoto, Japan

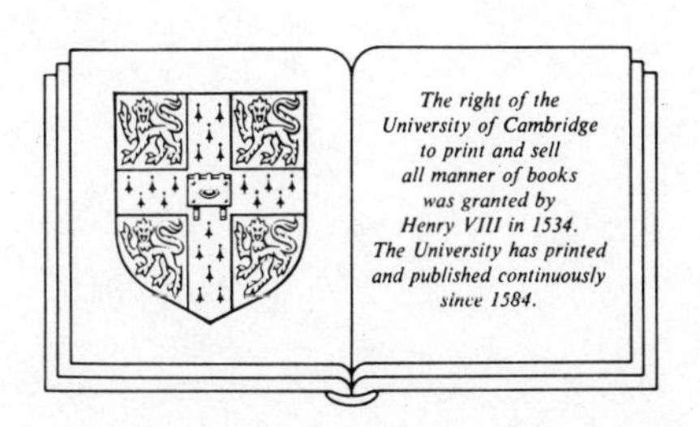

CAMBRIDGE UNIVERSITY PRESS
Cambridge
New York · Port Chester · Melbourne · Sydney

Published by the Press S
The Pitt Building, Trum
40 West 20th Street, New York, NY 10011, USA
10 Stamford Road, Oakleigh, Melbourne 3166, Australia

First published 1991

Printed in Great Britain by the University Press, Cambridge

British Library cataloguing in publication data
Theoretical studies of ecosystems: the network perspective.
1. Ecology
I. Higashi, M. II. Burns, T. P.
574.5

ISBN 0 521 36138 9 hardback

UP

Contents

Contributors

Timothy F. H. Allen
Dept. of Botany
University of Wisconsin
Madison, Wisconsin 53706

Joseph Bentsman
Dept. of Mechanical and Industrial Engineering
University of Illinois
Urbana, Illinois 61801

Thomas P. Burns
Dept. of Zoology
Kyoto University
Kyoto 606, Japan

Donald L. DeAngelis
Environmental Sciences Division
Oak Ridge National Laboratory
Oak Ridge, Tennessee 37831

Bruce Hannon
Dept. of Geography
University of Illinois
Champaign, Illinois 61820

Robert A. Herendeen
Illinois Natural History Survey
Champaign, Illinois 61820

Masahiko Higashi
Faculty of Science and Technology
Ryukoku University
Otsu 520–21, Japan

Kohkichi Kawasaki
Science and Engineering Research Institute
Dohshisha University
Kyoto 602, Japan

Ramon Margalef
Department de Biologie
Universitat de Barcelona
08028 – Barcelona, Spain

Donald C. Mikulecky
Dept. of Physiology/Biophysics
Medical College of Virginia
Richmond, Virginia 23298-0001

Hisao Nakajima
Dept. of Mathematics and Physics
Ritsumeikan University
Kyoto 603, Japan

Robert V. O'Neill
Environmental Sciences Division
Oak Ridge National Laboratory
Oak Ridge, Tennessee 37831

Bernard C. Patten
Dept. of Zoology and Institute of Ecology
University of Georgia
Athens, Georgia 30602

William M. Post
Environmental Sciences Division
Oak Ridge National Laboratory
Oak Ridge, Tennessee 37831

Nanako Shigesada
Dept. of Biophysics
Kyoto University
Kyoto 606, Japan

Ei Teramoto
Faculty of Science and Technology
Ryukoku University
Otsu 520–21, Japan

Robert E. Ulanowicz
University of Maryland
Chesapeake Biological Laboratory
Solomons, Maryland 20688-0038

Preface

This book is about recent efforts toward enrichment of ecosystem theory. Ecosystems, conceived of as more than simply a collection of organisms, and more than energy and nutrient flows, trophic webs and competition communities, are the full interactive *network* among coexistent living organisms and their non-living milieu. The diversity of viewpoints and approaches for theoretical study of ecosystems that are collected here share the perception of ecosystems as networks, and represent the state-of-the-art. We believe that exchange and interaction among these diverse approaches will generate further exciting research and a better and deeper understanding of ecosystems. With that belief, we gathered at two related symposia, entitled 'New Approaches to Ecological Networks' and 'Search for Ecosystem Principles: A Network Perspective', held during the joint meetings of the Ecological Society of America (ESA), the International Society of Ecological Modelling (ISEM) and the IV International Association of Ecology (INTECOL) Congress in August 1986 at Syracuse, New York. This book grew out of that opportunity, and reflects the results of exchange during and after the meeting. We sincerely hope this volume leads to new empirical research, theory and understanding of the nature of ecosystems.

Acknowledgements

The interactions culminating in this book began while the editors were at the Institute of Ecology, University of Georgia, Athens. We thank Dr Bernard C. Patten for providing there a stimulating and open environment for discourse. Professor Ei Teramoto and his Mathematical Biology Group, including Drs Shigesada, Nakajima and Kawasaki, gave to M.H. liberal time for study and readjustment on his return to Kyoto after six years in the United States. They were also very kind hosts to T.P.B. during his visits. We thank Drs W. E. Grant (ISEM) and A. P. Covich (ESA) for their assistance in organizing the symposia from which this book grew and Cambridge University Press for their patience and professionalism in seeing it into print. Financial support was provided through a grant-in-aid from the Japanese Ministry of Education, Culture and Science to M. H., NSF awards BSR 8215587 and 8114823 to B.C. Patten and US–Japan Cooperative Research Program 85-154201/BAMR-208 awarded to E. Teramoto and B. C. Patten. None of this would have been possible without the concerned efforts of the authors, both towards their own contributions and in review of others'. Drs W. Cale, S. Jørgensen, S. H. Levine, I. Aoki, Y. Imai and H. Hirata gave graciously of their time to review one or more chapters. And last, but not least, is the debt we owe to our wives and families for their strength and understanding during our absences and their love and support every day of our lives together.

Introduction

Enrichment of ecosystem theory

MASAHIKO HIGASHI AND THOMAS P. BURNS

I.1 Ecosystems

A standard definition of *ecosystem* is 'a community of organisms and their physical *environment* interacting as an ecological unit' (Lincoln, Boxshall & Clark, 1982). There are two extreme interpretations of this definition. One is the regional (habitat) concept, which restricts the ecosystem to a regional biota together with all its environmental factors. The other extreme is to interpret ecosystem as any biological entity (e.g. organism, population, community, and even possibly cells) together with its environment.

For the purpose of characterizing ecosystems, it seems useful and convenient to distinguish two concepts of ecosystem reflecting these two extremes:

1. ecosystem as a physical entity: a *dynamical system* consisting of a biological entity, typically a regional biota (community), together with its environment (Tansley, 1935), and
2. ecosystem as a paradigm for science: *entity-environment unit.*

I.1.1 *Ecosystem as dynamical system*

The concept of ecosystem as a physical entity consisting of a regional biota and its environment is the more usual usage today. The central idea is that an ecosystem as a physical entity is a dynamical system subject to relevant physical laws. To this concept, however, the concept of biological entity-environment unity is essential. Ecosystems are more than just energy and nutrient flows, trophic webs and competition communities, they are the full interrelations among coexistent living organisms and their non-living milieu (e.g. Tansley, 1935). Thus, the study of ecosystems naturally requires tools for investigating *complex systems.*

I.1.2 *Ecosystem as paradigm for science*

The concept of ecosystem as a scientific paradigm with entity-environment unity as the central concept has been less standard, but may have more profound implications. First, it has potential to bring ecological principles into a broader domain of science, in particular other areas of biological research, for example immunology (self-nonself relations), cell biology (cell as ecosystem), neurobiology. Also, recent research trends in information science and technology show a serious interest in system-environment unity, or environment-conscious thinking, as networking via communication cables increases among larger numbers of computers. This more liberal and general notion of ecosystem, the idea of unity between (biological or living) entity and its environment, is applicable to any level of the biological hierarchy, and thus, helps to connect concepts that have been restricted conventionally to use on their own levels. For example, it may be possible to generalize to any level of the biological hierarchy the concepts of natural selection (Darwin, 1859), fitness (Wright, 1969), and adaptability (Conrad, 1983), all of which concern the relationship between a biological entity and its environment. Burns, Patten & Higashi (Chapter 8) explore this possibility.

To understand ecosystem, as a complex physical entity or as a paradigm for science, requires formal means (methodologies) to represent the relationship between entity and environment.

I.2 The network perspective

I.2.1 *Modeling nature*

We argue that both theorists and empiricists do *modeling*; all scientific research begins with a model, regardless of the extent to which we are conscious of it, because only through a model (filter, framework) can we perceive nature, or rather some aspect of it. If this is so, then the question becomes what kinds of modeling methods are we actually using. As a field of science advances, and thus the complexity of the research object and problem increases, more systematic and well-founded methodologies for modeling should be of key interest. Research toward developing better methods and basic studies of the foundations of modeling itself become of great importance.

Mathematical modeling is a systematic methodology that has proved successful in discovering and understanding the underlying processes and causes in nature based on its observable parts and their relationships. This is true in many fields of science besides physics, and thus has shown its generality. Biological and life sciences, particularly during the last two decades, have applied mathematical modeling to a variety of fundamental problems ranging from biochemical reactions and other biophysical processes of macromolecules, to physiological processes (e.g. nervous systems, brain models) and morphogenesis of multicellular systems (e.g. L-system, spatial pattern formation, cell sorting models), to population interactions and other ecological processes.

Modeling is homomorphic mapping, i.e. constructing an image (a model) of reality by abstracting a particular aspect. It simplifies apparent *complexity* in nature so that we may comprehend. Thus, a model is perhaps the goal of any research in basic science; we work toward a model by which we can better understand and sometimes predict the natural phenomenon in question. Prerequisites to modeling include specification of the object system and focusing the question or problem. This eventually comes down to specifying what is the *state* of the system, i.e. identifying an optimal set of state variables, the macroscopic quantities that allow an effective description of a specific aspect of the system under study. Identifying the optimal state variables is thus a primary goal of all study. Indeed, the history of classic thermodynamics in the last century provides an example of a struggle to discover the most useful set of state variables, which almost entirely led to the discovery of thermal theory.

Margalef (Chapter 1) claims that ecosystems are anisotropic, that is exhibit different properties when measured along axes on different dimensions. This could result in models of the same ecosystem having different network structure depending on what type of interaction is being measured and modeled. Allen and O'Neill (Chapter 4) explore this fundamental issue of *system specification* for ecosystem study with a focus on issues of complexity, and provide ideas on how we can improve our models for complex large-scale systems like ecosystems using notions from hierarchy theory (Simon, 1957, 1962; Allen & Starr, 1982; O'Neill *et al.*, 1986; Lawton, 1987). Mikulecky (Chapter 3) suggests that thermodynamics provides the best set of state variables for ecosystems. If so, the most significant dimension on which to measure ecosystem processes at any scale may be that pertaining to entropy. Network models of entropy production and flow in ecosystems are feasible (Aoki, 1987, 1988).

I.2.2 *Modeling nature as networks*

The network perspective is not a theory, but rather a means of viewing or approaching nature or its parts through a particular formalism – network models. When we choose to recognize discontinuities or boundaries in space–time as entities, the world appears to be made up of discrete, but interacting entities – a network (for a further elaboration on this point see Mikulecky, Chapter 3). Nervous systems are perhaps the paradigm networks, being composed partly of interacting neurons 'connected' chemically, behaviorally and physically (at certain scales of resolution). Network models are not constrained to compartmental flow–storage models, linear equations or steady state conditions, as is clear from Chapter 9 by Hannon and Bentsman.

One benefit of the network perspective is that a large body of mathematics exists to help analyze network models of many forms. If a system (a part of nature) is modeled and quantified well, then it becomes possible to utilize relevant mathematical tools, such as graph theory (topology), probability theory, linear algebra and dynamical system theory, to better understand the way the world works. A key to this methodological resource is the isomorphism that exists between a geometric and topographic representation of a system as a network and a matrix of coefficients from the equations representing a dynamical system. To provide a concrete meaning to this relationship, we present, according to this parallelism between topological and algebraic representations, some mathematical concepts relevant to the network perspective (Appendix AI.1). Mikulecky in Appendix A3.1 of Chapter 3 provides additional mathematical background to the topology–algebra parallelism.

The mathematical representation and treatment of networks allows a formal and rigorous approach to complexity. This is generally the case for model-based formalisms. Ecologists' models of the world are usually left unspoken while they communicate instead a simplified caricature. This does not mean that some degree of simplification isn't always necessary when modeling, only that our inability to deal informally with complex relationships makes it desirable, if not necessary, for the science of relationships to have a formal means to represent its knowledge without danger of significant loss of information. The network perspective, by virtue of its explicit consideration of and ability to deal with interactive complexity, encourages the formal realization of our understanding of relationships between all the parts of any whole, no matter how complex. Therefore, it can deepen our understanding of the essential nature of complex systems. Mikulecky (Chapter 3) presents a new modeling methodology that is general enough to potentially model any complex system, including ecosystems.

We argued earlier that we model nature prior to studying it, and at the same time, modeling and its fundamental concern, system state specification, are themselves important goals of all research. Perhaps the fundamental or ultimate goal of basic science is understanding what causes a particular phenomenon. Ulanowicz in Chapter 2 makes a philosophical argument that the network perspective allows us to identify, or at least begin to address non-teleologically, formal and final causes: the two (of four) Aristotelian categories of cause that have been neglected in spite of their importance for fully understanding biological order. Ulanowicz associates the former with 'structural' features of the network, and the latter with causation from the environment of the whole network, that is upper levels in the hierarchy of networks.

This reminder suggests a possible expansion of the search for answers to questions about causality. It may be that we have been asking the wrong questions, that is, searching for answers in categories of causation that do not give an effective way to organize an answer. A useful state descriptor for many problems could be a macro-quantity that represents structural properties of the system, for example cycling, or positive and negative feedbacks. This line of thinking has philosophical connections to the discussion of trophic dynamics in Chapter 5 by Higashi *et al.* and evolution in hierarchical networks in Chapter 8 by Burns *et al.*, where in particular the question could be asked, what are formal and final causes in evolution. Also relating to this point, Chapter 6 by DeAngelis and Post shows how positive feedback acts as just such a structural (formal in Aristotelian terminology) cause of population dynamics. This holistic systems perspective on population dynamics is complementary to the Newtonian cause (material and efficient causes in Aristotelian terms) traditionally sought in population ecology. To generalize, cybernetic and information-theoretic studies of phenomena seem to have close linkages to this extension of the search for causality in networks. (Note: Several chapters in this volume utilize information theory as a mathematical tool, and Appendix AI.4 provides a brief introduction to Shannon information theory.)

I.3 The network perspective in ecosystem ecology

The ecosystem is the manifestation of nature that is of primary concern in this volume. We discuss below the relationship between the concepts of ecosystem and the network perspective. We begin with the ecosystem as a physical system.

I.3.1 *Ecosystem as a physical network*

Ecosystems have been conceived of as more than simply a collection of organisms, like butterflies pinned onto a museum tray. Rather they are sets of interacting entities. Whether these entities are abstract functional components of some process, or species-populations of organisms eating, being eaten, or competing, these sets of entities and their interactions can be perceived and modeled as networks. This view of ecosystems is at least as old as the term, although Tansley (1935) did not explicitly define ecosystem as a network. Odum (1968) credits nineteenth-century naturalists with an emerging network view of their world, culminating in Forbes' (1887) 'The lake as microcosm'. Later, geochemical cycles of the globe (Riley, 1944; Hutchinson, 1944) and of ecosystems (Cooper, 1937; Hutchinson & Bowen, 1948), and feeding relations in terrestrial (Shelford, 1913) and aquatic communities (Thienemann, 1926; Rawson, 1930; Lindeman, 1941) were described as networks of flows among pools or populations. Hutchinson (1948) distinguished two implicitly network approaches to ecosystems: biogeochemical and biodemographical.

> ...described in terms of the transfer of some substance through the system, without employing any purely biological enumeration, such as the size of a population, the mode of approach will be characterized as *biogeochemical*. When a circular causal system is described in terms of the variation in numbers of biological units or individuals, or, in other words, in terms of the variation in the sizes of populations, the mode of approach is characterized as *biodemographic*.

This dichotomy persists today as ecosystem ecology's schism into the process-functional and population dynamics schools (O'Neill *et al.*, 1986).

Until recently, there has been little concern with the explicit network properties of ecosystems considered as 'circular causal systems' (Hutchinson, 1948). Systems of Lotka–Volterra equations can be conceived of as causal networks of interacting populations, but network properties of these systems have been virtually ignored. A notable exception was the exploration of the effect of system order and connectivity on the stability of randomly constructed systems (Gardner & Ashby, 1970; May, 1972, 1973; DeAngelis, 1975), which led to the study of the relationship of structure to stability and development of ecosystems (Levins, 1975; Tansky, 1976; Puccia & Levins, 1985). Even simple networks of this type show network properties such as indirect determination of the quality of relationships (Levins, 1975; Levine, 1976; Vandermeer, 1980). These attributes of Lotka–Volterra systems are just now being explored in detail.

Margalef (1963) may have been the earliest advocate for the network perspective in ecology. He focused attention on the structure of ecosystems; he saw interactions among constituent elements as the basis of this structure. Later, taking a strong stand for the ecosystem as physical network being the embodiment of organism–environment relationship, he made the following assertion:

> Ecology, I claim is the study of systems at a level in which individuals or whole organisms may be considered elements in interaction, either among themselves, or with a loosely organized environmental matrix. Systems at this level are named ecosystems, and ecology, of course, is the biology of ecosystems.
>
> (Margalef, 1968, p. 4).

Ecosystems, in Margalef's view, are also cybernetic systems with feedback loops resulting from 'elements linked by reciprocal influences' (Margalef, 1983). Ecosystem as both physical network and paradigm are implicit in his words.

The existence of interactive networks, signaled by indirect effects propagating around the system, was empirically demonstrated to be a significant determinant of organismal function (Patten & Witkamp, 1967). Theoretical investigation of indirect effects (Patten, 1982; Higashi & Patten, 1986, 1989) constitutes a conceptual basis for understanding complex interactions in ecosystems and the indirect effects recently demonstrated empirically by field manipulations (e.g. Kerfoot & Sih, 1988).

Hannon (1973) introduced input–output flow analysis to ecology from economics (Leontief, 1966), where the view of human production–function systems as networks of flows of commodities was well established. Inspired by this, several network analyses have been developed in recent years (Odum, 1960, 1971; Patten *et al.*, 1976; Finn, 1976; Barber, 1978; Matis & Patten, 1981; Herendeen, 1981; Patten & Higashi, 1984; Ulanowicz, 1983; Higashi, 1986*a*, *b*). These have led the search for network properties of ecosystems. Recent developments along several of these lines of research are presented in this volume (Margalef, Chapter 1; Ulanowicz, Chapter 2; Higashi *et al.*, Chapter 5; Hannon and Bentsman, Chapter 9; Herendeen, Chapter 10; Patten, Concluding Remarks).

I.3.2 *Ecosystem paradigm networked*

The ecosystem as a scientific paradigm (base model, framework) has become structured through network models. Central to this is the specification of the structure of environment, one half of the key notion of entity–environment unity. This structured-environment ecosystem paradigm has evolved into the open (hierarchical) network paradigm, where the rest of the network serves as the environment of each member of the network. The whole notion of the environment becomes relativistic: the world appears differently to different members according to their unique positions in the interactive network (Patten, 1982). This new paradigm is quite suited to coevolutionary questions, because the biological environment is naturally this structured relativistic environment. In particular, it provides a framework to investigate the real nature of 'natural selection' in ecosystems (Burns *et al.*, Chapter 8).

This paradigm shift has occurred mainly in the study of ecosystems, in the more restricted physical sense that we discussed above. But, we should note that, because ecosystem as a paradigm has broad relevance and application, any elaboration of it may have a significant influence on many fields of science. In fact, it is interesting to observe this trend in other biological and life sciences. Even information science and technology has shown parallel development toward a more structured or network perspective, in which every entity appears to be part of the relative environment of the others.

I.4 Other perspectives and their relations to the network perspective

Other properties of living systems have and will continue to be studied profitably from other than the network perspective, for example cybernetic behavior from a black box perspective. The population growth (logistic) models of Lotka (1925) and others (Pearl, 1928; Verhulst, 1938) have been extremely successful in ecology. We currently see two major perspectives on ecosystems as important to complement the network perspective – the hierarchical and evolutionary perspectives.

I.4.1 *The hierarchical perspective*

The network model *per se* is by definition not hierarchical; network modeling does not make a prior assumption (restriction) on any direction in its structure. However, in its application to reality a network model must specify the level to which it belongs; boundaries must be drawn, objects identified and interactions observed (sampled) at some interval. As Allen and O'Neill (Chapter 4) make clear, the network perspective inherently involves hierarchy. In fact, it is possible and useful to explicitly link these two perspectives. For instance, several contributors (Ulanowicz, Chapter 2; Mikulecky, Chapter 3; Burns *et al.*, Chapter 8;

Patten: Concluding remarks) have integrated the network and hierarchical perspectives, and view ecosystems as hierarchical networks.

I.4.2 ***The evolutionary perspective***

I.4.2i *Biological evolution*

The relationship between a biological entity and its environment is central to ecosystem study. Ecology is thus linked ineluctably to biological evolution, which provides the main process to adapt organisms to their environment, including their biological environment. Nevertheless, ecosystem ecology has virtually ignored evolutionary considerations. Perhaps this is because of the great distance between macroscopic thinking, in terms of a common currency such as energy and nutrients, and thinking of individual species, which is central to evolutionary ecology. However, evolution can change the members (species) of the ecosystem, thereby affecting ecosystem processes (function). When the ecosystem is seen as an interactive network, the linkage of evolution to ecosystem processes appears clearer. Furthermore, the hypothesis that indirect effects are significant if not dominant in ecosystems (Patten, 1982; Higashi & Patten, 1986, 1989) makes this linkage even stronger, suggesting that local evolutionary change at one point in an ecosystem network may affect the entire ecosystem.

On the other hand, the notion that the structure and function of an ecosystem affect the evolution of the species in it is too obvious, being almost equivalent to the definition of natural selection. However, it may not appear so obvious if we examine more closely the actual meaning of the terms involved. When an ecosystem is studied from the network perspective, identifying actual *pathways* and extent of influence on one member of the ecosystem from other members, the meaning of those terms are clarified. Thus, we may conclude that ecosystem study and biological evolution have significant cross-linkages, which have not been satisfactorily investigated, but might be explored more effectively through the network perspective. Burns *et al.* (Chapter 8) try to develop this linkage.

I.4.2ii *Evolution as a general process of ordered change*

Recent research in physics and general systems theory see evolution as a process of self-organizing change common to many far-from-equilibrium systems. These theories, more so than Neo-Darwinism, recognize that the entities of interest are systems evolving within still larger interactive systems; entities with environments both modified by and constraining their evolution. Thus, evolution is naturally and directly linked to the ecosystem paradigm.

For the study of ecosystems in the more restricted sense as physical systems, evolution as ordered change should also have relevance. In fact, this broader concept of evolution may provide a more complete evolutionary perspective on ecosystems from a hierarchical view. This was suggested in Margalef's (1968) early work; ecosystems and constituent species both change in ordered ways (i.e. evolve) through time. Ecosystem alterations can involve species replacement and (biological) evolution, depending on the reference time scale. The question becomes what is the relationship between these two changes taking place on different levels (temporal scales) at the same location. We are again led to the species–ecosystem interplay mentioned in the previous section, but the ecosystem is now seen as a dynamically ordered entity in its own right.

I.5 Challenges to theoretical issues on ecosystems

Recent advances in ecosystem ecology based on the network perspective and related approaches have accelerated theory development, identified new theoretical issues pertaining to ecosystems and are beginning to develop testable hypotheses of empirical interest. The works collected in this volume are updated reports on these efforts. Theoretical issues the chapters deal with include:

1. diversity, connectance and succession (Margalef, Chapter 1; Kawasaki *et al.*, Chapter 7)
2. trophic-dynamics (Higashi *et al.*, Chapter 5)
3. cybernetic (control) processes (DeAngelis and Post, Chapter 6; Hannon and Bentsman, Chapter 9)
4. behavioral trends, optimality principles (Ulanowicz, Chapter 2; Herendeen; Chapter 10)
5. evolution (Burns *et al.*, Chapter 8)

There is a diversity of network approaches to theoretical study of ecosystems in this book. This was exactly our intention. We have attempted to reflect in this collection the diversity of the state-of-the-art in this field, and to make comparisons possible. To enhance comparisons, an energy-flow model of an intertidal oyster reef in South Carolina, USA (Fig. I.8, see Appendix AI.3 for description) is used where needed as an example ecosystem network. (Where an author has changed the model slightly for his own purpose, the altered figure is reprinted separately.) Some of the diversity apparent now may be necessary, but much of it seems to be transient in nature and may eventually converge. In fact, collaborative research has already begun.

Finally, we observe that as more new theoretical problems arise, they demand elaboration and innovation of approaches and methodologies, and in particular, integration of related approaches to produce a more advanced one. For example, to produce empirically testable hypotheses about the effect of energy and nutrient circulation on the stability of ecosystems almost demands two traditionally separate approaches, the process-function and population dynamic approaches to ecosystems. We speculate that the cyclic positive feedback that we observe to occur among theory, empirical examination, and the discovery of new problems is the major driving force in the enrichment of ecosystem theory.

Appendix

AI.1 Basic mathematics for the network perspective

The purpose of this appendix is to provide a simple informal exposition on several related mathematical notions that we believe are relevant and useful to the foundation and development of the network perspective in ecology. One specific intention is to demonstrate the parallelism between topological (graphic) and algebraic representations of system structure for general relations defined on sets (Preparata & Yeh, 1973; Bollobas, 1979) and for two special classes of systems: Markov chains, a type of stochastic process (Kemeny & Snell, 1960; Feller, 1968), and dynamical systems of difference or differential equations (Bellman, 1953; Lefschetz, 1957).

AI.1.1 *Graphs, binary relations and matrices*

Cartesian product. The *cartesian product* $X \times Y$ of sets X and Y is the set of all the ordered pairs for elements of X and Y, i.e.,

$$X \times Y \equiv \{(x, y) \mid x \in X \text{ and } y \in Y\}. \quad \text{(A1)}$$

Ordered pair here means a pair for which the order does matter, i.e. pair (x, y) is distinguished from pair (y, x) unless x and y are identical.

Example A1. If sets X and Y are represented, respectively, by the x-axis and y-axis in a two-dimensional space, the cartesian product $X \times Y$ corresponds to the entire two-dimensional space.

Binary relation between two sets. A binary relation R from set X to set Y is a subset of the cartesian product $X \times Y$:

$$R \subseteq X \times Y. \tag{A2}$$

If $X = \{x_1, x_2, \ldots, x_n\}$ and $Y = \{y_1, y_2, \ldots, y_m\}$, binary relation $R \subseteq X \times Y$ is represented by a *bipartite graph* b_R from *vertex* class (*node* set) $X = \{x_1, x_2, \ldots, x_n\}$ to vertex class $Y = \{y_1, y_2, \ldots, y_m\}$ in which there is an arc from x_j to y_i (denoted by $x_j \rightarrow y_i$) iff (if and only if) $(x_j, y_i) \in R$. In this finite case, binary relation $R \subseteq X \times Y$ (or the corresponding graph b_R) may be also represented by an $m \times n$ *matrix* $\mathbf{R} = [r_{ij}]$ such that $r_{ij} = 1$ if $(x_j, y_i) \in R$, i.e. if there is an arc from vertex x_j to vertex y_i and '0' otherwise.

Example A2. Relation $R = \{(x_1, y_2), (x_2, y_1), (x_3, y_1), (x_3, y_2)\}$ from $X = \{x_1, x_2, x_3\}$ to $Y = \{y_1, y_2\}$, whose corresponding graph is shown in Fig. I.1, is represented by matrix $\begin{bmatrix} 0 & 1 & 1 \\ 1 & 0 & 1 \end{bmatrix}$.

Fig. I.1. The graph of a relation, $R = \{(x_1, y_2), (x_2, y_1), (x_3, y_1), (x_3, y_2)\}$ from $X = \{x_1, x_2, x_3\}$ to $Y = \{y_1, y_2\}$, which is represented by matrix $\begin{bmatrix} 0 & 1 & 1 \\ 1 & 0 & 1 \end{bmatrix}$.

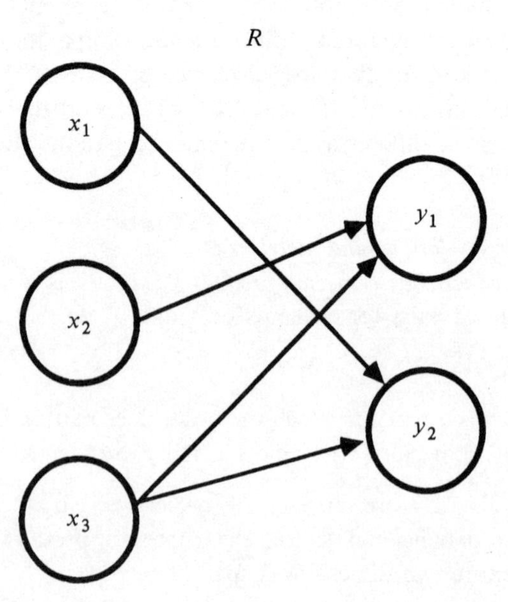

Composition of binary relations. The composition of two binary relations $R \subseteq X \times Y$ and $S \subseteq Y \times Z$, denoted by $S \circ R$, is a binary relation $S \circ R \subseteq X \times Z$ that consists of all pairs $(x, z) \in X \times Z$ such that $(x, y) \in X \times Y$ and $(y, z) \in Y \times Z$ for some $y \in Y$.

Let $X = \{x_1, x_2, \ldots, x_n\}$, $Y = \{y_1, y_2, \ldots, y_m\}$ and $Z = \{z_1, z_2, \ldots, z_k\}$, and let $R \subseteq X \times Y$ and $S \subseteq Y \times Z$ be the binary relations represented respectively by bipartite graph b_R from vertex class $X = \{x_1, x_2, \ldots, x_n\}$ to vertex class $Y = \{y_1, y_2, \ldots, y_m\}$ and bipartite graph b_S from vertex class $Y = \{y_1, y_2, \ldots, y_m\}$ to vertex class $Z = \{z_1, z_2, \ldots, z_k\}$. Then, the composition $S \circ R \subseteq X \times Z$ is represented by the bipartite graph from vertex class $X = \{x_1, x_2, \ldots, x_n\}$ to vertex class $Z = \{z_1, z_2, \ldots, z_k\}$ in which there is an arc $x_j \to z_h$ if and only if for some $y_i \in Y$ there is an arc $x_j \to y_i$ in graph b_R and an arc $y_i \to z_h$ in graph b_S. Let matrices $\boldsymbol{R} = [r_{ij}]$ and $\boldsymbol{S} = [s_{ih}]$ respectively represent relations $R \subseteq X \times Y$ and $S \subseteq Y \times Z$. Then, the matrix representing $S \circ R$, denoted by $\boldsymbol{S} \circ \boldsymbol{R}$, which defines one kind of product (multiplication) of matrices $\boldsymbol{S}$ and $\boldsymbol{R}$, is a $k \times n$ matrix whose element at the intersection of the hth row and the jth column [(h, j) element] $(\boldsymbol{S} \circ \boldsymbol{R})_{hj}$ is given as

$$(\boldsymbol{S} \circ \boldsymbol{R})_{hj} = \begin{cases} 0 & \text{if } \sum_{i=1}^{k} s_{hi} r_{ij} = 0 \\ 1 & \text{otherwise} \end{cases} \tag{A3}$$

or

$$(\boldsymbol{S} \circ \boldsymbol{R})_{hj} = (s_{h1} \wedge r_{1j}) \vee (s_{h2} \wedge r_{2j}) \vee \ldots \vee (s_{hm} \wedge r_{mj}) \tag{A4}$$

where '$\vee$' and '$\wedge$' respectively represent maximum and minimum operations.

Another kind of product (multiplication) $\boldsymbol{SR}$ of matrices $\boldsymbol{S}$ and $\boldsymbol{R}$ is defined as an $k \times n$ matrix whose (h, j) element $(\boldsymbol{SR})_{hj}$ is given as

$$(\boldsymbol{SR})_{hj} = \sum_{i=1}^{k} s_{hi} r_{ij}, \tag{A5}$$

and represents the number of paths from vertex x_j to vertex z_h found in the graph obtained by connecting two graphs b_R and b_S side by side overlapping the common vertex set $Y = \{y_1, y_2, \ldots, y_m\}$.

Example A3. Let $X = \{x_1, x_2, x_3\}$, $Y = \{y_1, y_2\}$ and $Z = \{z_1, z_2\}$, and let $R \subseteq X \times Y$ and $S \subseteq Y \times Z$ be binary relations represented by matrices

$$\boldsymbol{R} = \begin{bmatrix} 0 & 1 & 1 \\ 1 & 0 & 1 \end{bmatrix} \text{ and } \boldsymbol{S} = \begin{bmatrix} 1 & 1 \\ 1 & 0 \end{bmatrix}, \tag{A6}$$

respectively. Then, $S \circ R \subseteq X \times Z$ is represented by the following matrix:

$$\boldsymbol{S} \circ \boldsymbol{R} = \begin{bmatrix} 1 & 1 \\ 1 & 0 \end{bmatrix} \circ \begin{bmatrix} 0 & 1 & 1 \\ 1 & 0 & 1 \end{bmatrix} = \begin{bmatrix} 1 & 1 & 1 \\ 0 & 1 & 1 \end{bmatrix}, \tag{A7}$$

Fig. I.2. (Upper) The graph obtained by connecting two bipartite graphs b_R and b_S, respectively, for binary relation R from $X = \{x_1, x_2, x_3\}$ to $Y = \{y_1, y_2\}$, represented by matrix $\boldsymbol{R} = \begin{bmatrix} 0 & 1 & 1 \\ 1 & 0 & 1 \end{bmatrix}$, and binary relation S from $Y = \{y_1, y_2\}$ to $Z = \{z_1, z_2\}$, represented by $\boldsymbol{S} = \begin{bmatrix} 1 & 1 \\ 1 & 0 \end{bmatrix}$. (Lower) The graph for the composition $S \circ R \subseteq X \times Z$ of two relations $R \subseteq X \times Y$ and $S \subseteq Y \times Z$, which is represented by matrix $\boldsymbol{S} \circ \boldsymbol{R} = \begin{bmatrix} 1 & 1 \\ 1 & 0 \end{bmatrix} \circ \begin{bmatrix} 0 & 1 & 1 \\ 1 & 0 & 1 \end{bmatrix} = \begin{bmatrix} 1 & 1 & 1 \\ 0 & 1 & 1 \end{bmatrix}$.

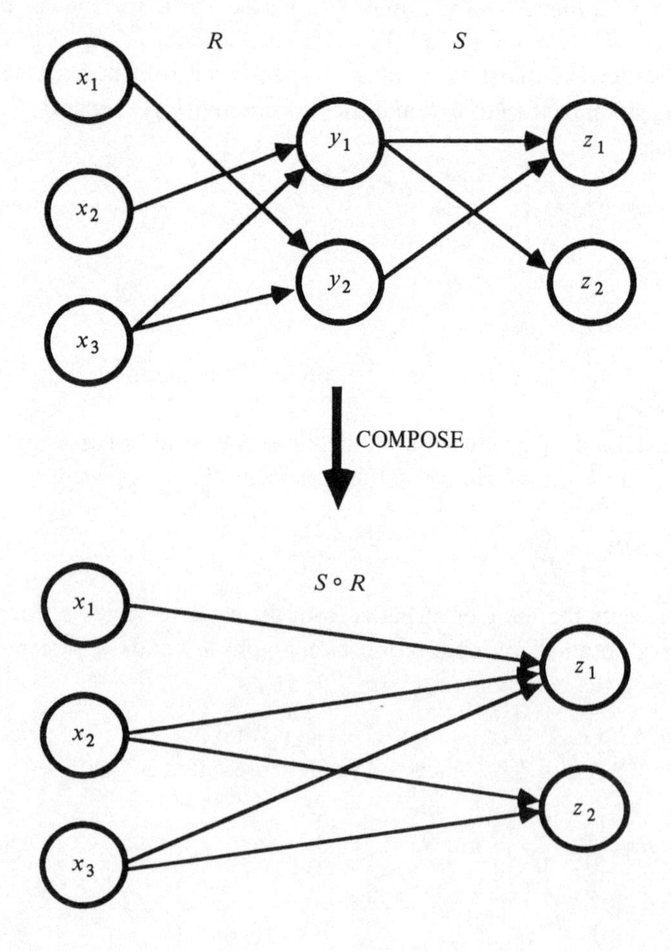

and the graph for $S \circ R \subseteq X \times Z$ is generated by composing the graphs for $R \subseteq X \times Y$ and $S \subseteq Y \times Z$ as illustrated in Fig. I.2. The other kind of matrix product is calculated as

$$\boldsymbol{SR} = \begin{bmatrix} 1 & 1 \\ 1 & 0 \end{bmatrix} \begin{bmatrix} 0 & 1 & 1 \\ 1 & 0 & 1 \end{bmatrix} = \begin{bmatrix} 1 & 1 & 2 \\ 0 & 1 & 1 \end{bmatrix}, \tag{A8}$$

whose element $(\boldsymbol{SR})_{hj}$ represents the number of paths from vertex x_j to vertex z_h in the graph (the upper figure in Fig. I.2) obtained by connecting two graphs b_R and b_S.

Binary relation on a set. A binary relation R on set XC is a subset of the cartesian product $X \times X$, i.e.,

$$R \subseteq X \times X. \tag{A9}$$

This is just a special case of binary relation between two sets in which two sets coincide.

If $X = \{x_1, x_2, \ldots, x_n\}$, binary relation $R \subseteq X \times X$ is represented by a bipartite graph b_R from vertex class $\{x_1, x_2, \ldots, x_n\}$ to the identical vertex class $\{x_1, x_2, \ldots, x_n\}$, or a directed *graph* (*digraph*) g_R on vertex class $\{x_1, x_2, \ldots, x_n\}$, in which there is an arc from vertex x_j to vertex x_i (or more simply from j to i) denoted by $x_j \to x_i$ (or $j \to i$) iff $(x_j, x_i) \in R$. In this finite case, binary relation $R \subseteq X \times X$ (or the corresponding graph g_R) may be represented by an $n \times n$ (square) matrix $\boldsymbol{R} = [r_{ij}]$ such that $r_{ij} = 1$ if $(x_j, x_i) \in R$, i.e., if there is an arc from j to i, and '0' otherwise.

Example A4. Relation $R = \{(x_1, x_2), (x_1, x_3), (x_2, x_1), (x_2, x_3), (x_3, x_1)\}$ on $X = \{x_1, x_2, x_3\}$, whose corresponding graphs b_R and g_R are shown in Fig. I.3, is represented by matrix $\begin{bmatrix} 0 & 1 & 1 \\ 1 & 0 & 0 \\ 1 & 1 & 0 \end{bmatrix}$.

Power of a binary relation on a set. The composition $S \circ R$ of two binary relations on set X, $R \subseteq X \times X$ and $S \subseteq X \times X$ may be derived as already shown for the more general case of relations between two sets. As a special case, if $S = R$, then we have the second *power* $R \circ R = R^2$. Further, if $S = R^2$, then we have the third power $R^2 \circ R = R^3$. By continuing this process, we have the sequence of powers $R, R^2, \ldots, R^k, \ldots$ Note that each of these powers R^k is a binary relation on set X.

Let $X = \{x_1, x_2, \ldots, x_n\}$. Then, in the digraph representing the kth power R^k of binary relation R on set X, there is an arc from vertex j to vertex i

if and only if there exists at least one *path* of length k (a set of connected k arcs) from vertex j to vertex i in the digraph g_R for binary relation R. In terms of matrices, we have two kinds of matrix power corresponding to the two kinds of matrix products defined above. The first kind of power $(\circ \boldsymbol{R})^k$ of the matrix $\boldsymbol{R}$ representing binary relation R on set X corresponds to the power R^k of the binary relation R. The second kind of power $\boldsymbol{R}^k$ should be related to the composite graph obtained by connecting k identical bipartite graphs b_R, and each element $(\boldsymbol{R}^k)_{ij}$ represents the number of paths from j in the left-most vertex set to i in the right-most vertex set within this composite graph, or equivalently, the number of paths of length k from j to i in the digraph g_R for relation R.

Example A5. For binary relation R on set X represented by matrix $\boldsymbol{R} = \begin{bmatrix} 0 & 1 & 1 \\ 1 & 0 & 0 \\ 1 & 1 & 0 \end{bmatrix}$, whose corresponding graphs b_R and g_R are shown in Fig. I.3,

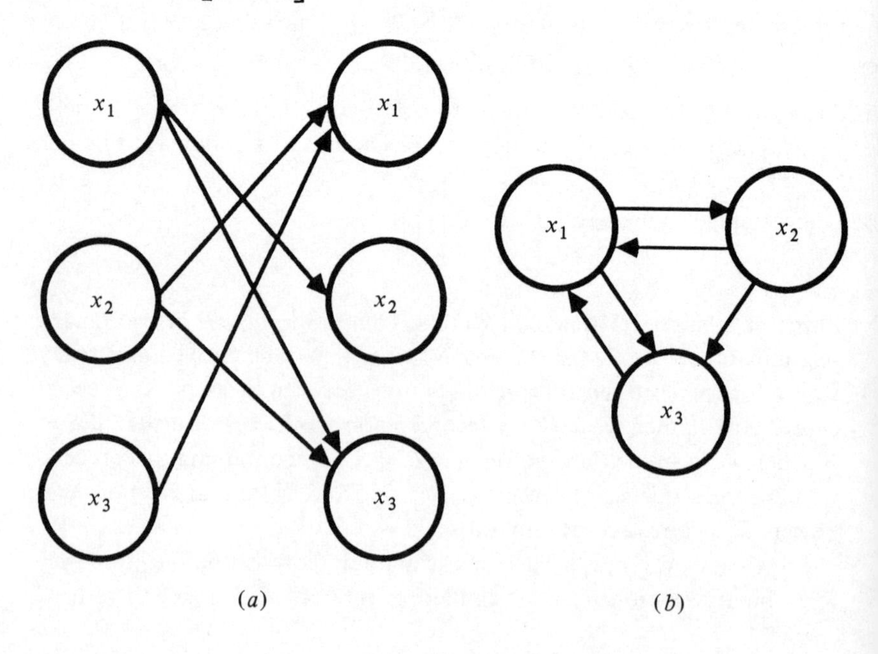

Fig. I.3. (*a*) The bipartite graph b_R and (*b*) digraph g_R of relation $R = \{(x_1, x_2), (x_1, x_3), (x_2, x_1), (x_2, x_3), (x_3, x_1)\}$ on $X = \{x_1, x_2, x_3\}$, which is represented by matrix $\begin{bmatrix} 0 & 1 & 1 \\ 1 & 0 & 0 \\ 1 & 1 & 0 \end{bmatrix}$.

Fig. I.4. *a–b*. For caption see p. 19.

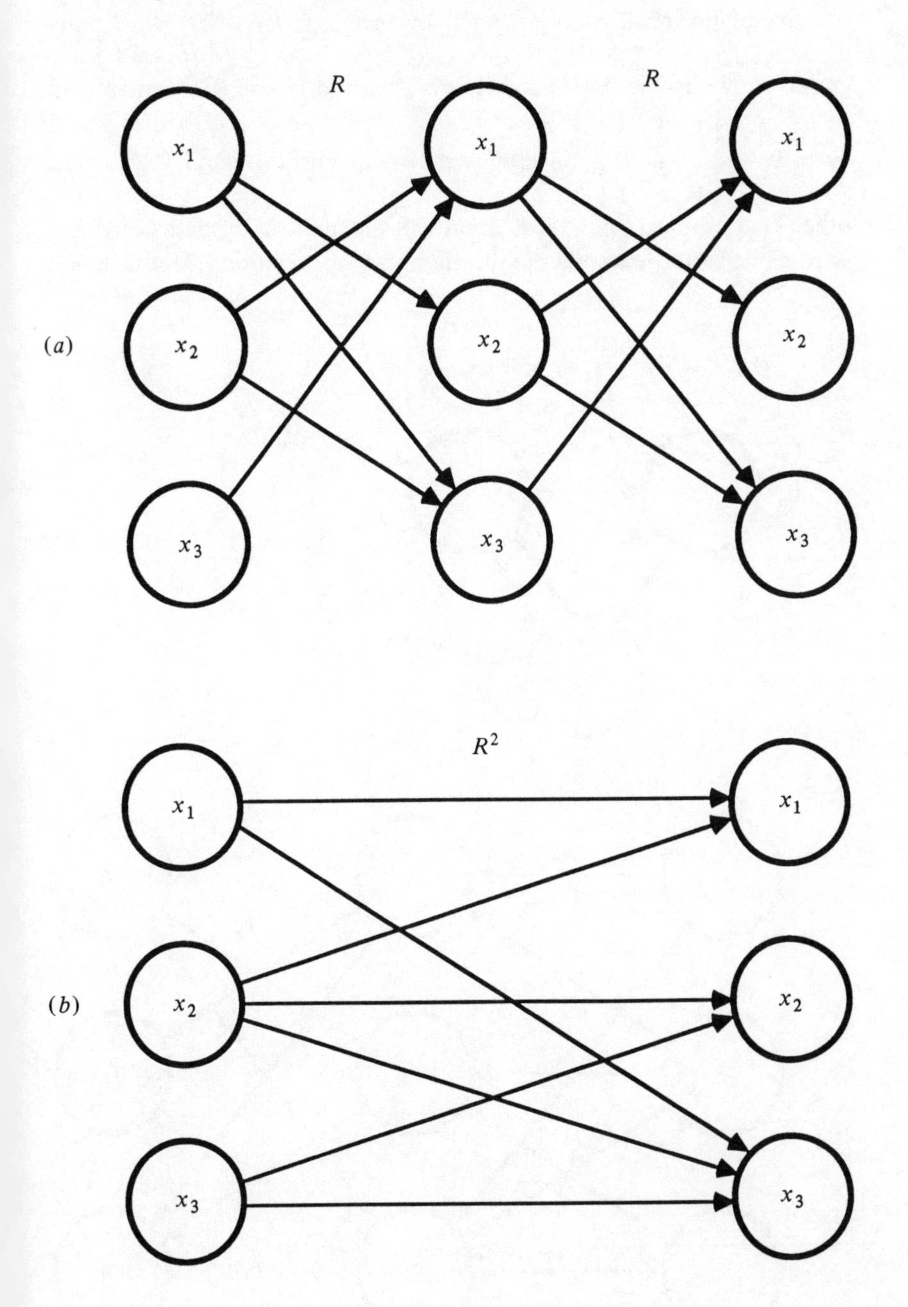

the second power R^2 is represented by matrix $(\circ \boldsymbol{R})^2 = \begin{bmatrix} 1 & 1 & 0 \\ 0 & 1 & 1 \\ 1 & 1 & 1 \end{bmatrix}$ and graphs shown in Fig. I.4(*b*) and I.4(*c*). The third power R^3 is represented by $(\circ \boldsymbol{R})^3 = \begin{bmatrix} 1 & 1 & 1 \\ 1 & 1 & 0 \\ 1 & 1 & 1 \end{bmatrix}$ and graphs shown in Fig. I.4(*e*)and I.4(*f*). The other kind of matrix powers $\boldsymbol{R}^k$ represent the numbers of paths of length k from one vertex to another in the digraph g_R for relation R; for instance,

Fig. I.4. *c–d*. For caption see opposite.

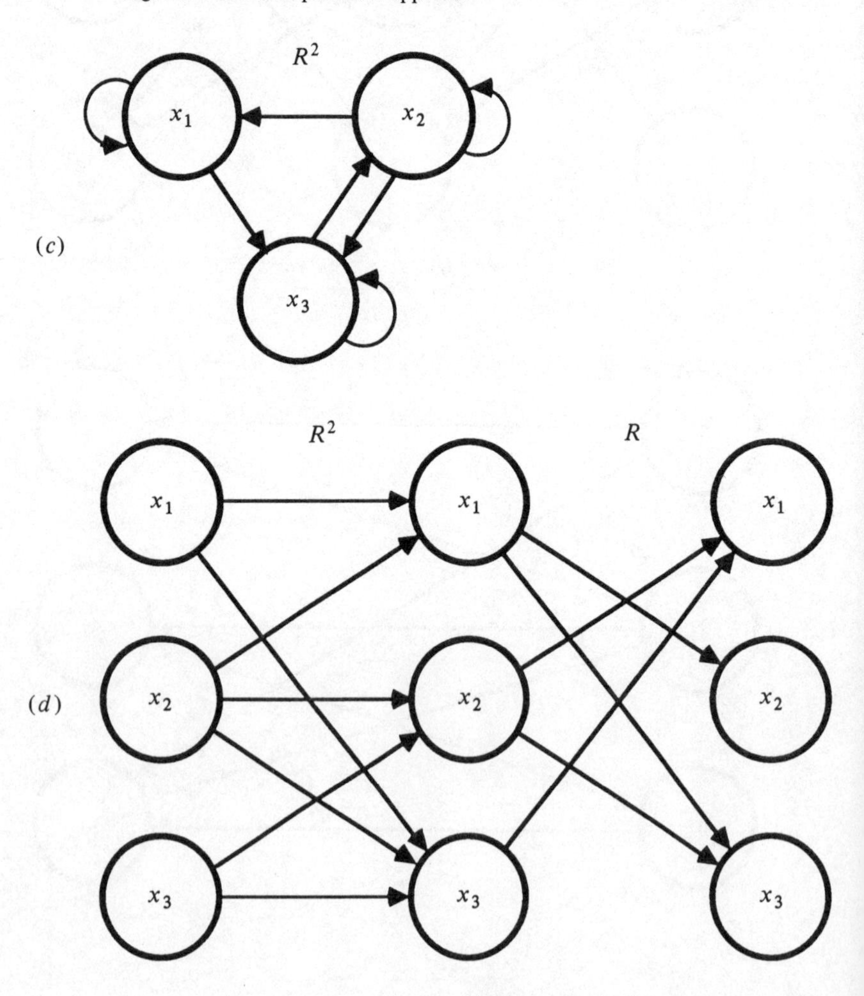

Fig. I.4. (*a*) The graph obtained by connecting two identical bipartite graphs b_R of the binary relation R on $X = \{x_1, x_2, x_3\}$ represented by matrix $\boldsymbol{R} = \begin{bmatrix} 0 & 1 & 1 \\ 1 & 0 & 0 \\ 1 & 1 & 0 \end{bmatrix}$; (*b*) the bipartite graph and (*c*) digraph for the second power R^2, which is represented by matrix $(\circ \boldsymbol{R})^2 = \begin{bmatrix} 1 & 1 & 0 \\ 0 & 1 & 1 \\ 1 & 1 & 1 \end{bmatrix}$, of binary relation R mentioned in (*a*); (*d*) the graph obtained by connecting the bipartite graph for the second power R^2 of binary relation R mentioned in (*a*) and the bipartite graph for R itself; and (*e*) the bipartite graph and (*f*) digraph for the third power R^3, which is represented by matrix $(\circ \boldsymbol{R})^3 = \begin{bmatrix} 1 & 1 & 1 \\ 1 & 1 & 0 \\ 1 & 1 & 1 \end{bmatrix}$, of binary relation R mentioned in (*a*).

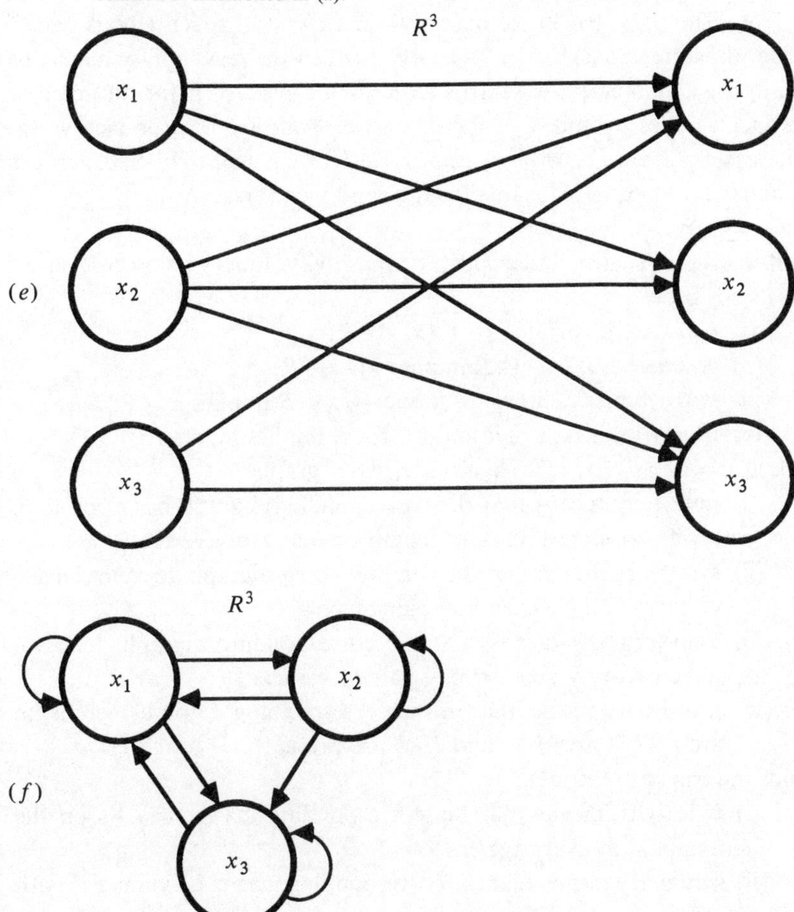

$\boldsymbol{R}^2 = \begin{bmatrix} 2 & 1 & 0 \\ 0 & 1 & 1 \\ 1 & 1 & 1 \end{bmatrix}$ indicates the number of paths from each vertex on the left-hand side to each vertex on the right-hand side in Fig. I.4(*a*), or equivalently, the number of paths of length 2 from column vertex to row vertex in digraph g_R.

Reachability and connectedness of digraphs. In a simple digraph g_R corresponding to a relation R on set $X = \{x_1, x_2, \ldots, x_n\}$, vertex x_i is said to be *reachable* from another vertex x_j if there is a path from j to i. Two necessary and sufficient conditions for x_i to be *reachable* from another vertex x_j are:

(i) The (i,j) element of $(\circ \boldsymbol{R})^k$ is one for at least one k $(1 \leqslant k \leqslant n-1)$.

(ii) The (i,j) element of maximum matrix $\vee_{k=1}^{n-1} (\circ \boldsymbol{R})^k$ equals one.

From the latter, $(\circ \boldsymbol{R})^{(k)} \equiv \vee_{k=1}^{n-1} (\circ \boldsymbol{R})^k$ is called the reachability matrix of digraph g_R. A digraph is said to be *weakly connected* if for any pair of distinct vertices x_i and x_j, either x_i is *reachable* from x_j or vice versa, whereas it is said to be *strongly connected* if for any pair of distinct vertices x_i and x_j, both x_i is reachable from x_j and vice versa.

Reflexivity, symmetry, antisymmetry, transitivity. Binary relation R on set X is said to be

(i) *reflexive*, if for any $x \in X, (x,x) \in R$;

(ii) *symmetric*, if $(x,y) \in R$ implies $(y,x) \in R$;

(iii) *antisymmetric*, if $(x,y) \in R$ and $(y,x) \in R$ implies $x = y$;

(iv) *transitive*, if $(x,y) \in R$ and $(y,z) \in R$ implies $(x,z) \in R$.

Let $X = \{x_1, x_2, \ldots, x_n\}$. Then, in terms of graphs,

(i) reflexivity means that the corresponding digraph has a self-loop (cycle, or closed path, of length one) at every vertex;

(ii) symmetry means that the corresponding digraph does not have a one-way arc between any pair of vertices;

(iii) antisymmetry means that the corresponding digraph does not have two-way arcs for any pair of vertices;

(iv) transitivity means that in the corresponding digraph there is an arc $j \rightarrow h$ if arcs $j \rightarrow i$ and $i \rightarrow h$ are present.

And, in terms of matrices,

(i) reflexivity means that the corresponding matrix has 1's on the diagonal at every vertex;

(ii) symmetry means that the corresponding matrix is symmetric with respect to the diagonal;

(iii) antisymmetry means that the corresponding matrix does not have 1s in both of the positions symmetric with respect to the diagonal;

(iv) transitivity means that the corresponding matrix $\boldsymbol{R}$ satisfies the condition $\boldsymbol{R} \supseteq \boldsymbol{R}^2$.

Transitive closure. The *transitive closure* of a binary relation R on a set is the smallest relation (in the sense of set inclusion) that contains R and is transitive. It can be shown that the matrix that represents the transitive closure of relation R on set $X = \{x_1 x_2, \ldots, x_n\}$ *equals the reachability matrix* $(\circ \boldsymbol{R})^{(n-1)}$ of digraph g_R.

Compatibility relations and classes. A *compatibility relation* on set X is a binary relation that is reflexive and symmetric. Consider a compatibility relation R on set X. Then,

(i) a *compatibility class* (or *compatible*) induced by R is a subset C of X such that for any members x and y of C, $(x, y) \in R$;

(ii) a *maximal compatibility class* (or *maximal compatible*, or *clique*) induced by R is a compatibility class induced by R that is not included by any other compatibility class induced by R;

(iii) the *complete cover* of X with respect to R is the collection (set) of maximal compatibility classes induced by R.

Equivalence relations and partitionings. *Equivalence relation* is a binary relation R on a set X that is reflexive, symmetric and transitive, (i.e. a compatibility relation on set X that is transitive). A maximal compatibility class of an equivalence relation R on set X is called an *equivalence class.* It can be shown that any two equivalence classes of an equivalence relation R on set X do not overlap each other (i.e. their intersection is empty). A *partition* π (X) of set X is a collection (set) of non-empty subsets of X such that any two of these subsets do not overlap each other, and these subsets together cover the entire set X (i.e. the union of these subsets equals set X). Thus, the complete cover, i.e. the collection (set) of all maximal compatibility classes (equivalence classes) of an equivalence relation R on set X is a partition of set X. Conversely, for each partition $\pi(X)$ of set X an equivalence relation R on set X can be defined as '$(x, y) \in R$ if and only if x and y belong to the same subset in the partition'. In this way, equivalence relations on set X correspond to partitions of X in *one-to-one fashion.*

Partial ordering and lattice. *Partial ordering* is a binary relation on a set that is reflexive, antisymmetric and transitive. *Poset* (or *partially ordered set*) is a set with a partial ordering $\geqslant$ defined on it. In a poset $(X, \geqslant)$, a lower (upper) bound of x and y is an element z of X such that $x \geqslant z$ and $y \geqslant z$ ($z \geqslant x$ and $z \geqslant y$), and a *meet* or *greatest lower bound* (*join* or *least upper bound*) z of x and y is a maximal (minimal) one among the lower (upper) bounds of x and y, i.e. a lower (upper) bound z of x and y such that there is no lower (upper) bound w of x and y satisfying $w \geqslant z$ ($z \geqslant w$). *Lattice* is a poset any two elements of which have a unique join and meet.

AI.1.2 *Finite Markov chains*

Temporally homogeneous finite Markov chains. A temporally homogeneous finite *Markov chain* (t.h.f. Markov chain) is a kind of stochastic process, and is defined as a process (motion or system) that takes n (a *finite* number of) distinct states and moves from state j to state i by probability p_{ij} ($0 \leqslant p_{ij} \leqslant 1$), where p_{ij} is constant (*time-invariant*). Note that the probability of the system's transition in the next step from the current state is independent of its past history, i.e. which states it had visited before it arrived in the present state. This property of a stochastic process is referred to as *Markov property*. The graph for a t.h.f. Markov chain is defined as a digraph in which each vertex j represents the corresponding state j of the chain and there is an arc from vertices j to i iff the

Fig. I.5. The graph for a hypothetical temporally homogeneous finite Markov chain, in which each vertex j represents the corresponding state j of the chain and there is an arc from vertices j to i if and only if the corresponding transition probability p_{ij} from state j to state i in the chain is not zero.

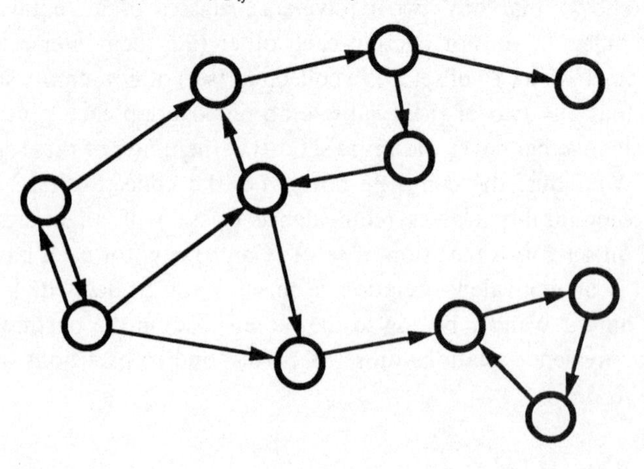

corresponding transition probability p_{ij} from state j to state i is not zero (Fig. I.5). The matrix $P = (p_{ij})$ is called the *transition matrix* of the chain, and satisfies the condition: for any j, $\sum_{i=1}^{n} p_{ij} = 1$.

Communication class of states. In a t.h.f. Markov chain with n states and transition matrix (p_{ij}), state j is said to communicate with state i iff vertices (representing states) j and i are reachable from each other in the graph for

Fig. I.6. (a) The partition of the state set of the Markov chain whose graph is depicted in Fig. I.5. Each shaded circle represents a communication class of states. (b) The partial ordering of reachability relation on the set of all the communication classes of states. C_4 and C_5 are ergodic classes, from which no other class is reachable, and the other classes are transient, which are not ergodic.

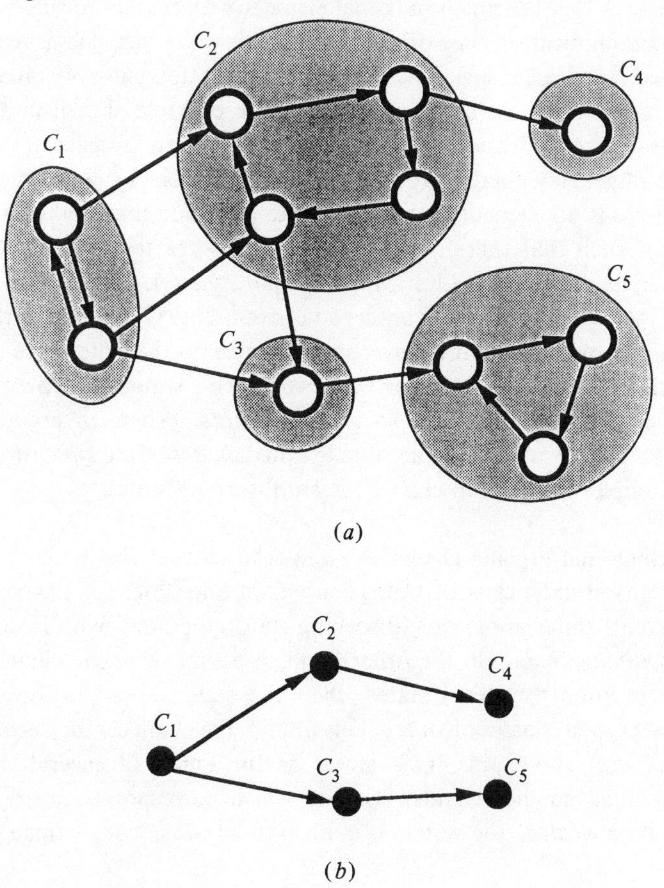

the chain. This *communication relation* between two states is an equivalence relation on the state set (the set of all the states), and thus induces the corresponding partition of the state set. Each equivalence class that constitutes this partition is called *communication class of states* (Fig. I.6(*a*) depicts the partition of the state set into communication classes of the Markov chain whose graph is shown in Fig. I.5).

Ergodic and transient classes and states. Further, in a t.h.f. Markov chain, communication class of states C is said to be reachable from communication class of states C' iff a state in C is reachable from a state in C' (then in fact every state in C is reachable from every state in C'). This reachability relation is a partial ordering on the set of all the communication classes of states (Fig. I.6(*b*) is the graph representing the partial ordering on the set of all the communication classes of states depicted in Fig. I.6(*a*)). A maximal element with respect to this ordering, i.e. a communication class of states from which no other class is reachable, is called an *ergodic class*, whereas a communication class of states that is not ergodic is called *transient class* (in the example shown in Fig. I.6, classes C_4 and C_5 are ergodic and the other classes are transient). In a finite poset, in general, there is at least one minimal element (an element such that there is no element that is smaller than it) and maximal element (an element such that there is no element that is greater than it; Birkhoff, 1967). Therefore, for a t.h.f. Markov chain, there is at least one ergodic class. A state belonging to an ergodic (transient) class has the property that the probability that the system will return to the state, once it leaves that state, is equal to (less than) one, and such a state is called an *ergodic* or *recurrent* (*transient* or *non-recurrent*) *state*. When an ergodic class consists of only one state, the state is called an *absorbing state* (in the Fig. I.6 example, the state in class C_4 is an absorbing state).

Absorbing and ergodic chains. As a special case, if for a t.h.f. Markov chain any ergodic class of states consists of one state, i.e. all the ergodic (recurrent) states are in fact absorbing states, then the chain is said to be an *absorbing chain*. On the other hand, if a t.h.f. Markov chain has no transient (non-recurrent) states, then the state set is partitioned into several ergodic classes of states. This implies the chain can be decomposed into several sub-chains (i.e. viewed as the union of several mutually independent smaller chains) each of which corresponds to an ergodic class, because once the system is in an ergodic class, it stays there forever.

A t.h.f. Markov chain whose state set consists of only one ergodic class is called *ergodic* or *irreducible chain.*

AI.1.3 *Dynamical systems*

Steady states of dynamical systems. A *dynamical system* is a system whose *state transition* is given by a set of differential equations

$$\frac{dx_i}{dt} = f_i(x_1, x_2, \ldots, x_n) \tag{A10a}$$

($i = 1, 2, \ldots, n$), or a set of difference equations

$$x_i(t+1) - x_i(t) = f_i(x_1, x_2, \ldots, x_n) \tag{A10b}$$

($i = 1, 2, \ldots, n$). The former is a *continuous-time* case, whereas the latter is a *discrete-time* case. In either case, the *steady states* (*equilibrium points, critical points, singular points, rest points,* or *fixed points*) of the system are given by the solutions $\boldsymbol{x}^* = (x_1^*, x_2^*, \ldots, x_n^*)$ for the set of algebraic equations

$$f_i(x_1, x_2, \ldots, x_n) = 0 \tag{A11a}$$

($i = 1, 2, \ldots, n$), or in the vector-matrix form,

$$\boldsymbol{f}(\boldsymbol{x}) = \boldsymbol{0}, \tag{A11b}$$

where $\boldsymbol{f}$ and $\boldsymbol{x}$ are the n-dimensional column vectors for f_i and x_i ($i = 1, 2, \ldots, n$) and $\boldsymbol{0}$ is the n-dimensional column zero vector (the vector whose elements are all zero). If x_i represents population (density), biomass or any positive physical quantity of species i in an ecological system (community), then positive solutions $\boldsymbol{x}^* = (x_1^*, x_2^*, \ldots, x_n^*)$, where $x_i^* > 0$ for $i = 1, 2, \ldots, n$, represent steady states of the system at which all species coexist.

Example A6. (Linear function case) If the response functions f_i are linear

$$f_i(x_1, x_2, \ldots, x_n) = \sum_{j=1}^{n} a_{ij} x_j + b_i \tag{A12a}$$

($i = 1, 2, \ldots, n$), i.e.,

$$\boldsymbol{f}(\boldsymbol{x}) = \boldsymbol{A}\boldsymbol{x} + \boldsymbol{b}, \tag{A12b}$$

where $\boldsymbol{b}$ is a column vector with b_i ($i = 1, 2, \ldots, n$) and $\boldsymbol{A} = [a_{ij}]$, and $\det \boldsymbol{A} \neq 0$ (the determinant of matrix $\boldsymbol{A}$ is not zero), then the steady state $\boldsymbol{x}^*$ of the system is given as

$$\boldsymbol{x}^* = -\boldsymbol{A}^{-1}\boldsymbol{b}, \tag{A13}$$

where A^{-1} is the inverse matrix of A (i.e. the matrix such that $A^{-1}A = AA^{-1} = I$, where I is the unit matrix).

Example A7. (Lotka–Volterra type) Consider a system whose response functions f_i are given as

$$f_i(x_1, x_2, \ldots, x_n) = x_i\left(\sum_{j=1}^{n} a_{ij}\, x_j + b_i\right) \tag{A14}$$

($i = 1, 2, \ldots, n$). If the system represents an ecological system comprising n species whose populations (or biomasses etc.) are represented by x_i ($i = 1, 2, \ldots, n$), matrix $A = [a_{ij}]$ is often referred to as *interaction matrix* in ecological literature, and the 'structure' of the interactions within the system can be illustrated by the sign digraph (digraph with sign '+' or '−' on each arc) corresponding to this matrix; in this digraph there is an arc from j to i if and only if $a_{ij} \neq 0$, and for each arc $j \to i$ sign (+) or (−) is assigned, according to $a_{ij} > 0$ or < 0. If $\det A \neq 0$, then the coexistence steady state $x^* = (x_1^*, x_2^*, \ldots, x_n^*)$, where $x_i^* > 0$ for $i = 1, 2, \ldots, n$, of the system is, if it exists, given also as

$$x^* = -A^{-1}b. \tag{A15}$$

Stability of the steady states. If any trajectory of a dynamical system starting at an initial state close enough to a steady state x^* stays close to it (i.e. for any $\varepsilon > 0$ there is some $\delta > 0$ such that, for any trajectory $x(t)$ of the system that satisfies $|x(0) - x^*| < \delta$, it holds that $|x(t) - x^*| < \varepsilon$ for $0 \leqslant t < \infty$), the steady state x^* is said to be (positively) *stable in the sense of Lyapunov*, or (*plus*) *Lyapunov stable*. If a steady state x^* is Lyapunov-stable and any trajectory $x(t)$ in a neighborhood of the steady state converges to the steady state as t goes to infinity ($|x(t) - x^*| \to 0$ as $t \to \infty$), x^* is said to be *asymptotically stable*. To study the local behavior around a steady state x^* of the dynamical system whose response functions are given by f, the linearization of the system around x^* is useful, which is given as

$$f(x) = A(x - x^*) + \mathbf{g}(x - x^*), \tag{A16}$$

where $A = (\partial f_i(x)/\partial x_j)$ and $\mathbf{g}(y) = 0(|y|)$ (i.e. $\mathbf{g}(y)/|y| \to 0$ as $|y| \to 0$). A criterion (sufficient condition) for the asymptotic stability of steady state x^* is given through this linearization; if the real parts (the absolute values) of the eigenvalues of matrix A, in the case of a continuous-time (discrete-time) dynamical system, are all negative (less than one), then steady state $\boldsymbol{x}^*$ is asymptotically stable. If the system represents an ecological system (community) consisting of n species whose populations (or biomasses etc.) are represented by x_i $(i = 1, 2, \ldots, n)$, then each element a_{ij} of matrix $\boldsymbol{A}$ in the linearization of the system around steady state $\boldsymbol{x}^*$ represents the *direct* interaction (impact) from species j to i, and matrix $\boldsymbol{A}$ is often in ecological literature referred to as a *community matrix* of the system at steady state $\boldsymbol{x}^*$. The interaction structure of the community can be illustrated by the sign digraph corresponding to its community matrix, just as in the case of the interaction matrix for a system of Lotka–Volterra type.

Example A8. (Linear function case) For the case of the linear functions $f(\boldsymbol{x}) = \boldsymbol{A}\boldsymbol{x} + \boldsymbol{b}$, the steady state $\boldsymbol{x}^* = -\boldsymbol{A}^{-1}\boldsymbol{b}$ is asymptotically stable if the real parts (the absolute values) of the eigenvalues of matrix $\boldsymbol{A}$ are all negative (less than one), in the case of a continuous-time (discrete-time) dynamical system.

Example A9. (Lotka–Volterra type) For the case of a system of Lotka–Volterra type (A14), the linearization of function $f(\boldsymbol{x})$ around steady state $\boldsymbol{x}^* (= -\boldsymbol{A}^{-1}\boldsymbol{b})$ is

$$f(\boldsymbol{x}) = \boldsymbol{A}\boldsymbol{D}[\boldsymbol{x}^*](\boldsymbol{x} - \boldsymbol{x}^*) + \boldsymbol{g}(\boldsymbol{x} - \boldsymbol{x}^*), \qquad \text{(A17)}$$

where $\boldsymbol{D}[\boldsymbol{x}^*]$ represents the diagonalized matrix of vector $\boldsymbol{x}^*$ (i.e. the diagonal matrix with the elements vector $\boldsymbol{x}^*$ as its diagonal elements). The steady state $\boldsymbol{x}^*$ is thus asymptotically stable if the real parts (the absolute values) of the eigenvalues of matrix $\boldsymbol{A}\boldsymbol{D}[\boldsymbol{x}^*]$ are all negative (less than one), in the case of a continuous-time (discrete-time) dynamical system. Note in ecological contexts the community matrix of this system at $\boldsymbol{x}^*$ is not $\boldsymbol{A}$, which is the interaction matrix of the system, but $\boldsymbol{A}\boldsymbol{D}[\boldsymbol{x}^*]$.

Sensitivity of steady states. Consider a dynamical system with n state variables $x_1, x_2, \ldots, x_n$ and m parameters $p_1, p_2, \ldots, p_m$ whose response functions f_i are given as

$$f_i(x_1, x_2, \ldots, x_n; p_1, p_2, \ldots, p_m), \qquad \text{(A18)}$$

$(i = 1, 2, \ldots, n)$. The steady state $\boldsymbol{x}^*(p_1, p_2, \ldots, p_m)$ determined for a set of parameter values $p_1, p_2, \ldots, p_m$ is given as the solution for the set of algebraic equations

$$f_i(x_1, x_2, \ldots, x_n; p_1, p_2, \ldots, p_m) = 0 \qquad \text{(A19)}$$

($i = 1, 2, \ldots, n$). By differentiating these equations with respect to p_k, we have

$$\sum_{j=1}^{n} \frac{\partial f_i}{\partial x_j} \frac{\partial x_j^*}{\partial p_k} + \frac{\partial f_i}{\partial p_k} = 0 \tag{A20}$$

for $i = 1, 2, \ldots, n$, that is,

$$\boldsymbol{A} \frac{\partial \boldsymbol{x}^*}{\partial p_k} = -\frac{\partial \boldsymbol{f}}{\partial p_k} \tag{A21}$$

where $\boldsymbol{A} = \left(\frac{\partial f_i(\boldsymbol{x}^*)}{\partial x_j}\right)$. The *sensitivity* of the steady state $\boldsymbol{x}^*(p_1, p_2, \ldots, p_m)$ with respect to the change in value of parameter p_k is thus given by

$$\frac{\partial \boldsymbol{x}^*}{\partial p_k} = -\boldsymbol{A}^{-1} \frac{\partial \boldsymbol{f}}{\partial p_k} \tag{A22}$$

if the inverse matrix $\boldsymbol{A}^{-1}$ exists. In ecological contexts, a change in value of a parameter p_k, which represents some external force affecting the growth f_i of each species, causes some change in steady state value $x_i{}^*$ of each species. Equation (A22) shows that in the evaluation of the total (direct and indirect) impacts from a change in parameter values to the population (or biomass etc.) levels of species, community matrix $\boldsymbol{A}$ is used again, but this time in the negative inverse form $-\boldsymbol{A}^{-1}$.

Example A10. (Linear function case) If b_i are considered as parameters in the case of linear functions f_i

$$f_i(x_1, x_2, \ldots, x_n) = \sum_{j=1}^{n} a_{ij} x_j + b_i \tag{A23}$$

and the steady state $\boldsymbol{x}^*$ is considered to be a function of $(b_1, b_2, \ldots, b_n)$, then the sensitivity of steady state $\boldsymbol{x}^*$ with respect to the change in values of b_i ($i = 1, 2, \ldots, n$) is given by

$$\frac{\partial \boldsymbol{x}^*}{\partial b_j} = -\boldsymbol{A}^{-1} \frac{\partial \boldsymbol{f}}{\partial b_j} = -\boldsymbol{A}^{-1} \begin{bmatrix} 0 \\ \vdots \\ 0 \\ 1 \\ 0 \\ \vdots \\ 0 \end{bmatrix} \begin{matrix} \\ \\ \\ \leftarrow j\text{th row} \\ \\ \\ \\ \end{matrix} \tag{A24a}$$

that is,

$$\left[\frac{\partial x_i{}^*}{\partial b_j}\right] = -\boldsymbol{A}^{-1} \tag{A24b}$$

Example A11. (Lotka–Volterra type) If b_i are considered as parameters in the case of the Lotka–Volterra type functions f_i

$$f_i(x_1, x_2, \ldots, x_n) = \left(\sum_{j=1}^{n} a_{ij}\boldsymbol{x}_j + b_i\right)x_i \tag{A25}$$

and the positive steady state $\boldsymbol{x}^*$ is considered to be a function of $(b_1, b_2, \ldots, b_n)$, then the sensitivity of steady state $\boldsymbol{x}^*$ with respect to the change in values of b_i $(i = 1, 2, \ldots, n)$ is given in terms of $[c_{ij}] \equiv -\boldsymbol{A}^{-1}$ as:

$$\left[\frac{\partial x_i{}^*}{\partial b_j}\right] = -(\boldsymbol{D}[x^*])^{-1}\boldsymbol{A}^{-1}\boldsymbol{D}[\boldsymbol{x}^*] = \left[\left(\frac{x_j^*}{x_i^*}\right)c_{ij}\right] \tag{A26}$$

AI.2 Mathematics of Shannon information theory

The purpose of this appendix is to provide a simple informal exposition on basic mathematics from the Shannon information theory. These notions were developed originally in the context of the mathematical theory of communication (Shannon, 1948) but have been found to be very general in their application (e.g. Ashby, 1965, 1969; Watanabe, 1969). They have been used in ecology and are used in the chapters of the present volume.

Measure of uncertainty. Shannon (1948) based his theory of information on the probabilistic description of uncertain situations. Suppose we are given a set of alternative outcomes $x_1, x_2, \ldots, x_n$ none of which is certain to occur and all of which are possible. Assume the likelihood of these possible outcomes is given by a probability distribution $p(1), p(2), \ldots, p(n)$, where $p(i) \geqslant 0$, $i = 1, 2, \ldots, n$, and $\sum_{i=1}^{n} p(i) = 1$. Is there a way to objectively measure how uncertain we are about the outcome?

The Shannon entropy. Shannon (1948) showed that the unique measure of uncertainty (i.e. the function of probability distributions that satisfies a certain set of properties which any reasonable measure of uncertainty should possess) is given by

$$H(\boldsymbol{x}) = -\sum_{i=1}^{n} p(j)\log p(j), \tag{A27}$$

where 'log' represents the logarithm to the base 2. To gain some idea about the properties of this function $H(\boldsymbol{x})$, take two extreme cases; first, consider the case in which $p(1) = 1$ and the other probabilities are all zero, i.e. among n alternative outcomes one particular case is certain to occur. Then,

$$H(\boldsymbol{x}) = -\log 1 = 0, \tag{A28}$$

indicating there is no uncertainty about the outcome. On the other hand, if all the alternative outcomes have an equal likelihood, i.e., $p(1) = p(2) = \ldots = p(n) = 1/n$, then

$$H(\boldsymbol{x}) = -\sum_{i=1}^{n} (1/n) \log(1/n) = \log n \tag{A29}$$

It can be shown that the function $H(\boldsymbol{x})$ takes the values between these two extreme cases:

$$0 \leqslant H(\boldsymbol{x}) \leqslant \log n \tag{A30}$$

for any probability distribution $p(1), p(2), \ldots, p(n)$; in particular, it attains its maximum with the equal probability (uniform) distribution $p(1) = p(2) = \ldots = p(n) = 1/n$, indicating that we have the most uncertainty about the outcome when all the alternative outcomes are equally likely to occur and cannot be discriminated in their likelihood. The function $H(\boldsymbol{x})$ decreases as the probability distribution $p(1), p(2), \ldots, p(n)$ becomes more uneven (less uniform) and more sharply distinguishes fewer cases out of the n alternatives, which implies that the outcome becomes more certain.

For example, in some communication system (e.g. telephony, telegraphy, oral speech, etc.) the alternative outcomes $x_1, x_2, \ldots, x_n$ may be a set of message symbols, one of which is chosen and transmitted through a 'channel' in some signal form. If the probability distribution for the choice of these symbols is given by $p(1), p(2), \ldots, p(n)$, then the uncertainty of this information source, i.e. the uncertainty of which symbol is actually chosen and transmitted, is measured by the Shannon entropy, Eq. (A27).

Degree of diversity. The concept of uncertainty captured by the Shannon entropy based on probabilistic notions involves a sense of the degree of *freedom in choice* or the degree of *diversity* (variety) in a system composition. In fact, the entropy function has been used in ecology mainly as a measure for the diversity of community composition.

Amount of information. The notion of *amount of information* is derived through that of degree of uncertainty. When the uncertainty about the outcome of an event is reduced, through an experiment, observation or some other process, we say that we gain information about the outcome.

The amount of information gained is measured as the reduction in the value of the uncertainty measure $H(\boldsymbol{x})$. Thus, for example, if one comes to know the actual outcome of an event $\boldsymbol{x}$ for which there were n alternative possible outcomes with likelihoods, $p(1), p(2), \ldots, p(n)$, after the event actually occurs and is observed, then the gain of information amounts to $H(\boldsymbol{x})\,[= H(\boldsymbol{x}) - \log 1]$.

Conditional entropy. Suppose now we have two events in question, $\boldsymbol{x}$ and $\boldsymbol{y}$, with n possible alternative outcomes for the former and m for the latter. Let $p(i,j)$ be the probability of the joint occurrence of x_i for $\boldsymbol{x}$ and y_j for $\boldsymbol{y}$. For example, in the communication system case $\boldsymbol{x}$ may be the message to be chosen by the transmitter, and $\boldsymbol{y}$ the signal to be received at the receiver. The entropy of the joint event $(\boldsymbol{x}, \boldsymbol{y})$

$$H(\boldsymbol{x}, \boldsymbol{y}) = -\sum_{i=1}^{n} \sum_{j=1}^{m} p(i,j) \log p(i,j) \tag{A31}$$

represents the uncertainty of which pair (x_i, y_j) out of $n \times m$ possibilities actually occurred.

The *conditional entropy of* $\boldsymbol{x}$ *given* y_j,

$$H_j(\boldsymbol{x}) = -\sum_{i=1}^{n} p_j(i) \log p_j(i), \tag{A32}$$

where $p_j(i)$ denotes the conditional probability of x_i given y_j, and $H_j(\boldsymbol{x})$ represents the uncertainty about the outcome for $\boldsymbol{x}$ given the occurrence of y_j for $\boldsymbol{y}$. The *conditional entropy of* $\boldsymbol{x}$ *given* $\boldsymbol{y}$ is defined as the average of these $H_j(\boldsymbol{x})$, i.e.,

$$\begin{aligned} H_y(\boldsymbol{x}) &= \sum_{j=1}^{m} p(j)\, H_j(\boldsymbol{x}) \\ &= -\sum_{j=1}^{m} \sum_{i=1}^{n} p(j)\, p_j(i) \log p_j(i) \\ &= -\sum_{i=1}^{n} \sum_{j=1}^{m} p(i,j) \log p_j(i) \end{aligned} \tag{A33}$$

where $p(j) = \sum_{i=1}^{m} p(i,j)$ represents the (marginal) probability of y_j. Eq. (A33) thus represents the average uncertainty regarding the outcome for $\boldsymbol{x}$ given the outcome for $\boldsymbol{y}$. In the communication system example, the conditional entropy $H_y(\boldsymbol{x})$ represents the uncertainty of which message $\boldsymbol{x}$

had been actually sent by the transmitter when we have received its corresponding signal, i.e. the uncertainty in inferring x from the knowledge of y; this is referred to as *equivocation*.

Mutual information. $H(x)$ and $H_y(x)$, both representing the uncertainty about the outcome for x, are proved (Shannon, 1948) to always satisfy the relation

$$H(x) \geqslant H_y(x). \tag{A34}$$

The difference, $H(x)-H_y(x)$, represents the reduction of uncertainty about (the outcome for) x by obtaining knowledge about (the outcome for) y; that is, it can be interpreted as the amount of information about x gained by the knowledge of y. The conditional entropy $H_y(x)$ can be rewritten (Shannon, 1948) as

$$H_y(x) = H(x,y)-H(y). \tag{A35}$$

This relation says that the uncertainty $H(x,y)$ about which pair (x_i, y_j) actually occurs can be considered in two steps: first, the uncertainty in choosing the outcome y_j, and second, the uncertainty in the choice of x_i given y_j. Thus, the previous difference can be written as

$$H(x)-H_y(x) = H(x)+H(y)-H(x,y). \tag{A36}$$

Fig. I.7. An illustration of the information conservation law represented by relation $H(x) = M(x,y)+H_y(x)$, where $H(x)$, $M(x,y)$ and $H_y(x)$, respectively, represent the amount of information contained in a message transmitted, that of information (on the original message x) transmitted through the channel to the receiver and preserved in the received signal, and that of information (on x) dissipated in transmission and missing in the received signal y.

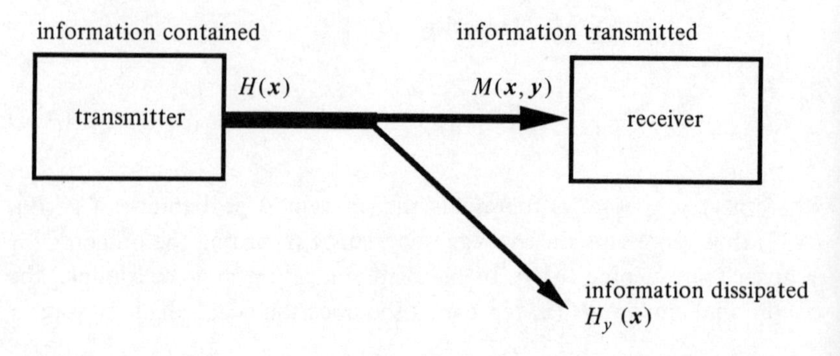

This quantity is symmetric with respect to x and y, as is obvious from the right-hand side of (A36) because $H(x,y) = H(y,x)$; thus, we have

$$\begin{aligned} H(x) - H_y(x) &= H(x) + H(y) - H(x,y) \\ &= H(y) + H(x) - H(y,x) \\ &= H(y) - H_x(y). \end{aligned} \tag{A37}$$

Therefore, the amount of information on x gained by the knowledge of y equals the amount of information on y gained by the knowledge of x. This equal amount of information is called the *mutual information* between x and y; let $M(x,y)$ be this mutual information.

In the communication system example, $M(x,y)$ represents the amount of information on the sent message x that is gained through the corresponding received signal y, i.e. the information amount (on the original message x) transmitted through the channel to the receiver; it is referred to as the *transmission rate* of the channel. The equivocation $H_y(x)$, i.e. the ambiguity in identifying (inferring) the specific message x actually sent from the knowledge of the received signal y, may be also regarded as the amount of information on x that is missing in the received signal y; for there is of course no uncertainty on the sent message x on the transmitter side, whereas the uncertainty on x amounts to $H_y(x)$ in the received signal. Thus, the equation, which is derived from the definition of $M(x,y)$,

$$H(x) = M(x,y) + H_y(x), \tag{A38}$$

may be regarded as the information conservation law, i.e.,

$$\begin{bmatrix} \text{information} \\ \text{contained} \\ \text{in the} \\ \text{message } x \\ \mathrm{H}(x) \end{bmatrix} = \begin{bmatrix} \text{information} \\ \text{transmitted} \\ \text{to the receiver} \\ \text{in the signal } y \\ \mathrm{M}(x,y) \end{bmatrix} + \begin{bmatrix} \text{information} \\ \text{dissipated in} \\ \text{transmission} \\ \text{due to noise} \\ \mathrm{H}_y(x) \end{bmatrix}$$

AI.3 Network model of an oyster-reef ecosystem

An energy-flow model of an intertidal oyster reef in South Carolina, USA (Fig. I.8) is used where needed as an example ecosystem network. We chose it because it is quantified and small, yet complex enough to represent generally the interactive or 'causal' structure of ecosystems. The model is described in detail elsewhere (Dame & Patten, 1981), so we give only a brief description here.

The oyster reef as modeled has six compartments: filter feeders (oysters and mussels), deposited detritus (particulate organic matter in sediments), microbiota (bacteria, yeasts and fungi), meiofauna (animals passing through a 1.0 mm sieve and retained on a 0.063 mm sieve), deposit feeders (macrofauna in sediments) and predators (mud crabs and other predators). Standing stock (biomass) of the compartments are given in the boxes of Fig. I.8 in terms of kcal m^{-2}. There is only one input into the system; 41.5 kcal m^{-2} d^{-1} (15137 kcal m^{-2} y^{-1}) of organic matter is consumed by the filter feeders. The 13 intercompartmental flows of energy (kcal m^{-2} d^{-1}) are feeding and egestion. Outputs exist from all compartments as respiration, export, or both. Exports are due to resuspension of sediments, or predation by transients. The oyster reef ecosystem as modeled is a cyclical network; 11 per cent of the energy moving through the system cycles. The model represents a steady state energy budget for the oyster reef based on an average annual water temperature of 20 °C. Ten years of observations supported the assumption of steady state.

The network serves as a basis for exemplifying and comparing analytical methods in four chapters. Higashi *et al.* (Chapter 5) use it to demonstrate

Fig. I.8. Intertidal oyster reef compartmental model, reprinted from Patten (1985). Numbers within rectangles indicate steady state standing crop energy storages (kcal m^{-2}), and those associated with arrows denote energy flows (kcal m^{-2} d^{-1}).

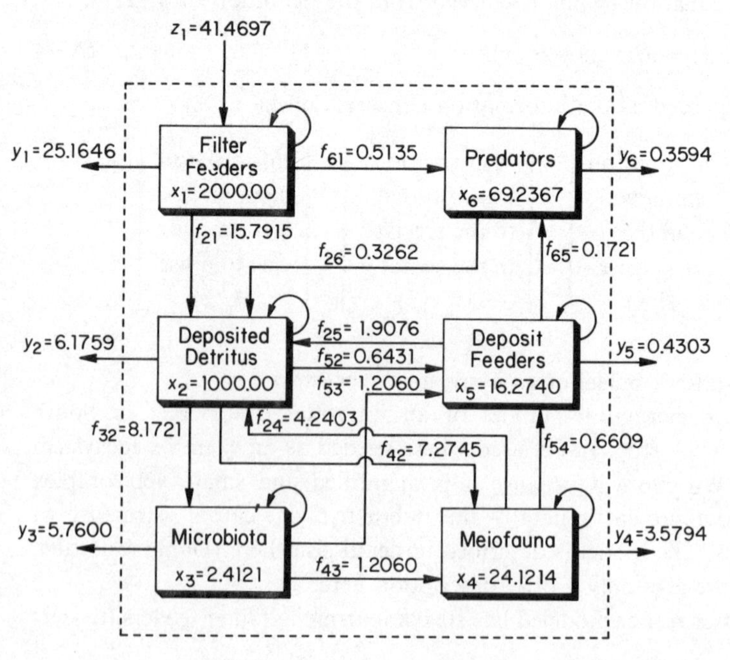

a new network analysis of trophic dynamics and structure. Mikulecky (Chapter 3) represents this energy-flow model as an electrical circuit, which can be simulated using SPICE, and Hannon and Bentsman (Chapter 9) adapt it to Hannon's flow-accounting format to demonstrate how their control theory can be implemented through manipulations of stocks and flows. Ulanowicz (Chapter 2) calculates the ascendancy (Ulanowicz, 1986) of the Fig. I.8 network, and there is sufficient information to calculate the exergy (Jørgensen, 1982) and Herendeen's proposed maximand (Chapter 10), if readers should care to do so.

References

Allen, T. F. H. & Starr, T. B. (1982). *Hierarchy: Perspectives for Ecological Complexity*. Chicago: University of Chicago Press.

Aoki, I. (1987). Entropy balance in Lake Biwa. *Ecol. Mod.*, **37**, 235–48.

Aoki, I. (1988). Entropy laws in ecological networks at steady state. *Ecol. Mod.*, **42**, 289–303.

Ashby, W. R. (1965). Measuring the internal information exchange in a system. *Cybernetica*, **1**, 5–22.

Ashby, W. R. (1969). Two tables of identities governing information flows within large systems. *ASC Comm.*, **1**, 3–8.

Barber, M. C. (1978). A Markovian model for ecosystem flow analysis. *Ecol. Mod.*, **5**, 193–206.

Bellman, R. (1953). *Stability Theory of Differential Equations*. New York: McGraw-Hill.

Birkhoff, G. (1967). *Lattice Theory*, 3rd edition. American Mathematical Society Colloquium Publications, Vol. 25.

Bollobas, B. (1979). *Graph Theory: An Introductory Course*. Berlin: Springer-Verlag.

Conrad, M. (1983). *Adaptability*. New York: Plenum Press.

Cooper, L. H. N. (1937). The nitrogen cycle in the sea. *J. Mar. Biol. Assoc. United Kingdom*, **22**, 183–204.

Dame, R. F. & Patten, B. C. (1981). Analysis of energy flows in an intertidal oyster reef. *Mar. Ecol. Prog. Ser.*, **5**, 115–24.

Darwin, C. E. (1859). *On the Origin of Species by Means of Natural Selection, or the Preservation of Favoured Races in the Struggle for Life*. London: John Murray.

DeAngelis, D. L. (1975). Stability and connectance in food web models. *Ecology*, **56**, 238–43.

Feller, W. (1968). *An Introduction to Probability Theory and Its Applications*. Vol. I, 3rd edition. New York: John Wiley.

Finn, J. T. (1976). Measures of ecosystem structure and function derived from analysis of flows. *J. Theor. Biol.*, **56**, 363–80.

Forbes, S. A. (1887). The lake as a microcosm. *Bull. Sc. A. Peoria*. Reprinted in *Ill. Nat. Hist. Surv. Bull.*, **15**, 537–50.

Gardner, M. R. & Ashby, W. R. (1970). Connectance of large dynamical (cybernetic) systems. Critical values for stability. *Nature*, **228**, 784.

Hannon, B. (1973). The structure of ecosystems. *J. Theor. Biol.*, **41**, 535–46.

Herendeen, R. (1981). Energy intensities in ecological and economic systems. *J. Theor. Biol.*, **91**, 607–20.

Higashi, M. (1986*a*). Extended input–output flow analysis for ecosystems. *Ecol. Mod.*, **32**, 137–47.

Higashi, M. (1986*b*). Residence time in constant compartmental ecosystems. *Ecol. Mod.*, **32**, 243–50.

Higashi, M. & Patten, B. C. (1986). Further aspects of the analysis of indirect effects in ecosystems. *Ecol. Mod.*, **31**, 69–77.

Higashi, M. & Patten, B. C. (1989). Dominance of indirect causality in ecosystems. *Am. Nat.*, **133**, 288–302.

Hutchinson, G. E. (1944). Nitrogen and biogeochemistry of the atmosphere. *Am. Sci.*, **32**, 178–95.

Hutchinson, G. E. (1948). Circular causal systems in ecology. *Annals New York Acad. Sci.*, **50**, 221–4.

Hutchinson, G. E. & Bowen, V. T. (1948). A direct demonstration of the phosphorous cycle in a small lake. *Proc. Natl. Acad. Sci. USA*, **33**, 148–53.

Jørgensen, S. E. (1982). Energy and buffering capacity in ecological systems. In *Energetics and Systems*, ed. W. Mitsch, R. Ragade, R. Bosserman & J. Dillon, pp. 61–72. Ann Arbor: Ann Arbor Science.

Kemeny, J. C. & Snell, J. G. (1960). *Finite Markov Chains*. New York: D. Van Nostrand.

Kerfoot, W. C. & Sih, A. (1988). *Predation: Direct and Indirect Impacts on Aquatic Communities*. Hanover: University Press of New England.

Lawton, J. (1987). Problems of scale in ecology. *Nature*, **325**, 206.

Lefshetz, S. (1957). *Differential Equations: Geometric Theory*. New York: Wiley-Interscience.

Leontief, W. W. (1966). *Input–Output Economics*. London/New York: Oxford University Press.

Levine, S. H. (1976). Competitive interactions in ecosystems. *Am. Nat.*, **110**, 903–10.

Levins, R. (1975). Evolution in communities near equilibrium. In *Ecology and Evolution of Communities*, ed. M. L. Cody and J. Diamond, pp. 16–50, Cambridge, MA: Belknap Press.

Lincoln, R. J., Boxshall, G. A. & Clark, P. F. (1982). *A Dictionary of Ecology, Evolution and Systematics*. Cambridge: Cambridge University Press.

Lindeman, R. L. (1941). Seasonal food-cycle dynamics in a senescent lake. *Am. Midl. Nat.*, **26**, 636–73.

Lotka, A. J. (1925). *Elements of Physical Biology* (Edition of 1956, under title *Elements of Mathematical Biology*). New York: Dover Publ., Inc.

Margalef, R. (1963). On certain unifying principles in ecology. *Am. Nat.*, **97**, 357–74.

Margalef, R. (1968). *Perspectives on Ecology*. Chicago: University of Chicago Press.

Margalef, R. (1983). *Limnologia*. Barcelona: Ed. Omega.

Matis, J. H. & Patten, B. C. (1981). Environ analysis of linear compartmental systems: the static, time invariant case. *Bull. Int. Stat. Inst.*, **48**, 527–65.

May, R. (1972). Will a large complex system be stable? *Nature*, **238**, 413–14.

May, R. (1973). *Stability and Complexity in Model Ecosystems.* Princeton: Princeton University Press.

Odum, E. P. (1968). Energy flow in ecosystems: a historical review. *Am. Zool.*, **8**, 11–18.

Odum, H. T. (1960). Ecological potential and analogue circuits for the ecosystem. *Am. Sci.*, **48**, 1–8.

Odum, H. T. (1971). An energy circuit language for ecological and social systems, its physical basis. In *Systems Analysis and Simulation in Ecology*, vol. 2, ed. B. C. Patten, pp. 139–211. New York: Academic Press.

O'Neill, R. V., DeAngelis, D. L., Waide, J. B. & Allen, T. F. H. (1986). *A Hierarchical Concept of Ecosystems.* Princeton: Princeton University Press.

Patten, B. C. (1982). Environs: relativistic elementary particles for ecology. *Am. Nat.*, **119**, 179–219.

Patten, B. C. (1985). Energy cycling, length of food chains, and direct versus indirect effects in ecosystems. In *Ecosystem Theory for Biological Oceanography*, ed. R. E. Ulanowicz & T. Platt, *Can. Bull. Fish. Aquat. Sci.*, **213**, 119–38.

Patten, B. C., Bosserman, R. W., Finn, J. T. & Cale, W. G. (1976). Propagation of cause in ecosystems. In *Systems Analysis and Simulation in Ecology*, vol. 4, ed. B. C. Patten, pp. 457–579. New York: Academic Press.

Patten, B. C. & Higashi, M. (1984). Modified cycling index for ecological applications. *Ecol. Mod.*, **25**, 69–83.

Patten, B. C. & Witkamp, M. (1967). Systems analysis of ^{134}cesium kinetics in terrestrial microcosms. *Ecology*, **48**, 813–24.

Pearl, R. (1927). The growth of populations. *Quart. Rev. Biol.*, **2**, 532–48.

Preparata, F. P. & Yeh, R. T. (1973). *Introduction to Discrete Structures.* Reading: Addison-Wesley.

Puccia, C. J. & Levins, R. (1985). *Qualitative Modeling of Complex Systems: An Introduction to Loop Analysis and Time Averaging.* Cambridge, MA: Harvard University Press.

Rawson, D. S. (1930). The bottom fauna of Lake Simcoe and its role in the ecology of the lake. *Univ. of Toronto Stud., Publ. Ontario Fish Res. Lab.*, **40**, 183.

Riley, G. A. (1944). The carbon metabolism and photosynthetic efficiency of the earth. *Am. Sci.*, **32**, 132–4.

Shannon, C. E. (1948). The mathematical theory of communication. *The Bell Syst. Tech. J.*, **27**, 379–423, 623–56.

Shelford, V. E. (1913). Animal communities in temperate America as illustrated in the Chicago region. *Bull. Geogr. Soc. Chicago* **5**, Reprinted (1977). New York: Arno Press.

Simon, H. A. (1957). *Models of Man.* New York: John Wiley.

Simon, H. A. (1962). The architecture of complexity. *Proc. Amer. Phil. Soc.*, **106**, 467–82.

Tansky, M. (1976). Structure, stability, and efficiency of ecosystem. *Prog. in Theor. Biol.*, **4**, 205–62.

Tansley, A. G. (1935). The use and abuse of vegetational concepts and terms. *Ecology*, **16**, 284–307.

Thienemann, A. (1926). Der nahrungkeislauf in wasser. *Verh. deutsch. Zool. Ges.*, **31**, 29–79.

Ulanowicz, R. E. (1983). Identifying the structure of cycling in ecosystems. *Math. Biosci.*, **65**, 219–37.

Ulanowicz, R. E. (1986). *Growth and Development: Ecosystems Phenomenology.* New York: Springer-Verlag.

Vandermeer, J. (1980). Indirect Mutualism: Variations on a theme by Stephen Levine. *Am. Nat.*, **116**, 441–8.

Verhulst, P. F. (1938). Notice sur la loi que la population suit dans son accroissement. *Corresp. Math. Phys.*, **10**, 113–21.

Watanabe, S. (1969). *Knowing and Guessing*. New York: John Wiley.

Wright, S. (1969). *Evolution and the Genetics of Populations*. Chicago: University of Chicago Press.

Part I

Perspectives on the network approach to ecosystems

1

Networks in ecology

RAMON MARGALEF

1.1 A thousand kinds of ecological nets

The concepts of conductor, channel, and tree, and their combination into webs or nets, which act as pathways in an organized system of transportation of materials, energy, and information, could have originated from a consideration of the circulation of blood, or the combined functions of support and transportation in vascular plants. The notion of flows external to organisms, like the hydrological cycle and the cycles of elements in the biosphere as they are increasingly understood in this century, together with a consideration of marine currents and river flow, introduce a second type of net which is less precise and much less controlled by life. Water, gases, and a number of particulate and dissolved substances travel along paths which both diverge and converge. Changes in speed along these paths lead to dilution or concentration. Concentration causes a temporary sequestering of material from the flow: material stored in the sediments, or in biologically less active compartments such as the high altitude atmosphere. Ecological science must study both and eventually combine them.

Most of the energy that supports this flow comes from the sun; only a fraction of it comes indirectly through organisms and their activities. Nevertheless, living organisms are both dependent on flows and cycles, and effective in moving and directing them. Transport mediated by plants, from roots to leaves and back, has inspired a dynamical picture of the ecosystem. Movement in animals and their feeding activities complement the same dynamic perspective, and have introduced the vivid image of discrete connections or links between stations or controlling points in a transport system, i.e. the food chain. Although grounded on pre-ecological conventional knowledge ('large fish eat small fish', 'a flea hath smaller fleas that on him prey, and these have smaller fleas'), the web of trophic relations extended among animals has grown to be one of the most effective page fillers in recent ecological theory.

Ecology and ecologists follow the flows of matter and energy through pathways that continue, externally to the body, the metabolic pathways of organisms. Fluxes are driven by forces, so the comparative study of metabolism soon led to considerations of efficiency, optimality, and thermodynamics (Lotka, 1925; Lindeman, 1942; Odum & Pinkerton, 1955; H. T. Odum, 1960, 1983), as well as to discussions of the observed stability and persistence of ecological networks. As an aside, an emphasis on energy helped introduce the debatable practice of converting inputs and outputs from grams of organic carbon, and the like, into calories, joules, or watts. A number of similarities with electrical and electronic circuitry – life has, in general, a semi-conductor quality – provided supplementary impetus for the development of mathematical models of ecosystems, and led to the easy acceptance that computers were well designed to simulate ecosystems.

It is debatable that large input–output models were thought of first by economists or by ecologists, but a cross-fertilization led to compartment models with a network of fluxes among a number of boxes. These models may be too fixed, too conventional, for the taste of the naturalist. When looking to economic models for inspiration, countercurrent flows of goods (matter and energy) and money introduce themselves. The money cycle and the energy flow have been associated by H. T. Odum in a way that is often convincing; this writer when pushed for analogies tends to identify money with information. The ratio money/goods is a price, and its differences along a path generate a flow.

The concept of network has been imposed on ecology from outside and 'sold' as a necessary reference to construct models. Is it worthwhile, therefore, to spend time on networks? Perhaps only to learn not to care about nets that are too simple. Maybe just to have some fun; the biosphere is actually a very complicated and strange network. Paths linking the dominant and large species can be visualized, and the ratio between costs of procuring goods and the value of the goods inspires a large fraction of modern ecological theory. But the network, when passing from animals to plants, ends with the network inside organisms, in very small capillaries (roots, hyphae) resting over a large store of dead organic matter more or less resistant to decay. The geometry imposed on the network is consistent – a small number of trophic levels along the vertical and an indefinite extension of the network over the horizontal plane, with the ancillary problems of defining connections for individuals, consolidating them for species or other selected sets, and drawing this over a background of bacteria and fungi.

We can imagine the functional structure of the biosphere as an aggregate of approximately vertical fibers, more or less laterally linked and matted. The horizontal distance between them is an effective device for isolation or for limitation of exchanges. This situation reintroduces, through the discussion of networks, the old problem in ecology of drawing boundaries between functional units (biocoenoses). A small island, or a lake could be modeled reasonably as an entity or a system. But, a large surface covered by a relatively uniform vegetation gave origin to the concept of zonal biocoenoses, with neighboring points obviously interacting and with an intensity of interaction diluted with distance. The consideration of ecological networks poses the same problem again, perhaps in the following form. How is it possible to idealize and then to consolidate actual nets in the form of paths between individuals, then between species, between trophic levels, etc., without disregarding in an unacceptable way the fact that each one of the entities, boxes, or pools, extends over a large space?

1.2 How real, needed, and helpful are networks?

Networks are used as a background for most exercises in ecological modeling. The network describes the paths followed by matter, single chemical elements, or energy. They are drawn for a selected problem, and sometimes to satisfy a selected expression. The segments of the paths start and end usually in organisms or groups of organisms defined in an appropriate way. Not only organisms can be added together into larger categories, but also organisms plus volumes of the environment, or volumes of the environment considered as pools of definite components. In expressions of the form $dN_i/dt = (b_i + \sum a_{ij} N_j) N_i$, where N_i is the number of individuals of type i, the interactions are stated and flows are supposed, but space is forgotten. We are left with the disquieting image of millions of animals dancing on the head of a pin. Actual exchange happens between individuals occupying defined positions in space, which for the occasion can be close, but not always. Interactions are not simple, so one should take into account the (statistical) distribution of the instantaneous values of the distances between all pairs of potential reactants relative, of course, to their motility.

Flows and exchanges happen between individuals, but if all the individuals of one kind behave in a similar way, it is possible to aggregate them (Ulanowicz, 1983, 1984; Hirata & Ulanowicz, 1984). Statistical properties of the networks can be followed through a process of aggregation, which does not need to stop at species, but can proceed to larger ecological entities (e.g. phytoplankton, zooplankton, meiobenthos, detritus) in an attempt to produce a model that is both simple and satisfactory. But the result is a highly idealized net; real flows are much more criss-crossed. In forgetting the very dense real web of relations, we are apt to lose essential qualities or properties, like diffusion or turbulence, attached to the real physical background of all exchanges. In considering information, there is no aggregation loss when elements to be aggregated have only one inflow from a common element and one outflow to a common element, that is when they are parallel (Hirata, 1990). If relationships do not change with aggregation, both are robust.

Forces driving fluxes are based on organisms, and appear more active in animals or phagotrophic organisms than in primary producers and osmotrophes, those organisms gaining their nutrition through diffusion (e.g. bacteria and fungi). The degree of connectivity among individuals and species, the fact that flow links are also feedbacks, and the variety of organisms involved in the fabric of the ecosystem might lead to considerations of the relations between available organisms, organization of fluxes, and degrees of persistence of the systems – I do not believe in equilibrium points, because the very nature of ecosystems is to shift continuously (see below). In a network, redistribution of fluxes over the network might be associated with the waxing and waning of populations. Obviously, at any time, fluxes must comply with a number of constraints, like the rules that apply to electrical circuitry, which often has provided useful inspiration.

Odum (1960) and others have proposed electrical circuits as analogues of the network organization of the ecosystem. It is more than an analogy, since the flow of electrons characterize ecosystem networks and lead naturally to a useful discussion of the organization of lakes (Margalef, 1983) or of any ecosystem. It can be added, continuing the comparison for didactic purposes, that the nodes or junctions in networks (organisms) are not simple resistors, rather semiconductors in which the current depends on and influences internal structure. The analogies for biomass, production, and memory in life are charge (amperes), intensity (volts), and magnetism. An image that might be inspiring has information and memory expanding orthogonally to the main flow. The pulsating character of life and the combination of alternating current (AC) with induction devices suggests, perhaps, the convenience of using AC for electrical analogues of ecosystems, and making use of induction devices, like the ferrite memories of the 1950s. The constraints associated with multiple coils winding around an iron may prove realistic in the exploration of constraints in the connectivity of ecological networks. Being wary of going astray with overextended analogies, electrical models can be relegated to a didactic and ancillary purpose.

Odum (1971) developed a symbolic representation of ecosystems and ecosystem function in terms of energy fluxes that makes considerable sense, although it may need some complementary description. The main network is embedded in another network of flows of energy and materials (e.g. air, water) around the organisms, which is essential for the support of life and must be included in any picture of the biosphere. To the external flows are associated forces of two kinds: (1) those dependent on external or exosomatic energy that is doing work in the fluid machine of the atmosphere and hydrosphere, and (2) those generated by the organisms, such as movement by animals. From my point of view, external energy (rainfall, upwelling, plowing, etc.) acts as a throttle or regulator of the flow of energy involved in primary production, and also at other levels, for instance when it facilitates the accumulation of potentially reactive organisms. It goes without saying that there might be a cascade of switches acting as multipliers: muscular work that lets fossil fuel energize earth-moving machines, that help build a dam, that changes availability of water, that increases primary productivity over a given land area. Odum introduces the concept of embodied energy, which is practically equivalent to this idea; according to him (Odum, 1983) the energy of one type that generates a flow of another can be accounted as embodied in the second flow.

Nobody doubts that flows in ecosystems are discontinuous, but the implications, in modeling generally, are rarely taken into account. Plants assimilate in the daytime, and animals have discontinuous contacts within the species (sexual contacts) or between species (predators chasing prey), but persistent connections do exist in the case of parasites and symbionts. Discontinuities in function (think of the circadian and seasonal rhythms) have developed together with the capacity for storage and a lengthening of the life span with the result that life bridges time and internalizes and assimilates short term environmental fluctuations. The anticipation of change, as well as the intermittent contacts among animals, may enhance stability not only by spreading the influence of an animal over many others, but also, more theoretically, through a restriction of the connectivity inside the system.

The analysis of the distinct food links involving animals has fired attempts at generalization and typification in the frame of evolutionary theory and natural selection with an emphasis on subjects like efficiency, foraging strategies, trade-offs between different requirements, and especially, the distinction between *r*-, *k*-, and other evolutionary strategies. This has filled many pages in ecology journals during the recent decades. As often happens, an insight produces first a classification, then an ordination. The desirable stage is the identification of a generating mechanism built around definite relations between turnover, individual size, and amount and quality of the information stored and used.

The next steps – carried out simultaneously – are the combination of food links into food chains and then food webs. This can be approached in different ways. One very general approach is to assemble many species at random and explore their dynamics, whether they collapse or simplify – optimally as believed – around a simple food-chain. Some function computable on the properties of the pertinent set could exist and act as a guide. Selection at the level of the ecosystem is still an anathema to most biologists, and in proposing food-chain theory, separate optimizations for each link are usually considered (Pimm, 1982, 1984). Nevertheless, some regularities could be gathered and even tentatively explored from the comparative study of a number of food chains, both in the laboratory and in nature. For instance, the number of links in a chain of species related as predator and prey for each successive pair, usually does not exceed five. But this sequence can be continued upwards by five more links with a chain of parasites. The definition of parasites should emphasize less the 'nasty' character than the fact that the relations of size and turnover are inverted in the parasite chain, in comparison with a 'normal' predator–prey chain. Parasites that are closer to the top, in what concerns the conduction of matter and energy up from the primary producers, have shorter lives and faster turnovers than organisms closer to the source of energy. They can also be permanently connected to their source of food.

The current belief that ordinary food chains are the most needed pieces for constructing ecological models should be deflated or, at least, moderated through the recognition that most energy involved in the ecosystem network (around 80%) flows through bacteria and fungi, and goes through a reservoir of organic matter of different degrees of persistence, from the easily metabolizable compounds to highly recalcitrant molecules (e.g. cellulose, lignin, polyphenols, fats, ectocrine messengers). Two facts emerge, animals have conquered their place against a background of osmotrophs, and many lifeless and quasi-lifeless materials play an important role in the organization of the ecosystem network.

Plants offer the extremely interesting example of networks that are not completely external to the organisms, although external to the cells, and not as internalized as in animals. Shoots and roots in flowering plants are support and conduction systems made of hard, durable materials. They develop differently from place to place depending on local solicitations. They are adaptive and 'cultural', and they can be compared to the highways and communication networks of human civilization. There are gradations, of course. In fungi, the living part of the organism is a very effective transporter over long distances. Macrophytes are in an intermediate position; wood loses biological activity with time, remains as a transportation link in the ecological network and later, as an organizer of space only. Social insects and human societies share with plants the capacity of constructing an external transport system, which provides a template for ecological networks, not only for the species principally responsible, but also for many more organisms. This is another complication to consider when establishing a protocol for the study of ecological networks.

We have thus far run into several problems: the possible aggregation of the components of a network by their affinities into species, or trophic levels, or by any other acceptable criteria; the disregard for real complication and criss-crossing of real paths; the convenience of overlooking the intermittent nature of the interactions and averaging flows, with the result that they appear steady; and forgetting that organisms are embedded in a fluid matrix which may be important in the organization and regulation of flows.

1.3 Back to the beginning: the complementary aspects of ecological networks

Two different approaches to the study of networks are possible. They are complementary and should be considered simultaneously, although it is usual practice to prefer one and give only scant consideration to the other.

First, there is the analysis of flow organization. Here one tries to determine the network inside a volume of fluid, for example the flow of sea water around an island-rich or fjord-rich coast, by condensing the flow around a number of trajectories, which will have a number of junctions. This can be compared conceptually to the flow of energy bound to organic carbon through a poorly structured or artificially simplified food web. The networks extend in a three-dimensional space, and allow for temporary storage in places. The model assumes continuity, and allows for the ordinary relations between flows, pressures, resistances, storages and the more generalized association in the distribution of forces and fluxes. From the point of view of ecological models, the sum of the inputs to a point i ($\sum f_{ji}$; $j = 0, 1, \ldots, n$) less the outputs from the same point i ($\sum f_{ij}$; $j = 0, 1, \ldots, n$) is equal to storage in i (X_i) plus the dissipation from i (D_i). That is:

$$\sum_{j=0}^{n} f_{ji} - \sum_{j=0}^{n} f_{ij} = X_i + D_i. \tag{1}$$

This provides a convenient starting point for much theory on networks (Ulanowicz, 1983, 1984; Hirata & Ulanowicz, 1984; Hirata, 1990). A synoptic appreciation of the composition of the net can lead to the aggregation of the components in a gradual way, and to the development of associated concepts that have to do with macroscopic properties of the whole system, such as ascendency (Ulanowicz, 1986). This approach has been the most common background for ecological modeling, where the contents of the compartments were considered as black boxes, or to have fixed properties.

The second approach is to develop an interest in the nodes or junctions of the network, where centers of interaction that operate with information (e.g. organisms, neurons) might be as important or more important than the pathways. Both individuals and neurons are information gathering and preserving centers subjected to historical change, and possessing directive or decision making power, which often depends on very complex responses to specified or selected patterns of external stimuli extended over large segments of the system.

The information processing function can appear more important than the organismic functions of supporting material life, although it is dependent on it. Here the discontinuity of ecosystems is most important; compare the actual biosphere with an hypothetical biosphere made of a continuous film around the planet. The most important result of the work by Ramon y Cajal on the nervous system has been the recognition of the individuality of the neuron. Proper understanding of behavior can only be based in the discrete properties of the subjacent mechanism. My unhappiness with the models proposed by Patten (1982, 1985) about indirect effects and the possibilities that recycling offers to the ecosystem, arise from my conviction that 'knots' or nodes – the organisms in the network – are terminal stations: beginnings and ends to all flow and all action, where effects stop and begin again anew. If each step is independent, are indirect effects dubious?

As a criticism against present day trends in modeling, especially concerning aquatic ecosystems, it has to be stressed that organisms are not only carriers of matter and energy, but also subjects of natural selection. This involves a capacity for decision making which is not incorporated in the usual models that specify only flows. Organisms act also as temporary reservoirs of a mix of materials. In the usual formulation of population change:

$$dN_i/dt = \left(b_i + \sum_{j=1}^{n} a_{ij} N_j\right) N_i, \tag{2}$$

if the change is measured in energy, this change has to be equal to or less than X_i (Eq (1)) and perhaps of the same order or less than D_i.

If the far-from-equilibrium thermodynamics of open systems applies, a trend of lower and lower D_i should be predicted, and the increase in entropy has to be evaluated against preserved information, which is proportional to biomass and special properties of biomass. In ecological terms this means a slower turnover, larger size, and better brains (Margalef, 1980; Matsuno, 1978; Tansky, 1976). Speed of evolution can be associated with the flow of energy and the corresponding possibilities of working with information. However, natural selection is free, because the flow happens in the maintenance of the system anyway; it works with deaths which are inevitable.

It is urgent to achieve a satisfactory understanding of the relations between the complementary aspects of networks, centered respectively on the links and the nodes, blending both in a common model. It is unfortunate that consideration of networks as links, or else nodes, junction points, or centers of interaction have been used as independent approaches, each one sufficient in itself. Not surprisingly, the need for a complementary approach is urgent when we face the reality that networks are evolving structures, which change deeply in the relations between links and nodes.

In my opinion, thinking in terms of networks is just one, more or less appropriate, way to look at nature and does not qualify as a scientific theory, as it can neither be proved nor disproved. But inside the network approach, special theories can be formulated and discussed in the frame of the same networks. We can discuss, as an example, the possibility for a steady state in a system. In a network representation, the hypothesis that energy exchanged pays for the accreted information leads to the acceptance that the center of gravity around which the entropy increases is closer to the primary producers, because entropy is produced with energy exchanged and work done locally. This center of gravity must be more or less distant, in the organization of the ecosystem, from the center of gravity around which information is gained in the system. This information may be ecological, in terms of species richness and interactions, genetic or cultural. The decay of energy and the increase of information is an historical process, and each transaction generates the scenario for the next move. As in all biological function and natural selection, a continuous shift is to be expected, and in consequence we might conclude that the persistence of the structure of a net is unlikely. This is ecological theory. If it is correct, the consequence may be doom for many models that have been proposed.

1.4 Connectivity in networks

A network consists of a number of links (L) and of nodes or junctions (S). A net requires continuity extended to all components. The minimum number of links is $L = S-1$; below that the network is cut into two smaller nets. The maximum number of links is $L = S(S-1)/2$. More generally, $L = \alpha S^k$, where k is between 1 and 2. These considerations are helpful in an aggregated network, in which nodes stand for species, and links stand for actual interactions existing for each pair of species. It is expected that k is somewhere between 1 and 2. In any hierarchical structure inside a system (e.g. individuals linked as species, species in trophic levels, or in any other convenient subsystem), we may conjecture that structures are inside other structures in an homogeneous way and that missing links are evenly distributed. Perhaps we can describe such persistence of style or pattern in the different levels as a fractal property, and quantify it, for instance, through the fraction of missing or interrupted links at each level. The repetition of the pattern at different scales could be a justification for making reference to fractals (Mandelbrot, 1983). But it also may happen that no uniformity prevails and that the values for k may be lower in the 'peripheric' areas of very specialized and strongly dependent organisms (e.g. parasites, parasitoids).

Comparison with other networks might be helpful. In a fisherman's net, $L = 2S$, in a tri-dimensional lattice $L = 3S$, with $k = 1$ always. In a more 'organized' net k is larger, where organization is size dependent, like electronic circuitry and ecological networks. But k should be less than 2, so far as these are 'flexible' systems and not short-circuited.

Relative connectivity, or connectance, C, can be defined as:

$$C = \sum_{i \neq j} a_{ij} p_i p_j / \sum_{i \neq j} p_i p_j, \qquad (3)$$

where p_i is the probability associated with the numerical abundance of each species ($\sum p_i = 1;\ i = 1, \ldots, n$) and a_{ij} are the probabilities of interaction when i and j meet. Since the expression $D = \sum p_i p_j / \sum p_i^2$ is suitable for diversity, connectivity is contained inside diversity. Diversity sets a limit to the maximum connectivity in a system. The product $C \times S$, where S is the number of species, is also limited, and several authors have suggested that this product falls in the range between 2 and 12, and in most cases remains around 4 (Gardner & Ashby, 1970; DeAngelis, 1975; DeAngelis *et al.*, 1978; Yodzis, 1980; McNaughton, 1978; Rejmanek & Stary, 1979; Cohen & Briand, 1984; Briand, 1985; see also Lane & Collins, 1985). Better precision is achieved if one writes $C \times S^E$, or $C \times 2^H$, in which $E = H/\log_2 S$ is evenness, and H is diversity (Shannon's entropy). The product $C \times S^E$ is related to k and has to remain between 2 and 4. In a large number of electric circuits, it was around 3.5 (Margalef & Gutierrez, 1983). The expression 2^k is valued between 2 and 4, and is 3.5 for $k = 1.8$.

The limits and regularities that have been observed in connectivity are not surprising, and are a consequence of the actual values which tend to fall around the center of the available range. However relatively narrow the range, differences can be significant. Actual values of connectivity may be relevant for the functionality of the system, and as an expression of the internal stresses and interactions that configure the ecosystem. This poses problems and the first one is what links should be counted. Fish nets have strings, and electronic circuits have wires; obviously the contacts and exchanges of matter and energy in organisms (among individuals, species, or 'groups', according to the degree of aggregation) also qualify as links. But what about exchange of signals (information) and how do we 'count' the indirect links that Patten recognizes?

The subject is inspiring in relation to neglected fields in ecology, and no dogmatic statements should be accepted. As suggested before (in the use of a_{ij}), degree of connection or interaction between two elements is not necessarily 0 or 1, rather any value between them. Probably, 'true' links have the quality of feedback cycles arising through natural selection. The same happens in economic networks. They work, of course, in an asymmetrical manner. Between prey and predator, matter and energy flow together with information, but more information flows from predator to prey. This information is responsible for the evolution of the prey. The asymmetry in evolution – the prey tends to evolve properties of immediate defensive utility, the predator tends to evolve the capacity to learn and in general to manipulate more information – is grounded in the actual properties of the network, and comparable statements can apply to any network, for instance to nervous systems and social systems. If the systems are hierarchical systems (Allen & Starr, 1982; Salthe, 1985), development as a hierarchy requires cutting some of the links; another plea to recognize the partial connectivity that goes with an organization that is internally flexible, and not a rigid object. Convention has it that everything in an ecosystem is connected to everything else: an assertion that makes little sense.

The subject of the limitation and of the organization of the links inside networks can be discussed with profit by considering triangular relations or ternary systems. This leads us back to traditional ecology. Links among species can be characterized as feedback cycles: p = positive, n = negative, or non-existent. I am aware that others can think of transfer of matter and energy without feedback, but I am not so sure if this actually happens in things related to life and organisms. A standard situation, involving competition, is the ternary system n–n–p; in standard ecological theory, the system should cut out the positive link, with the net loss of one link in three. Other possibilities, like p–p–p, n–n–n, and n–p–p are inherently unstable (this reminds me of the construction of syllogisms). We have here a reasoning and a frame in which to understand the intermediate values that connectivity attains. A number of strings should remain, and remain loose.

1.5 Anisotropy in dynamic and historic networks

I come to perhaps the only testable hypothesis that can be hatched out of these considerations about ecosystem networks, and it is one that may play against the general views. It runs against a trend opposed rather vociferously to giving much importance to the hypothetical strictures of natural selection (Strong *et al.*, 1984). If I disagree with the blanket application of the theory of natural selection in standard ecological theory, it is because of the systematic disregard of further constraints based on history and thermodynamics. If we assume that disturbances and the time that has been available for organization qualify the workings of natural selection, then I could agree that a generous measure of randomness can enter into the apparent organization of ecosystems, but not always in the same amount. This depends on the time and the resetting of ecological succession, a function of the local spectrum of disturbance. And it depends also on the location or region in an anisotropic network.

Anisotropic means 'exhibiting properties with different values when measured along axes in different directions' or 'assuming different positions in response to external stimuli' (Webster's New Collegiate

Dictionary) and I want to apply this quality to an ecosystem, or to an ecosystem network, to mean that the whole expanse of the net is not homogeneous in relation to properties concerning matter, energy, and information, and that the whole dynamics of the ecosystem works toward increasing the differences, until the whole system is reset (catastrophically) to a state comparable to a long passed one. Resetting follows a sudden input, usually in the form of external energy, unpredictable to the actual organization of the system and its components.

This means that network organization has something to do with succession, and implies that essentials of succession theory have less to do with 'orderly' development of ecosystems than with the basic asymmetry: fast change toward simplicity and slow change toward complexity. Traditional ecology is too flat, like laminar flow; we have to inject the analogue of vorticity.

The conjecture about anisotropy can be bounded inside the two limits afforded by single links and whole networks. In what concerns single links, remember the comments on the relations between predator and prey. All the energy available to this binary system goes through the prey, and from its perspective, dissipation is absolutely most important. The quality of information, which accretes through natural selection and also through cultural mechanisms, is heterogeneous: more genetic information for the prey, a larger fraction of cultural information for the predator. In addition, top organisms may have learning to learn, or deuterolearning in the parlance of G. Bateson. Individual size, life span, turnover, allocation of energy to maintenance, growth, and other needs appear subjected to constraints and driven toward expected forms by selection and evolution. Over a whole network of relations in the ecosystem, the 'center of gravity' in the amount of exchanged energy – and increase of entropy – is closer to the places of external input into the system, that is, closer to the primary producers. If we are allowed to speak of a 'center of gravity' for the increase and preservation of information in all forms, from DNA to trees, anthills, soil, and towns (not excluding wood and books), it is obviously located around the large animals, especially large predators with big brains.

It should be possible to find the same polarity in the nature of the junction between links and nodes, and this might be of great help, as it would result in a network theory free from the schizophrenia of dealing either with links or with nodes. Links are more important as carriers of energy, conductors without memory while subjected to noise. Nodes or junctions work like semiconductors and support, accrete, and reorganize information. Links may be a constraint necessarily imposed by space, but evolution surely can make them very short, as in the case of symbionts and parasites, in a way that has to be evaluated from new points of view. Certainly, the strategies of evolution are not uniformly enhanced in the different domains of the ecosystem network.

One might feel rather skeptical about the direct utility of networks in ecology: To produce beautiful sketches? To program computers? Checking them against nature suggests that actual networks are transient (ecosystems, while persistent, are changing), and thus cannot be a serious basis for very general models. Weather states not only move over the Earth, they change and evolve as well. No machine can turn twice remaining identical. A new dimension of change has to be introduced in the ecological models, that until now have been too flat. Dynamics of ecosystems cannot be understood as results of fixed interactions between persistently 'well behaved' components, generating regular cycles. All this behavior can be traced down to properties of the system of nodes and links, and of networks generally.

1.6 Summary

At least two different approaches are possible in the study of ecosystem networks: (1) The analysis of the organization of the flows. Here one starts with a volume of 'fluid' subjected to flow and eddies, and then takes out a known volume, leaving a network of flows. Examples are the circulation of water along a coast with many islands and fjords, and the flow of carbon bond energy through a poorly structured food web. A number of conditions concerning flows, pressures and resistances must be met. (2) Development of interest in the junctions, knots, or nodes of nets, where centers of interaction (e.g. neurons, organisms) may be more important than pathways. Organisms are not simply carriers of atoms, but also subjects of natural selection. Interaction stops and starts anew at each station, node, or reservoir, so that through them indirect actions across the net are minimized.

Ecology must combine the two complementary aspects: flows and nodes. Flows of matter and energy going through an organized system (in a node or knot of the net) experience a decay in energy which is partially matched by an increase of information (or organization). In consequence, the properties of the net shift or evolve. The nodes evolve faster, and the links can be reorganized after that.

Ecosystems and languages are internally flexible systems made of subsystems (organisms, words) that are replicable. Particular relations among these components fall into two groups: the more generalized relations between classes of components (organization of food webs, syntax), leading to concepts like style and organization, and secondly, the more particularized relations among individual components giving a unique picture (particular organization or particular constraints, 'what is said' in language). A number of constraints are related to space (two or three dimensions in organization, forced one dimensional chain in language).

A consideration of the ecosystem in terms of networks, applying both complementary points of view, leads to the conclusion that the ecosystem network is anisotropic: the 'center of gravity' of the function of energy dissipation is closer to the region of the primary producers, and the 'center of gravity' of the function of potential accumulation of information is closer to the region of the top carnivores. A careful analysis of each link of the prey–predator type provides useful hints to explain the anisotropy of the entire net. It explains also the selection for large size, long life, learning ability, and, in man at least, the ability to learn to learn, leading to reflection and cultural evolution.

A more detailed analysis of the network's links as feedback cycles leads to a classification of theoretical ternary systems into those altogether unstable and others that are dynamic and temporarily persistant (e.g. competition communities). This is the reason why ecosystem networks never can be totally connected and rigid. More or less for every three possible links, one is cut. There is also a lower limit to connectivity; that is, when the whole net is cut in two. In such areas, interesting considerations can be made in relation with diversity and with an allometric or fractal approach to the disposition of links. Of course, to cut potential links is a contribution toward the development of a hierarchy, and here disposition of the links rejoins the subject of the lack of coincidence between the places where energy is dissipated, and where information becomes apparent.

References

Allen, T. F. H. & Starr, T. B. (1982). *Hierarchy. Perspectives for Ecological Complexity*. Chicago and London: University of Chicago Press.

Briand, F. (1985). Structural regularities in freshwater food-webs. *Verh. intern. Verien Limnol.*, **22**, 3356–64.

Cohen, J. E. & Briand, F. (1984). Trophic links of community food webs. *Proc. Natl. Acad. Sci., USA*, **81**, 4105–9.

DeAngelis, D. L. (1975). Stability and connectance in food web models. *Ecology*, **56**, 238–43.

DeAngelis, D. L., Gardner, R. H., Mankin, J. B., Post, W. M. & Carney, J. H. (1978). Energy flow and the number of trophic levels in ecological communities. *Nature*, **273**, 406–7.

Gardner, W. R. & Ashby, W. R. (1970). Connectance of large dynamic (cybernetic) systems. Critical values for stability. *Nature*, **228**, 784.

Hirata, H. & Ulanowicz, R. E. (1984). Information theoretical analysis of ecological networks. *Int. J. Systems Sci.*, **15**, 261–70.

Hirata, H. (1990). Information theory and ecological networks. In *Complex Ecology: The Part-Whole Relation in Ecosystems*, ed. B. C. Patten & S. E. Jørgensen. The Hague: SPB Academic Publishing. (In press.)

Lane, P. A. & Collins, T. M. (1985). Food web models of a marine plankton community: an experimental mesocosm approach. *J. Exp. Mar. Biol. Ecol.*, **94**, 41–70.

Lindeman, R. L. (1942). The trophic dynamic aspect of ecology. *Ecology*, **23**, 399–418.

Lotka, A. J. (1925). *Elements of Physical Biology* (Edition of 1956, under title Elements of Mathematical Biology). New York: Dover Publ. Inc.

Mandelbrot, B. B. (1983). *The Fractal Geometry of Nature*. New York: W. H. Freeman & Co.

Margalef, R. (1980). *La Biosfera entre la termodinamica y el juego*. Barcelona: Ed. Omega.

Margalef, R. (1983). *Limnologia*. Barcelona: Ed. Omega.

Margalef, R. & Gutierrez, E. (1983). How to introduce connectance in the frame of an expression for diversity. *Am. Nat.*, **121**, 601–7.

Matsuno, K. (1978). Evolution of dissipative system: A theoretical basis of Margalef's principle on ecosystem. *J. Theor. Biol.*, **70**, 23–31.

McNaughton, S. J. (1978). Stability and diversity of ecological communities. *Nature*, **274**, 251–3.

Odum, H. T. (1960). Ecological potential and analogue circuits for the ecosystem. *Am. Sci.*, **48**, 1–8.

Odum, H. T. (1971). An energy circuit language for ecological and social systems, its physical basis. In *Systems Analysis and Simulation in Ecology*, vol. 2, ed. B. Patten, pp. 139–211. New York: Academic Press.

Odum, H. T. (1983). *Systems Ecology: An Introduction*. New York: John Wiley & Sons.

Odum, H. T. & Pinkerton, R. C. (1955). Time's speed regulator: the optimum efficiency for maximum power output in physical and biological systems. *Am. Sci.*, **43**, 321–43.

Patten, B. C. (1982). On the quantitative dominance of indirect effects in ecosystems. In *Analysis of Ecological Systems*, ed. W. K. Laueroth, G. V. Skogerboe & M. Flug, pp. 27–37. Amsterdam: Elsevier.

Patten, B. C. (1985). Energy cycling, length of food chains, and direct versus indirect effects in ecosystems. *Canad. Bull. Fish. Aquat. Sci.*, **213**, 119–38.

Pimm, S. L. (1982). *Food Webs*. London: Chapman and Hall.

Pimm, S. L. (1984). The complexity and stability of ecosystems. *Nature*, **307**, 321–6.

Rejmanek, M. & Stary, P. (1979). Connectance in real biotic communities and critical values for stability of model ecosystems. *Nature*, **280**, 311–13.
Salthe, S. N. (1985). *Evolving Hierarchical Systems. Their Structure and Representation.* New York: Columbia University Press.
Strong, D. R., Simberloff, D., Abele, L. G. & Thistle, A. B. (eds.) (1984). *Ecological Communities: Conceptual issues and the Evidence.* Princeton: Princeton University Press.
Tansky, M. (1976). Structure, stability and efficiency of ecosystem. *Prog. Theor. Biol.*, **4**, 205–62.
Ulanowicz, R. (1983). Identifying the structure of cycling in ecosystems. *Math. Biosci.*, **65**, 219–37.
Ulanowicz, R. (1984). Community measures of marine food networks and their possible applications. In *Flows of Energy and Materials in Marine Ecosystems*, ed. J. R. Fasham, pp. 23–47. New York: Plenum.
Ulanowicz, R. (1986). *Growth and Development.* New York: Springer-Verlag.
Yodzis, P. (1978). Environmental randomness and the tenacity of equilibria. *J. Theor. Biol.*, **62**, 185–9.
Yodzis, P. (1980). The connectance of real systems. *Nature*, **284**, 544–5.

2

Formal agency in ecosystem development

ROBERT E. ULANOWICZ

2.1 Introduction

Perhaps the greatest reservation that most scientists and philosophers have against the concept of holistic or 'top-down' causality is the apparent absence of a robust exposition of just how such influence might arise. True, those interested in hierarchy theory have made great strides of late in clarifying the role of higher-level constraints in generating order at the finer scales (Allen & Starr, 1982). But the idea of a constraint usually conjures up a mental image of a wall or a boundary – rather passive elements in the scheme of things. How can a wall ever hope to compete as an agent of cause against the exciting imagery evoked by the story of magical dancing molecular genes at work *directing* the assembly of the mature phenome?

If top-down causality is to achieve any credibility, it is necessary that its proponents point to a plausible *agent* behind such influence. As I have done elsewhere (Ulanowicz, 1986), I wish to focus here upon the role of positive feedback as an agent that helps to order events at the micro levels. What I wish to achieve in this narrative is to enhance the case for positive feedback by more accurately mapping its domain in the Aristotelian scheme of causality, by better enumerating its attributes, and by demonstrating that it affects both the extensive and intensive properties of flow networks.

2.2 Aristotelian causality

One of the unfortunate aspects about much of biological discourse is that the concept of causality is often used in generic fashion without any mention of the truly intricate nature of causal linkage. Aristotle was quite aware that cause was not a simple notion and defined four categories to describe the phenomenon: (1) material, (2) efficient, (3) formal and (4) final. Moreover, a single event can have more than one type of cause. The familiar example is the building of a house. The material cause lies in the bricks, mortar and wood used in the construction; the efficient cause is embodied in the laborers who assembled the materials; the formal cause is usually considered to derive from the architectural principles or blueprints according to which the house was put together; and the final cause was the desire for a house on the part of those who ordered the building, or more generally, the need for housing in the area where the residence was built.

For most of its prosperous existence, science has been confined to the quantification of material and the elucidation of efficient causes. However, in order to fully apprehend the nature of biological order, it appears necessary to entertain the possibility of and to attempt to quantify formal, final and non-proximate causes (Rosen, 1985; Patten *et al.*, 1976).

Recently, Salthe (unpubl. ms) has offered the very provocative insight that the appearance of efficient, formal and final causes appears to be correlated to the hierarchical context in which they are being observed. Thus, those agents of cause observed at finer scales are more likely to be classified as efficient, whereas final cause (if it is considered at all) usually derives from outside the system in question, i.e. at some higher level. Formal cause, as its name implies, is intimately tied up with the structure of the system itself and thus is a feature of the focal level. One should not attempt to draw the trichotomy too strictly, but the distinctions will prove useful in the discussion of agency presented in the following sections.

2.3 Material and efficient causality

Networks of material and energy flows are particularly well-suited to the exploration of causality in ecology (Platt *et al.*, 1981). Each arc is the palpable result of temporal sequences of similar arcs in the past, and it contributes to the material cause of like arcs at future times. Whence, the flow network portrays the aggregate of material causes in its underlying ecosystem. Furthermore, the topology of the network permits the quantitative exploration of non-proximate material causes. When Hannon (1973), Patten *et al.* (1976) and others employ linear algebra to assess the indirect or non-proximate causes for ecological events, they are pioneering a long-overdue form of scientific inquiry.

Unfortunately, flow networks are often ill-suited to the elaboration of efficient causality. As mentioned earlier, efficient causes lie predominantly at finer scales, i.e. within the nodes of the flow graph. Thus, genetic and behavioral mechanisms, and adaptive 'strategies' do not always lend themselves to depiction in terms of networks – at least as networks cast at the ecosystems level. However, environmental perturbations constitute one class of efficient causality that can be represented in the network format. In any event, the description of these efficient causes is the crux of most conventional biology and ecology. There is no dearth of activity nor paucity of discoveries in this realm. The chief goal here is to demonstrate how the study of flow networks constitutes a complementary endeavor that allows one to incorporate less familiar aspects of causality into ecology and biology.

2.4 Positive feedback as a formal agent

Of the four causes cited in the familiar example of constructing a house, perhaps the least explicit is the formal cause. Whereas one is content with the other three kinds, one is impulsively driven to look beyond the blueprint toward the architect. Why? It appears the blueprint alone does not possess sufficient *agency*. It does not generate any action on its own. Its power is entirely derivative of the architect. A true agent should be autonomous to some degree of the causes which engender it.

That the blueprint is a weak excuse for the formal cause of a house is more an inadequacy of the example (house building) than it is a flaw in the underlying concept (formal cause). Formal cause is an essential element of practically all biological phenomena. It often happens in biology that the present *form* of a system mediates what will subsequently appear and transpire. Hence, Weiss (1969) cites the intriguing example of early blastular formation. During the first few cellular divisions, when the diameter of the cell is not greatly exceeded by the blastular diameter, practically all the component cells communicate equally with the external environment. Not too long into the process of growth, however, geometrical relations dictate that a fundamental dichotomy appears between 'internal' and 'external' cell members, which distinction may induce corresponding differences in cell morphology and function.

A situation that is more pertinent to ecology occurs during succession, where the juxtaposition of the component processes and their rates may result in a change in the environment (the build up of detritus or waste products, a change in light regime, etc.) that sets the stage for the next system configuration.

The difficulty with these two examples is that the source of agency is not very explicit. In the example by Weiss, the emerging geometrical constraint appears almost as a passive element of the system.

Positive feedback is an example of a formal cause whose agency is much more apparent. In its simplest form positive feedback occurs when the activity of a given element increases the activity of one or more other elements that in turn increase the activity of the original element still more. Positive feedback is usually represented graphically in the form of a unidirectional closed cycle of influence as in Fig. 2.1. Biogeochemical cycles of material or energy in ecosystems are examples of positive feedback configurations of material causality (Ulanowicz, 1983). Hence, positive feedback is explicitly formal in character.

Positive feedback is imbued with at least five attributes, all of which contribute toward its inherent agency. It is (a) semi-autonomous, (b) emergent, (c) growth enhancing, (d) selective and (e) competitive.

The autonomous element in positive feedback is evident in the example shown in Fig. 2.1. At the focal level the cycle has no external cause. All causality, direct or indirect, originates within the system. Of course, no real system can correspond exactly to Fig. 2.1, for then it would violate logical (Goedel's hypothesis) and thermodynamic (second law) constraints. Real cycles communicate with the external world as in Fig. 2.2. But it likewise would be a mistake to label the system in Fig. 2.2 as entirely non-autonomous. A measurable fraction of each component's material causality is seen to originate in itself. Real systems with feedback are partially autonomous.

The appearance of cyclical causality in a system is contingent upon the position of the focal level in the hierarchical scheme. Suppose, for example, that at a given level of observation one sees only a subset of the elements in a particular cycle, as illustrated by the solid boundary in Fig. 2.3. The pathway of causality connecting this subset is strictly non-

Fig. 2.1. A closed, autonomous cycle of influence.

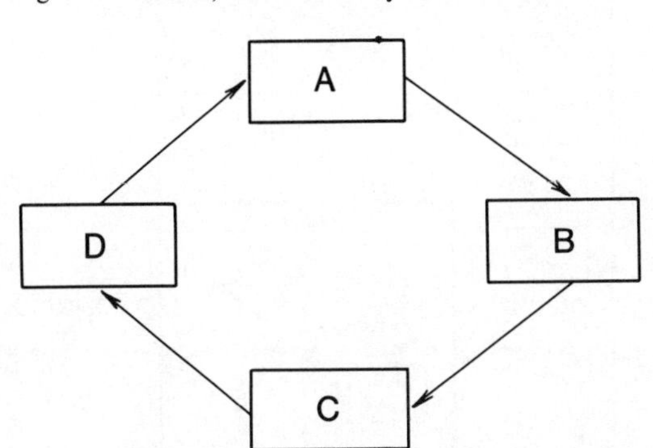

autonomous. However, if one expands the domain of observation (the dotted boundary in Fig. 2.3), then one might become aware of a degree of system autonomy associated with the newly perceived cycle that has *emerged* from increasing the scale of observation. Autonomous behavior becomes more apparent as one increases the field of observation.

That positive feedback is growth-enhancing is virtually tautological. In the absence of overwhelming constraints, an increase in activity anywhere in the cycle serves to engender greater activity everywhere else in the loop.

Fig. 2.2. A semi-autonomous cycle of influence. (Units are arbitrary.)

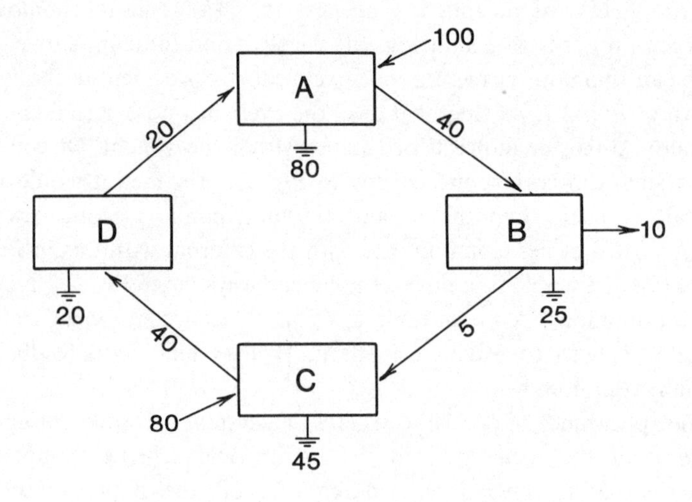

Fig. 2.3. An autonomous cycle emerges from an enlargement of the scope of observation from the solid square to the dotted.

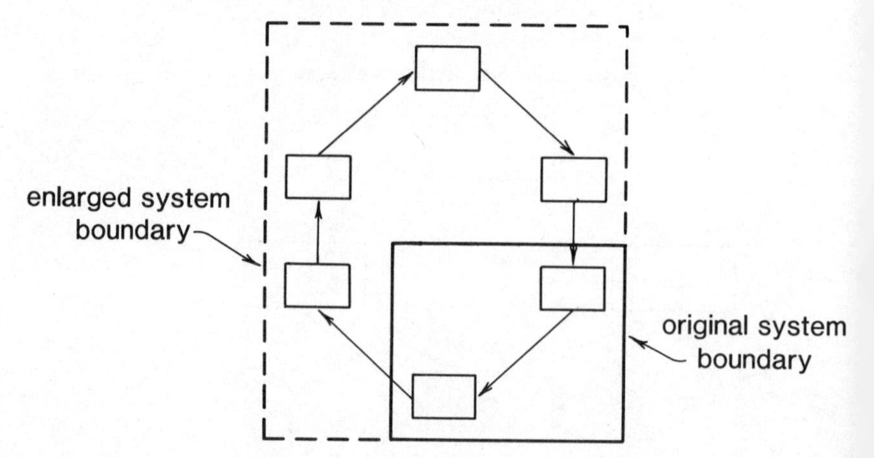

The activity level of the cycle is progressively elevated until it is restrained in some way from further increase.

Most important as regards network development is the possibility that positive feedback can serve as an agent of selection. To see how selection might arise one only needs to observe what happens when a perturbation changes the activity of any compartment in a cycle. If the change diminishes the outputs of the given node, then the negative result will propagate around the cycle upon itself. Conversely, if the change is incremental, it will be reflected positively upon itself (rewarded). Hence, by its very nature, positive feedback discriminates among the perturbations occurring in its cycle. The persistence of the characteristics of component elements are directly influenced by the feedback structure in which they occur.

The selection pressure engendered by feedback acts not only to change the features of the components, but also may help to alter the list of participating components. To better grasp this possibility, one may imagine that through some mechanism, stochastic or otherwise, a new element enters the system in Fig. 2.2 giving rise to the new configuration in Fig. 2.4. The new species, E, is seen to be more efficient at conveying a small amount, ε, of material cause from A to C. The pathway through E is progressively rewarded, and, if the whole system is acting near its limits (as it eventually must), the continued growth of the pathway A–E–C will occur at the expense of the activity of B. After a while B is displaced

Fig. 2.4. The cycle in Fig. 2 acquires a new element, E, that is more efficient than B at transporting medium from A to C.

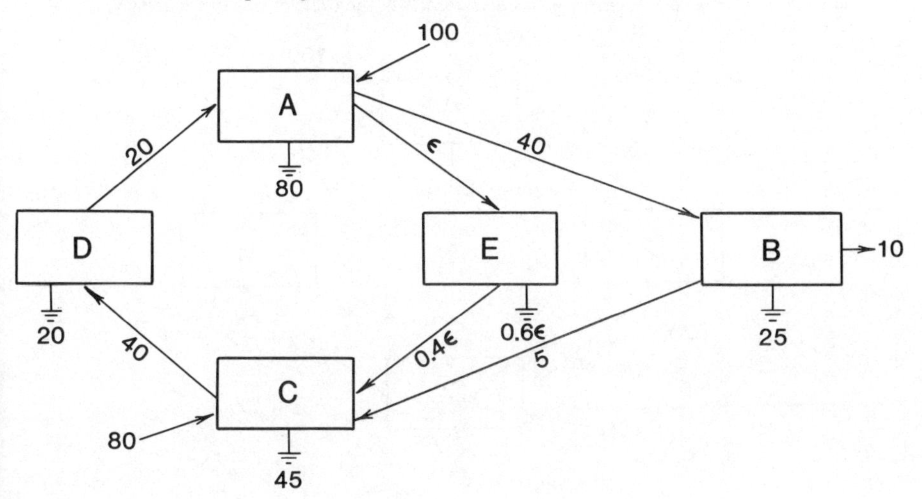

by E in the cycle, as shown in Fig. 2.5. It is possible to imagine that ultimately all of the original components are replaced by more efficient members, so that an identifiable structure may persist beyond the lifetimes of its constituents, all the while playing an active role in guiding its eventual make-up.

Selection in the face of limitations inevitably results in competition, such as occurred between B and E in Fig. 2.4. One also can expect competition between feedback loops whenever common elements appear in both circuits. But direct overlap of the cycles is not absolutely necessary. For example, a single resource can (directly or indirectly) contribute to more than one positive feedback loop, setting the stage for competition between non-overlapping cycles. Thus, each of the various feedback loops is seen to take on its separate agency and appears as a center into which material or energy is inexorably drawn, much in the fashion of Denbigh's 'chemical imperialism'.

The emerging picture of positive feedback as a formal agent is both a dynamic and an intriguing one. This single formal process has both extensive and intensive consequences. Its growth enhancing characteristics impel the system toward ever greater levels of flow activity, and the aggregate level of flow in a system is a common measure of a system's 'size' or extent (e.g. the GNP, or Gross National Product in economics). But flow is not being enhanced uniformly in the network. Rather, a greater proportion is being more narrowly channeled along those feedback pathways of higher transfer efficiencies. In the absence of any mechanisms

Fig. 2.5. Element E has displaced B in the cycle of positive feedback.

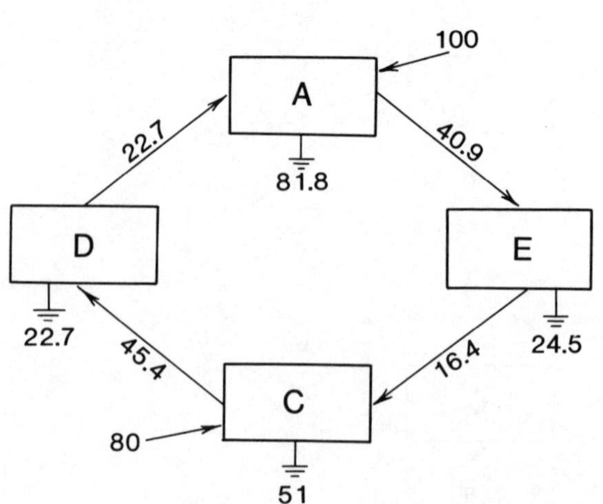

generating new components and/or pathways (which are always present in non-senescing systems), the evolving network topology would appear increasingly more simplified, or better articulated. This progressive articulation depends only on the ratios of the various flows and is an intensive feature of the development of the system.

Thus, the growth and development of a system network are seen to be separate outward manifestations of an underlying unitary process.

2.5 Quantifying the effects of formal agency

The foregoing examples serve as heuristic tools for understanding formal agency. But in order to make such agency more amenable to objective scrutiny, it becomes necessary to outline how one might measure the effects of positive feedback in ecological and other living systems. The network format serves this purpose admirably. As remarked earlier, Patten (1982) has used network representations of ecosystem kinetics to assay non-proximate causality in ecosystems. This same format allows one to trace quantitatively the effects of formal and (as will be discussed in the next section) final agencies.

To demonstrate how to calculate measures of size and organization in ecological networks, I will employ the structure of the oyster reef community shown in Fig. I.8. However, it should be stressed that this configuration is only an instantaneous snapshot of the ecosystem, so that one is limited to calculating the static attributes of size and organization. Growth and development, on the other hand, are temporal increases in these static measures, and to quantify these dynamic processes requires that the network be fully described at two or more distinct times.

Earlier, system size was assumed to be the total amount of flow activity in the system. Most simply put, the total system throughput of the oyster reef community is the sum of the magnitudes of all the arrows depicted in Fig. I.8. This amounts to 125.05 kcal m^{-2} day^{-1}.

The organization of the oyster community is not as simply determined. As a prelude, it is helpful to redefine slightly the flows as labeled in Fig. I.8, such that $f_{i0} = Z_i$ and $f_{n+1,i} = Y_i$. Then the total systems throughput, T, may be written more conveniently as

$$T = \sum_{i=0}^{n} \sum_{j=1}^{n+1} f_{ji}.$$

All other things being equal, a strongly organized community is highly articulated in the sense discussed in the last section. Knowing where a quantum of energy or mass is located at any time in a highly articulated system yields a great amount of information as to where the same quantum will be after its next transition. When averaged over all the components in the system, the amount of such information, A, may be quantified as

$$A = K \sum_{i=1}^{n+1} \sum_{j=0}^{n} (f_{ij}/T) \log \left[\frac{f_{ij} T}{\left(\sum_{k=1}^{n+1} f_{kj} \right) \left(\sum_{m=0}^{n} f_{im} \right)} \right].$$

This complicated looking formula is a version of the well-known average mutual information adapted for network theory by Rutledge *et al.* (1976) and Hirata & Ulanowicz (1984). The quantity K is the arbitrary scalar factor which is inherent in all information variables.

The interested reader should verify for him/herself that A is largest when the network for which it is being calculated is maximally articulated, as in Fig. 2.1. Conversely, A is identically zero when no articulation is evident, i.e. when each compartment exchanges equal amounts of medium with all compartments (and with itself). All real networks have values of A intermediate to these extremes.

Growth and development are purported to be separate manifestations of a single underlying agent, positive feedback. Consonant with this observation is the assumption that size and organization are co-factors in a single system attribute. Now, the scalar factor, K, in the formula for mutual information may be chosen at the discretion of the observer. Its dimensions are usually set by the base of the logarithmic operator. Tribus & McIrvine (1971), however, urge that K be chosen so as to impart actual physical dimensions to the information being measured. What better choice is there than to set $K = T$, making size and organization literal and numeric co-factors in a single network attribute, called the ascendency? A rise in ascendency, therefore, represents an increase in system size or organization (or both), i.e. growth and development. As positive feedback is a formal cause of growth and development, the results of its agency will be reflected in the changes in network ascendency.

The ascendency of the oyster reef network is calculated by substituting the various flows into the formula for A, setting $K = T$ and the logarithmic base to 2. The result is that $A = 166.35$ kcal-bits m^{-2} day^{-1}.

2.6 Quantifying final cause

Previously, it had been difficult to conceive of an example of final cause in physical or biological terms, much less to hope to quantify its agency. However, if one accepts the suggestion that the notion of final cause is a consequence of the hierarchical view of nature, assigning a

Table 2.1. *Components of the ascendency of the oyster reef community (Fig. I.8).*

	0	1	2	3	4	5	6	7
0	.0	.0	.0	.0	.0	.0	.0	.0
1	66.0368	.0	.0	.0	.0	.0	.0	.0
2	.0	17.3193	.0	.0	6.3166	3.9940	0.4626	.0
3	.0	.0	20.3457	.0	.0	.0	.0	.0
4	.0	.0	16.5011	1.3529	.0	.0	.0	.0
5	.0	.0	0.3377	3.4712	1.2934	.0	.0	.0
6	.0	0.6036	.0	.0	.0	0.6272	.0	.0
7	.0	21.9374	−1.5914	6.2657	1.2455	−0.4096	0.2374	.0

Values in kcal-bits m^{-2} day^{-1}. Component in row i and column j was generated by the flow from j to i. The 0 represents external inputs; the 7, the combined exports and respirations.

number to such top-down influence might not be an impossible task. Toward this end, it is noted that one may rewrite the formula for ascendency as the double sum of $(n+2)^2$ terms,

$$A = \sum_{i=1}^{n+1} \sum_{j=0}^{n} f_{ij} \log \left[\frac{f_{ij} T}{\left(\sum_{k=1}^{n+1} f_{kj} \right) \left(\sum_{m=0}^{n} f_{im} \right)} \right].$$

The 49 terms generated by the oyster reef network are arrayed in Table 2.1. One sees that corresponding to each non-zero f_{ij} is a logarithmic term that is a function of the configuration of the *entire* network. When the medium in circulation is energy (as it is in the oyster reef network) and the base of the logarithms is 2, then A has the dimensions of power-bits. Traditionally, power functions in irreversible thermodynamics are written as the sums of products of conjugate pairs of fluxes and forces (Onsager, 1931). If this analogy is applied to the terms in the last formula, one sees that the logarithmic terms correspond to the forces. That is, the logarithmic terms might be said to express the whole-system level pressure (formal cause at the level of the system, final cause at the level of the flow) upon their corresponding flows. These 'factors' may be positive or negative (although A is always non-negative, the particular components of A may be negative) and the degree to which the flows do *not* respond to their conjugate forces is an indication of the severity of the constraints to which the system is subject.

As an example, the contribution to A from f_{54} is read from Table 2.1 to be 1.2934 kcal-bits m^{-2} day^{-1}. Dividing this value by $f_{54} = 0.6609$ kcal m^{-2} day^{-1}, assigns a value to the conjugate 'force' of 1.957 bits. Because the system is in steady state, the tendency of this system-level pressure to further increase f_{54} is being balanced by unwritten constraints, such as energy limitations, mass balance, environmental limits, perturbations, etc.

This analogy of a system-level force apparent in the network is highly speculative and tenuous at best. However, it was drawn to illustrate how it might be possible to measure the effect of formal cause operating at the system level when manifested as a final cause of events (flows) at a finer scale. Elsewhere (Ulanowicz, 1986), I show how influence from even higher level configurations upon the given system is quantified via another network information variable called the 'tribute'.

2.7 Conclusions

Practically all the progress associated with the last 300 years of the scientific age has been made by elucidating material, efficient and proximate causes. Formal, final and non-proximate causes have not figured much in the body of scientific narrative. The idea that such causes exist has been considered fanciful by many; that they might be measured has, to the best of my knowledge, not received serious consideration.

Portraying complex evolving systems in terms of their constitutive networks is thereby seen as an enormous aid to viewing new perspectives on reality. With some input from hierarchy and information theories, network analysis now allows one to quantitatively track the effects of formal, final and non-proximate causal agencies that, for all practical purposes, have been heretofore neglected.

What progress as has been made thus far in elucidating these non-Newtonian forms of causality has been largely phenomenological in nature. But such has been the case with most beginnings in the history of science. There is every reason to hope that further research into the network analysis of living systems will lead to radically clearer insights into the life process.

2.8 Summary

1. Formal cause is associated with the structure of ecological systems.
2. Positive feedback, represented as closed cycles in ecological networks, helps order events at the hierarchical level below. Positive feedback is an agent of formal causality.
3. Positive feedback is imbued with at least five attributes, all of which contribute to its inherent agency; it is semi-autonomous, emergent, growth enhancing, selective and competitive.

4 Growth and development are separate manifestations of a single underlying agent, positive feedback.
5 Flow networks serve for the measurement of formal cause operating at the system level and manifested as a final cause of events at a lower level.

Acknowledgements

The author would like to thank Stanley Salthe for an enriching exchange of letters on hierarchy and causality. Helpful comments on the draft manuscript from Thomas Burns, Ron Pilette and Bruce Hannon are also gratefully acknowledged. Travel and publication preparation were underwritten by the US Office of Naval Research, grant number N00024-86-C-5188.

References

Allen, T. F. H. & Starr, T. B. (1982). *Hierarchy*. Chicago: University of Chicago Press.

Hannon, B. (1973). The structure of ecosystems. *J. Theor. Biol.*, **41**, 535–46.

Hirata, H. & Ulanowicz, R. (1984). Information theoretical analysis of ecological networks. *Int. J. Systems Sci.*, **5**, 261–73.

Onsager, L. (1931). Reciprocal relations in irreversible processes. *Phys. Rev.*, **37**, 405–26.

Patten, B. C. (1982). On the quantitative dominance of indirect effects in ecosystems. In *Analysis of Ecological Systems: State of the Art in Ecological Modelling*, ed. W. K. Lauenroth, G. V. Skogerboe & M. Flug, pp. 27–37. Amsterdam: Elsevier.

Patten, B. C., Bosserman, R. W., Finn, J. T. & Cale, W. G. (1976). Propagation of cause in ecosystems. In *Systems analysis and simulation in ecology*, vol. 4, ed. B. G. Patten, pp. 457–79. New York: Academic Press.

Platt, T., Mann, K. H. & Ulanowicz, R. E. (1981). *Mathematical Models in Biological Oceanography*. UNESCO Monographs on Oceanography Methodology 7. Paris: UNESCO Press.

Rosen, R. (1985). Information and complexity. In *Ecosystem Theory for Biological Oceanography*, ed. R. E. Ulanowicz & T. Platt, pp. 221–33. Ottawa: Canadian Bulletin of Fisheries and Aquatic Sciences 213.

Rutledge, R. W., Basorre, B. L. & Mulholland, R. J. (1976). Ecological stability: an information theory viewpoint. *J. Theor. Biol.*, **57**, 355–571.

Tribus, M. & McIrvine, E. C. (1971). Energy and information. *Sci. Am.*, **225**(3), 179–88.

Ulanowicz, R. E. (1983). Identifying the structure of cycling in ecosystems. *Math. Biosci.*, **65**, 219–37.

Ulanowicz, R. E. (1986). *Growth and Development: Ecosystems Phenomenology*. New York: Springer-Verlag.

Weiss, P. A. (1969). The living system: determinism stratified. In *Beyond Reductionism*, ed. A. Koestler & J. R. Smythies, pp. 3–55. New York: MacMillan Co.

3

Network thermodynamics: a unifying approach to dynamic nonlinear living systems

D. C. MIKULECKY

3.1 Introduction: Network thermodynamics as a new approach to complex systems

3.1.1 *What is network thermodynamics?*

3.1.1i *Networks, systems, topology and geometry: connectedness and measurement*

Network thermodynamics is a relatively new field which uses both classical and non-equilibrium thermodynamics and kinetics in conjunction with graph/network theory to provide a unified analysis of highly structured systems. Its original purpose was to provide a tool to analyze living systems (Peusner, 1970; Oster, Perelson & Katchalsky, 1973), but it has also become a much more general approach to complexity in physical and engineering systems as well (Thoma, 1975; Chandrashekar & Wong, 1982; Breedvelt, 1984).

The basic principles of network thermodynamics follow very closely the development of modern electronics. The concepts of resistance, capacitance and inductance have their counterparts in all dynamic systems that involve material flows whether electrical or not, and therefore, the network approach seems very natural. That there is significant overlap with other approaches is one reason why network thermodynamics is a unifying formalism. The overall systematic structure provides a framework for formulating the solution to problems in a rigorous, consistent and illuminating manner.

One central contribution is that the formulation consistently distinguishes those aspects of a system that are due to the measurable, material properties (geometry) and those due to the organization or connectedness (topology) of the system. To make this distinction concrete, consider the fact that a television set has a lot of elements in it – resistors, transistors, picture tube, capacitors, etc. These are the geometric (remember that geometry means 'to measure the earth') properties. If they sit disconnected in a box, they do not make a picture on the tube. Once connected properly, they constitute the system we know as a television receiver. Thus, topology is also an essential part of a system.

We will see that this combination of metric properties with connectedness is a key component to network thermodynamics. It appears in other approaches as well, but almost always without the systematic and clearly identified structure.

3.1.1ii *Why networks?*

Most biological systems are organized to an impressive extent. If we observe a tissue and see it composed of a group of cells, we are forced to entertain the possibility that some of the materials in the tissue move among these cells. The morphology of the system suggests that we identify a number of compartments with natural barriers (membranes) between them. This is true at all levels of the organism's hierarchical organization, including the whole animal. It might seem sufficient to identify the system as a compartmental system and use compartmental analysis to analyze it. For many purposes, this is adequate, and much has been learned from this approach. If the flow we are interested in is coupled to other flows, however, or possibly some of the barriers in the system have special transport properties (carriers, channels, exchangers, etc.), the transition to these more complicated (and often nonlinear) problems becomes much easier to formulate and their analysis more simple using the network approach. Network thermodynamics and compartmental analysis lead to the same solution of the problem, but the former can be generalized to a broader class of problems and presented in such a way that the links with other approaches become obvious, requiring little additional manipulation to achieve.

3.1.1iii *Why thermodynamics?*

The thermodynamic constraints on any model can be incorporated at a number of stages in its development. Often, thermodynamic constraints on models are ignored, sometimes with disastrous consequences. Network thermodynamics requires that these constraints be incorporated throughout the modeling process. Because the approach leans heavily upon and contributes new insights into non-equilibrium and equilibrium thermodynamics, this interplay is very natural. It is no surprise that the elements chosen by electronic network analysts to represent various parts of an electrical system are distinguished by their thermodynamic properties, namely their manner of manipulating energy. The resistor is the energy dissipator, the capacitor an energy storage element of one kind, and the inductor the representative of inertial energy storage. This is a general and natural categorization in any dynamic system. Network thermodynamics is the next logical stage in the development of thermodynamic thinking (Mikulecky, 1984), broadening its scope to highly organized, nonlinear, dynamic systems and erasing the traditional gaps between thermodynamic and kinetic approaches.

3.1.2 *Contributions of network thermodynamics to the field of thermodynamics*

The ability to model and simulate large, highly organized systems is enough to make network thermodynamics an approach worthy of consideration. However, this extension of thermodynamic reasoning into the realm of complex, organized systems (Mikulecky, 1984) sheds a new light on the fundamentals of thermodynamics. These new findings are certainly worth our attention, but are beyond the scope of this work. A brief list of these new findings is all we can afford to present here. They are:

1. The discovery of a metric structure underlying thermodynamics (Peusner, 1983*a*, *b*, 1985*a*, *b*, 1986*a*, *b*).
2. A class of 'thermokinetic potentials' topologically connect state spaces belonging to distinctly different physical processes (Peusner, 1982, 1986*a*; Peusner *et al.*, 1985; Mikulecky, Sauer & Peusner, 1986).
3. The extension of Kedem and Caplan's energetics of non-equilibrium systems to a broader class of systems than those covered by the Onsager theory (Onsager, 1931*a*, *b*; Odum & Pinkerton, 1955; Peusner, 1983*a*, *b*, 1985*a*, *b*, 1986*a*; Mikulecky *et al.*, 1986; Caplan & Essig, 1986).
4. The generation of a unifying graph-theoretical approach to dynamic systems which can include flow graphs, King–Altman and Hill diagrams and others (King & Altman, 1956; Mason & Zimmerman, 1960; Hill, 1977; Peusner *et al.*, 1985; Peusner, 1986*a*; Mikulecky *et al.*, 1986).
5. The macroscopic and microscopic proof of Onsager's reciprocity for connected physical and biological networks using Kirchhoff's laws and Tellegen's Theorem (Peusner, 1970, 1982, 1983, 1985*a*, *b*, 1986*a*, *b*; Peusner *et al.*, 1986; Mikulecky *et al.*, 1986).

6 'Completes' the picture of systems provided by non-equilibrium thermodynamics and shows that a more holistic representation and analysis can lead to profound results.

The unifying quality of network thermodynamics goes beyond these contributions to thermodynamics. We will now examine the unifying qualities of network thermodynamics in more detail.

3.1.3 *Network thermodynamics as a unifying formalism*

Because network thermodynamics is a thermodynamic formalism, it has some obvious unifying attributes. First of all it can accomplish what any other approach to systems of material flows can. This means that it is sufficiently general to be universally applicable. Secondly, it is compatible with many other approaches and can be readily adapted to them. It is not difficult to extrapolate that *any* approach yielding solutions to complex, highly organized systems is a special case of network thermodynamics. Thus, the approach has obvious applications in ecology (Mikulecky, 1985). The network and graph theoretical aspects of the formalism lead to an immediate application to hierarchical systems. This will become an important qualification of the common distinction between thermodynamic and mechanistic models. It is worthwhile to list many other approaches that are naturally encompassed by network thermodynamics:

1 Signal flow graphs.
2 Electrical networks.
3 Compartmental analysis.
4 Dynamic systems models.
5 Flow analysis.
6 King–Altman and T. Hill diagrams.
7 Bond graph techniques.
8 Metabolic networks.
9 State transition models.
10 Equilibrium and non-equilibrium thermodynamics.
11 Chemical kinetics.

This partial list should justify the statement that this is a unifying theory.

3.2 The origins of network thermodynamics

Network thermodynamics began with Kirchhoff's seminal work on network and graph theory (Kirchhoff, 1976). As in many other instances, network thermodynamics came from a need to rigorously analyze and simulate complex living systems. In this section we give a brief summary of that development.

3.2.1 *Network and graph theory*

Kirchhoff (1976) recognized that there were two distinct, but necessary, properties of electrical circuits; the elements in the networks and their connections or topology. He then developed a systematic method for dealing simultaneously with both of these. Since his time, various ways of utilizing network and graph theory have occurred in biology, for example, enzymatic and transport systems (King & Altman, 1956; Mason & Zimmerman, 1960; Hill, 1977), molecular structure (Nahikian, 1964; MacDonald, 1983), food chains (MacDonald, 1983), population biology (Lewis, 1977), and many others. In none of these was the use of network and graph theory as systematic or all-encompassing as in electronics and electrical network theory. This explosion resulted in the computer age. In a way it is a curious fantasy, but it is now also obvious that it could have been worked out equally well to serve the needs of biology. This is only because the generality of the theory has now been established.

As in electrical networks, the more general networks we find in living organisms, flows of energy, substances, charge, etc. are best represented by linear graphical representation which allows us to utilize Kirchhoff's two conservation laws. These ideas are developed in detail through examples later. We will see that Kirchhoff's 'voltage law' as generalized by network thermodynamics, has a profound connection with certain fundamental concepts from physics, and in particular, thermodynamics (Peusner, 1986*a*). Tellegen's theorem arises as a consequence of Kirchhoff's laws holding in a system. This important theorem was not in the non-equilibrium thermodynamic paradigm, because this paradigm analyzes a part of any system while the network approach forces a holistic view.

Network and graph theory are much more than convenient diagrammatic or schematic representations of systems. They are also ways of embodying fundamental approaches. This is better seen by an example. In Fig. 3.1(*a*), a complex network of resistors is energized by a single potential source (battery or generator). This set of electrical resistances connected as a network could be an analog model for a set of membranes through which diffusion occurs or a set of pipes through which a volume of fluid flows. The potential difference due to the source would then be a concentration and pressure difference, respectively. The diagram in Fig.

3.1 (*b*) is a single resistor (membrane or pipe) connected to a source. Non-equilibrium thermodynamics was developed for such steady state systems (systems of resistances and sources *only*, no capacitors or inductors). Resistors completely determine the steady state; capacitors (storage) and inductors (inertia) furnish the dynamics. The important thing to note is that the flow *through* the resistor is equivalent to the flow through the source and the drop in potential *across* the resistor is supplied by the source. Hence, examination of either the resistor or the source reveals everything about the system. Tellegen was the first to show that the holistic description of systems resulted in very powerful relations between

Fig. 3.1. (*a*) A large resistive network connected to a single source. (*b*) A single resistor connected to a single source.

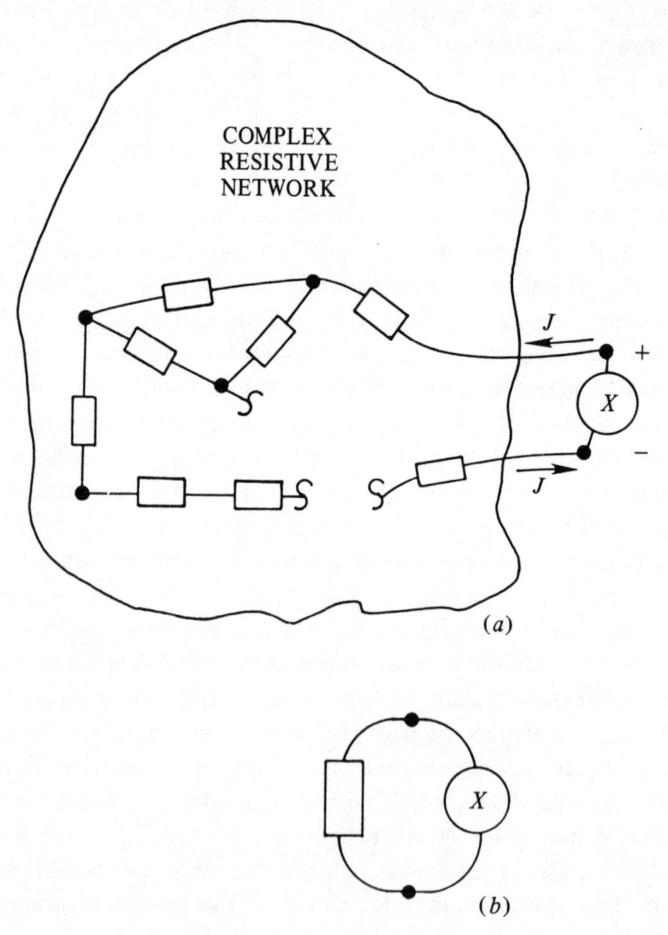

the flows and forces, which were true in dynamic states, as well as steady states, and which generalized to *all* systems connected in the same way (with equivalent topologies). Tellegen's theorem is a simple one, namely, if both load and source are included, the flow, $\boldsymbol{J}$, and force, $\boldsymbol{X}$, vectors are orthogonal:

$$\bar{J} \cdot \bar{X} = 0$$

or

$$J_1 X_1 + J_2 X_2 + \ldots + J_n X_n = 0.$$

This is also

$$\sum_{i=1}^{n} J_i X_i = \sum_{k=1}^{n} J_k X_k$$

where i indicates summation over the terminals (places to be experimentally manipulated) and k summation over all internal network branches.

In non-equilibrium thermodynamics, a portion of the above sum, namely the internal branches, was the focus of interest and this relation was never seen. The almost unbelievable result of Tellegen's theorem rests only on the fact that the flows and forces (differences in potentials) obey Kirchhoff's laws. The extension of Tellegen's theorem is even more of a jolt to intuition; if two systems of totally different nature obey Kirchhoff's laws and have identical connectedness (topology) the flows from one and the forces from the other also obey Tellegen's theorem! This is one of the features that sets the network thermodynamic paradigm apart from, and as a measure for, all others that attempt to provide holistic diagrammatic approaches to systems.

3.2.2 *The thermodynamics of highly organized systems*

At this time there really is no physics of highly organized systems. Rosen addressed this issue at length (Rosen, 1985*a*, *b*, *c*, 1986, 1988) and he suggested that the Newtonian paradigm fails for 'complex systems'. Only recently have topological and differential geometric approaches to physics (Harrary, 1967; Abraham & Marsden, 1978; Peusner, 1983; Nash & Jen, 1983; Curtis & Miller, 1985) been recognized as valuable on a large scale. No authors have made application to thermodynamics as systematically as those who have begun network thermodynamics (Peusner, 1970; Oster *et al.*, 1973; Chandrashekar & Wong, 1982; Peusner, 1986*b*).

Network thermodynamics uses all we know about simple systems and combines it with topology, mainly through graph theory, to treat more complicated systems. The results of this extension have been striking. The ideas have found widespread application in a number of areas of biology.

3.2.3 *Applications in biology and other fields*

Network thermodynamics has been applied to achieve more holistic representations of reductionist results. For example, studies involving transport into cells (Rapundalo & Feher, 1981) and subsequent metabolism of the transport substance (May & Mikulecky, 1983; White & Mikulecky, 1986; Huf & Mikulecky, 1986) or the sites of action of hormones (May & Mikulecky, 1983; Fidelman & Mikulecky, 1986) put together data obtained on components to model the simultaneous operation of these components in the whole system. Likewise, systems too complicated to study by other methods have been modeled using network thermodynamics (Thomas & Mikulecky, 1978; Mikulecky & Thomas, 1979; Plant & Horowitz, 1979; Atlan *et al.*, 1979; Oken, Thomas & Mikulecky, 1981; Mikulecky & Peusner, 1985; Walz & Caplan, 1986). Recently, the possible applications to ecology have been explored (Mikulecky, 1985) and this paper will pursue this idea.

Applications outside biology are also worth noting, for example, many applications have been made in engineering and physics (Thoma, 1975; Chandrashekar & Wong, 1982; Breedvelt, 1984). In this context we should acknowledge that the standard mathematical repertoire for a scientist is heavy in analysis, differential equations, and related topics, but extremely light in subjects like topology and differential geometry. A number of books introducing these subjects into physics, for example, are now on the market (Harrary, 1967; Abraham & Marsden, 1978; Nash & Jen, 1983; Peusner, 1983; Curtis & Miller, 1985). As more and more highly organized systems are studied, the need to know how to apply these topics will grow. Network thermodynamics brings the topology and the geometry in the measured properties of the elements together to treat both the organizational as well as the mechanistic aspects of systems. It does so in a way that pays full recognition to the results built upon the pioneering work of Kirchhoff in electrical network theory.

3.3 The network thermodynamic approach

3.3.1 *The conceptual framework of network thermodynamics*

Network thermodynamics is fundamentally founded in graph theory (Kirchhoff, 1976). A network is often distinguished from a simple graph by assigning to the nodes and branches special physical meaning and symbols to codify this meaning. No matter how this is done, the code must be unambiguous and ultimately result in a graph from which a solution to the network can be obtained. Furthermore, the distinction between the connectedness of flow pathways and the nature of those flows must be made as clear as possible.

3.3.1i *Representations of complex systems*

There are many ways to pictorially represent complex systems. Generally, each field produces a seemingly unique representation adapted to that field only to find that with slight modification, translation, or both it is equivalent to many others. In ecology the same is true. A superficial scan of ecological models, reveals compartment models (Platt, Mann & Ulanowicz, 1981), signal flow graphs (Patten *et al.*, 1976), energy circuits (Odum, 1972) and many other diagrammatic representations. In physiology, which is my field, there is also a proliferation of representations, including those already mentioned and bond graphs (Oster *et al.*, 1973; Atlan *et al.*, 1979; Gebben, 1979; Plant & Horowitz, 1979), equivalent circuits (Finkelstein & Mauro, 1963), Hill/King–Altman diagrams (King & Altman, 1955; Mason & Zimmerman, 1960; Hill, 1973; Peusner *et al.*, 1985), and the more traditional metabolic maps, etc., which now decorate the doors and walls of many laboratories and offices. Peusner chose to use a linear graphical representation taken from electronic networks and we have stuck to it.

At times it was worth comparing those representations and showing how they could be unified (Mikulecky & Thomas, 1979; Peusner *et al.*, 1985). The reason was to show that it is the *formalism* which really matters, and that various representations can be translated into others. In ecology this is no less true. One important example of how the choice of a representation could become misleading is that of signal flow graphs (Patten *et al.*, 1976). In his discussion of this representation, Patten says 'The signal content of a node is thus not affected by outgoing signals, that is it is distributive (non-conservative)'. Examples from his discussion are given in Fig. 3.2(*a*) and 3.2(*b*). In Fig. 3.3(*a*) and 3.3(*b*), these are redrawn as linear (pseudo-electrical) graphical networks. The mathematical relations depicted are preserved, as well as a conservation property which was not in the signal flow graph representation, namely that the flows and forces in the network obey Kirchhoff's law (and, therefore, Tellegen's theorem applies). Mason & Zimmerman (1960) devote an entire chapter of their book on electronic networks to signal flow graph representations of these Kirchhovian networks. The importance of conservation proper-

Fig. 3.2. (*a*) Signal flow graph for the relation $y = H, Z, H_2, Z_2$. (*b*) Signal flow graph for the relation $y_i = H_i Z$, $i = 1, \ldots, n$.

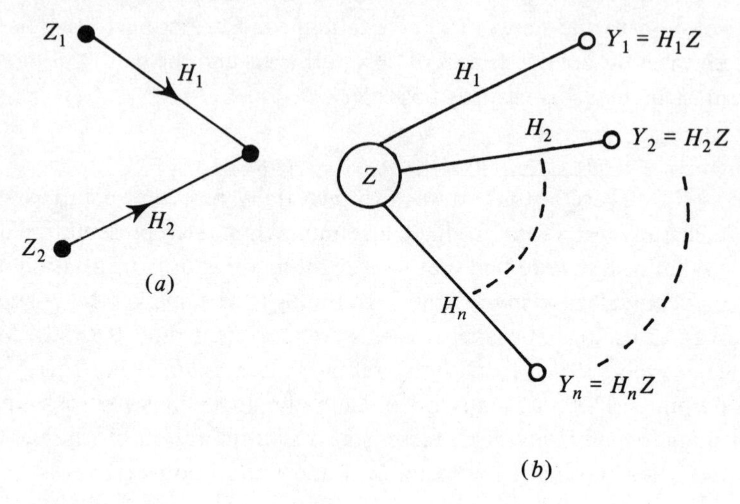

Fig. 3.3. (*a*) Network representation for the relation in Fig. 3.2(*a*). Notice that, in this case, all quantities are conserved. (*b*) Network representation for the relation in Fig. 3.2(*b*).

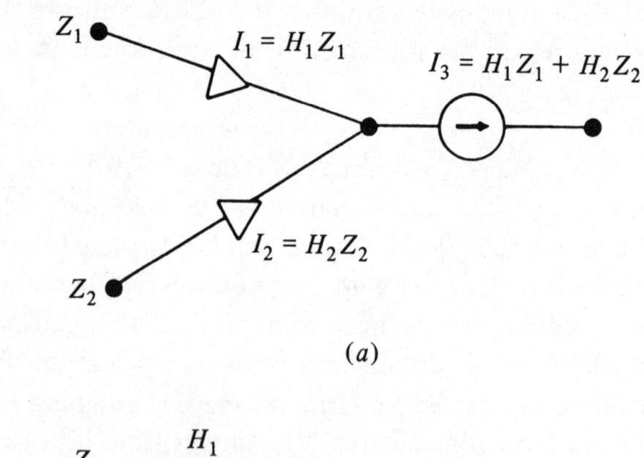

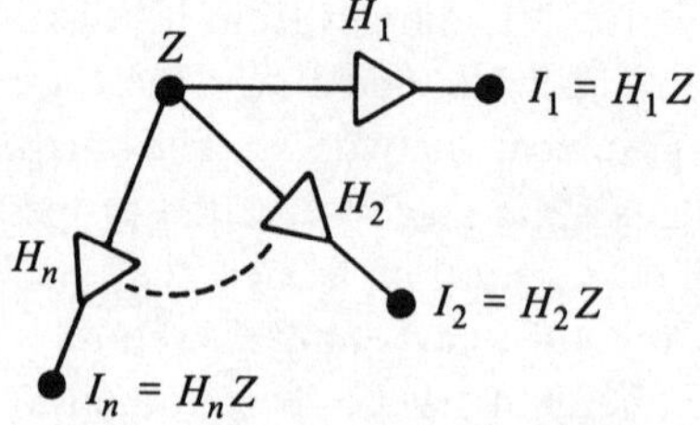

(*b*)

ties in network thermodynamics is central and therefore representations that connote this aspect of the formalism are preferred. Of course, if the system does not obey Kirchhoff's laws, the signal flow graph is certainly a correct representation and the Kirchhovian network is not. Signal flow graphs are more general, but they obscure the conservation relations in Kirchhovian systems. Another formalism that deals with Kirchhovian systems, but makes no explicit acknowledgement of their applicability is flow analysis (Patten *et al.*, 1976; Platt *et al.*, 1981). The relative roles of topology and geometry in the formalism needs to be made clear if a unification of ideas is to be achieved.

3.3.1ii *The relationship between thermodynamic and 'mechanistic' descriptions of systems*

(a) Simple systems – Traditionally, the distinction between a thermodynamic and a 'mechanistic' description of a system arises when one is forced to focus on a given level in the hierarchical structure. A model is then chosen to represent the 'mechanism' at this level, but this model must then conform to certain constraints imposed by more general thermodynamic considerations (often implicit in the position of this mechanism in the hierarchy). On the other hand, purely thermodynamic models focus on input/output relationships and should be independent of mechanism. In other words, a number of mechanistic models will fit a given thermodynamic model. It pays to look more carefully at the distinction between mechanistic and thermodynamic models in another context, namely, the hierarchical structure of a system. It is not as clear a categorization in that more complicated context.

(b) Hierarchical systems – As soon as we admit that our system is at a level of organization in a hierarchy, the distinction between mechanistic and thermodynamic becomes somewhat blurred. For example, in a compartmental system, the description of the system's connectedness (topology) coupled with the rate expressions for transitions between the compartments constitutes a 'mechanistic' model depending specifically on the topology and rate constants. Adding thermodynamic constraints immediately forces the model into the subclass of things that have a realization in the 'real' world. (We hope it is clear that not all models are valid representations of reality, e.g. perpetual motion machines.) However, a more detailed look at the rate constants themselves shows that they are very phenomenological in character and can be further assigned a mechanistic model to describe their relation to more fundamental parameters, and so on. Thus at each hierarchical interface we may try to bring together the reductionist view along with a more holistic synthesis of the parts of submechanisms. We will see that network thermodynamics does this neatly.

3.3.1iii *The roles of topology and geometry in the formalism*

Systems can be efficiently described with two kinds of information: topological and geometric (Maloney *et al.*, 1986). Each of these words has its origin in a branch of mathematics and each connotes something profoundly different about a system. Topology is the branch of mathematics that deals with the *connectedness* of a system. The relationships in an electronic device depend on the elements being connected in a certain way. The components of your TV set might be rewired to provide some very different devices. These specific connections, shown in some sort of schematic diagram, are independent of how close you draw one element to the other in that schematic, but very dependent on what is or is not connected to what. Geometric properties, on the other hand, have measurement distinctly connected with them. These, for example, are the phenomenological description of flow–force relationships in non-equilibrium thermodynamics. In network thermodynamics they are called *constitutive* relations.

In hierarchical systems, it is very easy to mistake one of these properties for another. At some level of our hierarchical description a complex network with a very special topology may become a simple 'black box' described only in terms of geometric properties expressed as input/output relationships. Network thermodynamics seeks to define the topological and geometric properties clearly and distinctly whenever possible. A very simple example from electronics should show the importance of the distinction.

Two circuit elements may be wired together in either series or parallel. Likewise, two flow pathways in a living system may be combined in either way. These are the topological choices. However, those elements may be resistors, capacitors, or inductors and the behavior of the whole depends very much on *both* kinds of information. (Resistances in series add whereas capacitances in parallel add.)

The network thermodynamic formalism produces a representation that associates the conservation laws with the topology. The geometry comes in through the constitutive relations between efforts and flows in the elements being connected together. This will now be illustrated in more detail.

3.3.1iv *Conservation laws and Tellegen's theorem*

The most fundamental constraints that operate at almost all, if not all, levels of the organizational hierarchy are those of conservation: energy, mass, charge, etc. Network thermodynamics, following Kirchhoff, establishes the fundamental link between these laws and the system's connectedness. This idea, which was discussed in Section 3.3.1, is illustrated mathematically in Appendix A3.2. What is important for this discussion is that the introduction of the connectedness of the system and the constraints imposed by conservation are one and the same act. This is true whether we use network thermodynamics to solve the problem or any other method. The main difference, however, is the manner in which network thermodynamics provides a holistic application of these constraints that makes them explicit and identifiable.

As an 'automatic' consequence of this set of topological properties, Tellegen's theorem results (Tellegen, 1952). This has the form of a power conservation, but in its quasi-power form it is much more than a restatement of the conservation of power (see Appendix A3.1.2 for a mathematical demonstration). The implications of Tellegen's theorem for systems has only begun to be explored, but certainly the scope of system properties it has a bearing on, including stability and identifiability considerations, make it significant.

The conservation laws *and* Tellegen's theorem come from the *topological* properties of the system alone. No measurement is necessary for us to be able to utilize this information. As a corollary, the ability to 'sort out' this information from the geometric (phenomenological) aspects at each hierarchical level is among the most important contributions of network thermodynamics. The contribution of the topology and geometry are carried through to a solution of the network in Appendix A3.3.

3.3.2 *Method of approach*

The approach to any system involves an identification of the hierarchical level at which its description will be attempted and then a recognition of the topological and geometric aspects of the system at that level. To be more specific, the level is determined by the flows, as flow analysis in ecology and economics readily recognizes. These flows are through something and this is how level is focused upon. It leads also to a search for the connectedness.

We will briefly describe the approach using the intertidal oyster reef ecosystem (Dame & Patten, 1981) as an example.

3.3.2i *Topological description of the system*

The system's topology is provided by our visualization of the system's connectedness, either implicit or explicit – in the case of the oyster reef example, we can arrange the elements of that system in an explicit topology as in Fig. 3.4.

The compartments (nodes in the figure) are 1 – filter feeders, 2 – detritus, 3 – microbiota, 4 – meiofauna, 5 – deposit feeders, and 6 – predators. All six compartments have outputs, while there is a single input to compartment 1. The linear digraph in the figure illustrates the system's topology or connectedness by having the following branches which connect the nodes (compartments) and represent paths for the flow of energy between the compartments. They are 1 → 2, 2 → 3, 3 → 4, 4 → 5, 5 → 6, 1 → 6, 6 → 2, 3 → 5, 2 → 4, 4 → 2, 2 → 5, 5 → 2. [The topology of the system shows how conservation (Kirchhoff's flow law) must be applied to the flows at each node and to any potential-like quantity (Kirchhoff's effort

Fig. 3.4. Digraph representing the intertidal oyster reef system.

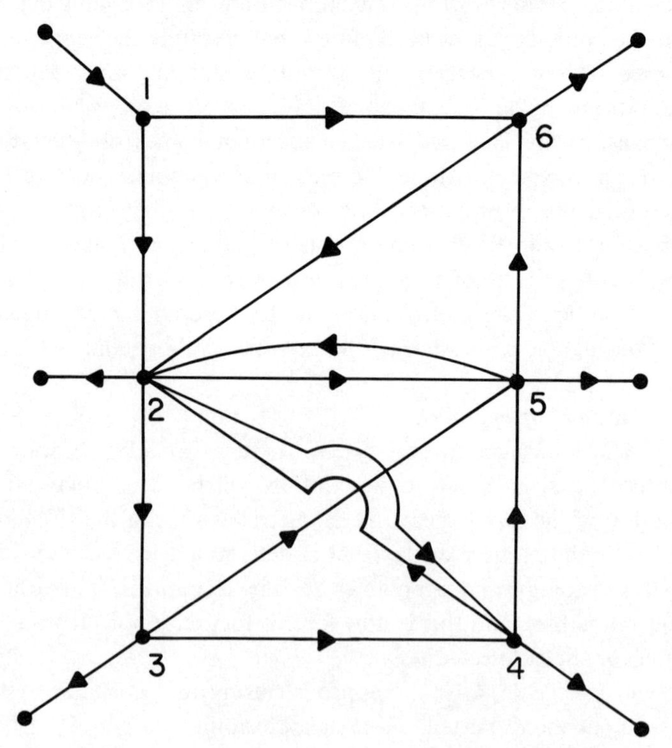

law) assigned to the compartments in the network (in this case the steady state energy storages).] These two laws are easily stated in terms of the various incidence matrices, which encode the connectedness of the digraph, and the vectors of flows and potentials, which are encodings of the two lists given above – the nodes and branches. This is shown in Appendix A3.2.

3.3.2ii *Geometric aspects*

A simple example of a system described by network thermodynamic reasoning is shown in Fig. 3.5. Two 'pools' or compartments or regions are separated from each other, but there is a path between them through which something of interest to us (biomass, numbers of individuals, energy, etc.) can move from one place to the other. This simple example captures the *mechanistic* essence of much of what goes on in an ecological system.

It seems popular and often expedient to measure a 'flow' in such a system. This flow is the rate at which our 'something' moves between the two pools and is expressed in units of the amount of the something (biomass, kilograms of carbon, etc.) per unit of time. The existence of this flow enables the definition of two geometric descriptors of the system – two 'constitutive' relations. The flow also follows a very specific topology (pool 1 $\rightarrow$ pool 2 via the path between them). This topology relates the two constitutive relations in a unique way. This is worth closer examination, because this simple example embodies the relationships between topology and geometry that any more complicated system will possess. The pools are actually capacitors. Their static definition relates a potential-like quantity (we will use concentration, but others would be equivalent, e.g. number density) to an extent or size for storage (in this case, volume) and to an amount entering or leaving the pool.

Concentration = amount/volume

or

$$C = N/V.$$

Fig. 3.5. A simple two pool system represented as a network.

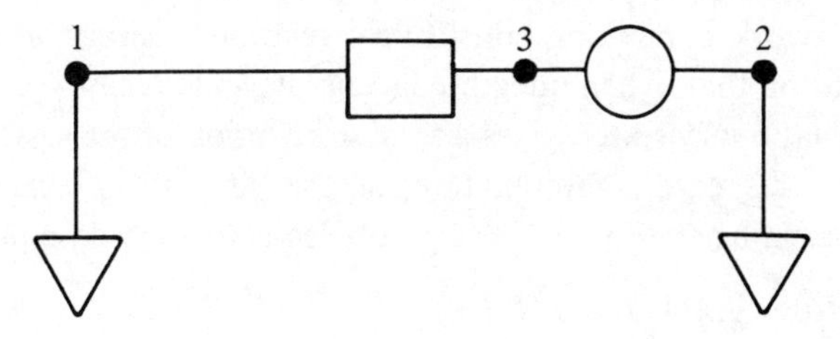

For a constant volume, this also has a dynamic form:

$$V\,\mathrm{d}C/\mathrm{d}t = (\mathrm{d}N/\mathrm{d}t) = J,$$

where $J = \mathrm{d}N/\mathrm{d}t$ is a flow into or out of the compartment, the rate of change of amount, N. The constitutive relation for the path between pools is often harder to come by, but for small enough flows, can often be assumed to be linear. In the linear case, we define a conductance as the proportionality constant between flow and force:

$$\text{conductance} = \text{flow}/\text{force}$$

or

$$L = J/X$$

where the force, X, in this case, is the difference between the pool concentrations,

$$X = C_1 - C_2.$$

In the nonlinear case, J is expressed as a *function* of X or vice versa. Notice that the use of the word 'force' here to describe a quantity that is merely a difference in two pool concentrations is in no way meant to imply causality. This force may or may not 'drive' the flow causally, but it is the conjugate variable to the flow and it is the variable that contains precisely those concentrations that are changing due to the flow and that enter into the capacitive constitutive relation. In fact, Kirchhoff's law tells us that for this topology the flows are all the same. Thus the three constitutive relations,

$$J_p = L(C_1 - C_2),$$

$$J_1 = V_1(\mathrm{d}C_1/\mathrm{d}t),$$

and

$$J_2 = V_2(\mathrm{d}C_2/\mathrm{d}t),$$

can be combined due to Kirchhoff's flow law:

$$J_p = -J_1 = J_2.$$

As in every network combining constitutive relations, Kirchhoff's laws lead to an equation that, when integrated (time-dependent case) or solved algebraically (stationary state), gives a total description of the system, i.e. a solution. In this case, manipulation of the capacitive constitutive relations and defining the force $X = C_1 - C_2$, leads to such an equation:

$$-\mathrm{d}X/\mathrm{d}t = LX(1/V_1 + 1/V_2).$$

Its solution by integration is:

$$X(t) = X_0 \exp(-t/\tau),$$

where the time constant, τ, is

$$\tau = V_1 V_2/[L(V_1 + V_2)].$$

An examination of the general treatment of the steady state case in Appendix A3.3 will show how these manipulations can be stated in completely general form. The time dependent general case is developed in detail elsewhere (Curtis & Miller, 1985).

It is worth summarizing what was done here. The pools themselves had a certain size or extent which may have been a volume of water or an area of forest, etc. In these pools were a given amount of the something that can flow to the other pool. The ratio of amount to extent or size is a potential-like quantity (by definition!). Such definitions may seem disturbing at first, however, they have some value. Their value lies in the fact that if they are accepted, the new formalism is then at our disposal. In this simple example, the result is almost too obvious: the two pools are governed by a capacitive constitutive relation and the flow between them acts as if it is going through a resistance! Empirically, we can establish a relationship between the instantaneous flow from one pool to the other and the instantaneous potential difference between the two pools. This relationship is geometric and can be shown to be a generalization of a resistance. Likewise, there is a relation between the flow into or out of a pool and the rate at which its potential changes. This defines a generalized capacitance.

After I introduced these definitions in a previous paper (Mikulecky, 1985) there were some evaluative working groups and a report was given by these working groups (see Mikulecky, 1985, for reference). One concern was the possibility that the constitutive relations may often be highly nonlinear. Yes, they may be multiple valued as well. The fact that the formalism has not yet been applied to systems with nonlinear constitutive relations has nothing to do with limitations of the formalism, but only reflects the stage of development of ecological models. Physiology, for example, is replete with nonlinear models. It is a strength that the formalism can deal with such complexity. Indeed, it shows that the formalism does distinguish between simpler and more complex systems, by recognizing classes of constitutive relations (linear vs. nonlinear). Since we base much of network thermodynamics on modern electronics, we also reap the fruits of that field's preoccupation with very nonlinear circuit elements. It is quite evident that any problem that can be shown to be isomorphic with a problem in electronics can be solved in the same way. Once again, a solution to the problem is a solution to the problem so that the essence is in the formulation of the problem. We have found that for most purposes it is easier to simulate the system on the circuit simulator SPICE to obtain a solution.

3.3.2iii *Computer simulation*

Modern electronics was forced to find methods for dealing with extremely nonlinear devices in networks. Their answer to dealing with these difficult problems was simulation on a digital computer. This eliminated much of the actual wiring together of elements into test circuits called 'breadboarding'. The chips contained networks too large and complicated to breadboard. The cheap, fast, and efficient replacement was computer simulation. To this day, the most effective simulator developed is the program SPICE (*S*imulation *P*rogram with *I*ntegrated *C*ircuit *E*mphasis).

SPICE is written to incorporate the mathematics of network theory in a systematic manner. The user feeds in the topology of a network by assigning node numbers to a network diagram (see Fig. 3.4 for example). The constitutive relations are chosen from a 'menu' that encodes the type of element by the first letter in its name. Resistors get R, capacitors get C, and so on. The program sets up and solves the algebraic (steady state) or differential equations (time-dependent case) by using a DC or a TRANSIENT analysis.

(a) Example: Simple 2 pool model (Fig. 3.5).

For example, the program for the example in Fig. 3.5 would be as follows:

```
FLOW BETWEEN TWO COMPARTMENTS
C1       1    0      1600     IC = 100
C2       2    0       500     IC =  10
RP       1    3      5E06
VFLOW         2         3     0
RINF          1         0     IE25
.TRAN         1       100     UIC
.PRINT        TRAN V(1)       V(2)      I(VFLOW)
.PLOT         TRAN V(1)       V(2)      I(VFLOW)
.END
```

The first line is the title which is a mandatory aspect of SPICE's format. Next comes a list of the circuit elements. The first is a capacitor attached between node 1 (pool 1) and ground (the reference concentration, which is zero, is assigned to the ground node, for which SPICE reserves the number zero). The order of the nodes listed after the name, C1, is 1 first and then 0. This is in conformance with SPICE's convention that the positive node always comes first. The next entry on the line is 1600, which is the volume of pool 1 in some prescribed set of units. Next the entry IC = 100 is the initial concentration in pool 1 in some appropriate set of units. The next line puts the capacitor for pool 2 into place in the same manner. Its volume is 500 and its initial concentration is 10. Finally, the resistance in the flow pathway between the pools is listed. It is connected to a flow meter named VFLOW at a 'dummy' node numbered 3. Node 3 has no physical significance in contrast to nodes 1 and 2 which represent the pools. In SPICE, flow meters are created out of potential sources, hence the 'V' in VFLOW, set to zero output. Thus, in the line for VFLOW a zero is listed after the node numbers denoting where it is connected. The resistor's value, 5×10^6 in appropriate units, is listed after its connection node numbers and is a separate line of the program, as in each distinct element. The first pair of numbers are the node numbers, with a polarity convention, and constitute the way SPICE creates an incidence matrix of the type discussed in the Appendix. Using the incidence matrices to force the relations in the Appendix, Kirchhoff's laws are forced to hold. In this manner, the mathematics of network theory come into play 'behind the scenes' if the user constructs the network and encodes it in SPICE properly.

(b) Intertidal oyster reef (Fig. 3.6).

The SPICE model for the intertidal oyster reef is very straightforward. An additional circuit element, the controlled source, is introduced. These objects have multiple inputs, either flows or forces and a single output, which can be either a flow or a force. With these combinations available, there are four kinds. For this model, the force controlled flow source is equivalent to Mason's unistor (Mason & Zimmerman, 1960). Its constitutive relation is:

Flow = constant × node concentration on feed side.

It is shown as a diamond with an arrow in it in the model. It is used when flow goes from a low node value (stock) to a high one. In any case where the observed flow is from high node value to low, a resistance can be used of value:

Resistance = (feed side node value − receiver side node value) /flow.

The values of the conductances for the unistors are calculated by the relation:

Conductance = flow/feed side node value.

The input/output flows are modeled by constant current (flow sources represented by circles with arrows in them). In the figure, the flow meters have been omitted for simplicity, but they do appear in the program listed in Appendix A3.4.

3.4 Summary

Network thermodynamics has now been established as a legitimate extension of thermodynamic reasoning into complex systems. By systematically separating topological from geometric properties, it helps identify at least some of the 'non-mechanistic' determinants of a

Fig. 3.6. A diagram of the SPICE model.

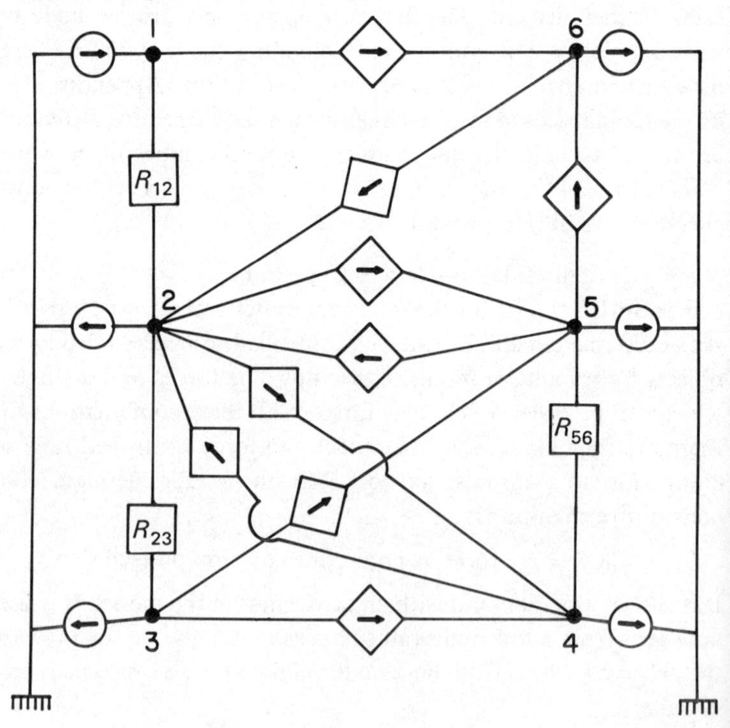

system's behavior. In this way, it clarifies some of Rosen's distinctions between 'mechanistic' and complex systems (Mikulecky, 1987*a*, *b*, 1988; Rosen, 1985*a*, *b*, 1986, 1988). Any systematic approach involving the possibility of a linear graphical representation of systems (or even others, such as bond graphs), that obeys Kirchhoff's laws is the subject of network thermodynamics. Electric circuits, pharmacological, biochemical and physiological systems, ecological systems, and many, many others are all special cases. Network thermodynamics systematizes these other approaches and unifies them by recognizing the necessary and sufficient conditions for the thermodynamically correct formulation of any system. These are the existence of a limited number of different types of elements (resistors, capacitors, inductors, sources, unistors and a few others) encompassing all known physical behaviors in networks and the need to specify a topology. From this formulation a rigorous answer is readily obtained by simulation or often, by straightforward mathematical analysis. Modern electronics has solved many problems involving some of the largest and most complicated nonlinear dynamic systems. We have no alternative at hand that is as powerful. It also tells us that anything that works is a variation of network thermodynamics. For these reasons, network thermodynamics can find use in ecological models. It can incorporate the information already being treated by flow analysis, in those cases where information on stocks is also available, or by environ analysis. As in all the other fields where it has been applied, the failing is never in the formalism but rather the inability of those who utilize the formalism to generate enough information to take advantage of its power. I offer that as a challenge to ecologists who wish to completely describe the objects of their studies.

Acknowledgement

I wish to thank Sheryl Warner and Isabel Sluder for a magnificent job of typing this manuscript.

Appendix

A3.1 Topological relations

A3.1.1 *Incidence matrices*

For this discussion only one set of incidence matrices will be presented, namely the node–branch set, although there are two other sets commonly used in network theory (Breedvelt, 1984). The topology of the system is encoded in a node–branch incidence matrix by translating the digraph in Fig. 3.4 into an array of ± 1 and 0 as follows:

The values of the elements of the incidence matrix are assigned by ordering the nodes vertically and the directed branches horizontally and then listing $+1$ if the node is an end point of a branch and the direction of the branch, as indicated in the digraph, is out of the node. If it is into the node, -1 is listed, and if the node and branch are not incident, a 0 is inserted. The incidence matrix for the intertidal oyster reef is then an array of the following kind:

	Branches								
Nodes	01	02	03	04	05	06	12	23	34
0	1	−1	−1	−1	−1	−1	0	0	0
1	−1	0	0	0	0	0	1	0	0
2	0	1	0	0	0	0	−1	1	0
3	0	0	1	0	0	0	0	−1	1
4	0	0	0	1	0	0	0	0	−1
5	0	0	0	0	1	0	0	0	0
6	0	0	0	0	0	1	0	0	0
Nodes	45	56	16	25	52	62	24	42	35
0	0	0	0	0	0	0	0	0	0
1	0	0	1	0	0	0	0	0	0
2	0	0	0	1	−1	−1	1	−1	0
3	0	0	0	0	0	0	0	0	1
4	1	0	0	0	0	0	−1	1	0
5	−1	1	0	−1	1	0	0	0	−1
6	0	−1	−1	0	0	1	0	0	0

The numbers across the top are the branches (ij is the branch from node i to node j).

As a code for the entire structure, the incidence matrix as written is adequate. Normally, it is helpful for computation to use the submatrix that results from choosing a reference node (here node '0') and eliminating all connections to it from the incidence matrix. We will call the entire incidence matrix A and the submatrix A^*.

A3.1.2 *Kirchhoff's laws and Tellegen's theorem*

If we define two flow vectors:

$$J = \begin{bmatrix} J_{01} \\ J_{02} \\ J_{03} \\ J_{04} \\ J_{05} \\ J_{06} \\ J^* \end{bmatrix} \qquad J^* = \begin{bmatrix} J_{12} \\ J_{23} \\ J_{34} \\ J_{45} \\ J_{56} \\ J_{61} \\ J_{25} \\ J_{62} \\ J_{24} \\ J_{35} \end{bmatrix}$$

the Kirchhoff flow law now takes the form:

$$A\bar{J} = \bar{0} \text{ or } A^*\bar{J} = \bar{0},$$

while the Kirchhoff effort (or force) law takes the form

$$A^{*T}\bar{c} = \bar{X},$$

where $\bar{c}$ is the vector of compartment storage values (potential), $\bar{X}$ is the vector of potential drops across the branches, and A^* is the incidence matrix above, after the reference node is removed.

It is a small step to use the structure of the incidence matrix to establish that

$$\bar{X} \cdot \bar{J} = 0,$$

which is Tellegen's theorem.

A3.2 The geometric relations

A3.2.1 *Resistors, one-port and multi-port*

The most common constitutive relation is the category of linear physical laws exemplified by Ohm's law (electricity), Fick's law (diffusion) and Poiseuille's law (bulk flow). They all have the form

$$\text{effort} = \text{conductance} \times \text{flow}$$

or

$$\text{flow} = \text{resistance} \times \text{effort}$$

where resistance = 1/conductance.

When these become nonlinear, the multiplication by a constant gets replaced by some function operator:

$$\text{flow} = R\,(\text{effort})$$

or

$$\text{effort} = L\,(\text{flow}).$$

In the nonlinear case R^{-1} and L^{-1} do not necessarily exist.

To extend the linear case to the realm of non-equilibrium thermodynamics the flow–effort relations are sets of linear equations:

$$J_i = \sum_{j=1}^{n} L_{ij} X_j.$$

These equations describe a network element called a 'multi-port'.

In the nonlinear case, these multi-ports become defined as functions of the form,

$$J_i = L_i(X_1, X_2, \ldots, X_n).$$

For compartmental systems and kinetics, the unistor introduced by Mason & Zimmerman (1960) is very useful:

$$\text{flow} = \text{rate constant} \times \text{donor side potential}.$$

This is especially true, as we will see, in cases where circuit simulation programs are adapted to the simulation of these more general networks. One important aspect of a constitutive relation is its holistic character with regard to the effort and flow. It is not necessary to assign a causality between these two aspects of the same event. From analogous situations elsewhere in science there is a tendency to speak of 'driving' forces and 'resultant' flows. In fact, even in electrical networks, the idea of an IR drop across a resistor connotes a different causal relationship than the idea of a source pushing a flow through a resistor.

A3.2.2 *Sources and controlled sources*

The extreme or limiting constitutive relations would be those cases where the conductance or resistance became infinite. These would be sources of flow independent of the effort across them or sources of effort independent of the flow through them. These are the 'independent' sources. The unistor mentioned above is the linear 1-port version of a class of sources which are more complicated. These are the so-called 'controlled' sources whose effort or flow output are dependent on inputs in a prescribed manner. Their presence in circuit simulation packages, such as SPICE, makes the modeling of solid state devices possible. It also allows us to model a variety of biological objects.

A3.3 Solving networks

The solution to our networks comes from the combination of the topological and geometric information. If we use a simple example of a linear resistive network powered by certain flow or effort sources, the geometry is embedded in a conductance matrix as:

$$\bar{J} = L(\bar{X} - \bar{X}_s) + \bar{J}_s.$$

Here the $\bar{X}$ vector is the branch efforts and the $\bar{X}_s$ vector is the effort sources. $\bar{J}_s$ is a vector of flow sources. These vectors possess a certain order consistent with that used to generate the incidence matrices. The matrix L here is diagonal and the entries along the diagonal are the branch conductances.

Using Kirchhoff's flow law:

$$A\bar{J} = 0 = AL\bar{X} - AL\bar{X}_s + A\bar{J}_s.$$

Using Kirchhoff's potential drop law:

$$0 = ALA^T\bar{c} - ALX_s + A\bar{J}_s.$$

Defining the node–branch admittance matrix

$$Y = ALA^T,$$

and realizing that for the linear case, it always has an inverse, the solution is:

$$\bar{c} = Y^{-1}AL\bar{X}_s - Y^{-1}A\bar{J}_s.$$

The main purpose in showing this solution is to illustrate the systematic manner in which the topological and geometric information are combined in the solution. Nonlinear and linear, non-steady state solutions can be constructed in similar, if not more complicated, systematic ways.

A3.4 SPICE program for intertidal oyster reef ecosystem

```
INTERTIDAL OYSTER REEF ECOSYSTEM
*From R. Dame and B. C. Patten,
*1981. Analysis of Energy Flows in an
*Intertidal Oyster Reef. Marine Ecology
*Progress Series 9:115–124
*Stock Units = KCAL/M2   Flow Units = S/DAY
VPOMO   1     0     2K
IRES    2     0     6.16
IMIC    3     0     5.76
IMEI    4     0     3.58
IDF     5     0     0.43
IPRED   6     0     0.35
R16     1     16    3785.9
V16     16    6     0
R12     1     12    63.33
V12     12    2     0
R45     4     45    11.89
V45     45    5     0
R23     2     23    122
V23     23    3     0
VK62    62    0     4.77M
VK56    56    0     10M
VK25    25    0     640U
VK52    52    0     117M
VK35    35    0     500M
VK24    24    0     7.27M
VK42    42    0     176M
VK34    34    0     500M
G62     6     26    POLY(2) 6 0 62  0 0 0 0 0 1
V26     26    2     0
G56     5     65    POLY(2) 5 0 56  0 0 0 0 0 1
V65     65    6     0
G25     2     125   POLY(2) 2 0 25  0 0 0 0 0 1
V25     125   5     0
G52     5     152   POLY(2) 5 0 52  0 0 0 0 0 1
V52     152   2     0
G35     3     53    POLY(2) 3 0 35  0 0 0 0 0 1
```

```
V35   53   5    0
G24   2    142  POLY(2)  2  0  24   0  0  0  0  0  1
V24   142  4    0
G42   4    124  POLY(2)  4  0  42   0  0  0  0  0  1
V42   124  2    0
G34   3    134  POLY(2)  3  0  134  0  0  0  0  0  1
V34   134  4    0
R1    1    0    1E25
R62   62   0    1E25
R56   56   0    1E25
R25   25   0    1E25
R52   52   0    1E25
R35   35   0    1E25
R24   24   0    1E25
R42   42   0    1E25
R34   34   0    1E25
R4    4    0    1E25
R5    5    0    1E25
*PROGRAM CNTL
.OPTIONS NOPAGE NUMDGT = 7  ITL1 = 1K
+GMIN = 1E-30  PIVTOL = 1E-30
+NODE LIST
.NODESET  V(2) = 10  V(3) = 10  V(4) = 10  V(5) = 10
+V(6) = 10
.END
```

Another SPICE simulation of an ecosystem has been made to duplicate the simulations of Maloney *et al.* (1986) of nitrogen fluxes associated with microbial interactions in plankton communities. This SPICE program is available upon request.

References

Abraham, R. & Marsden, J. E. (1978). *Foundations of Mechanics*. Reading, MA: Benjamin/Cummings.

Atlan, H., Panet, R., Sidoroff, S., Salomon, J. & Weisbuch, G. (1979). Coupling of ionic transport and metabolic reactions in rabbit reticulocytes: bond graph representation. *J. Franklin Inst.*, **308**, 297–309.

Breedveldt, P. C. (1984). *Physical systems theory in terms of bond graphs*. Ph.D. Thesis. Netherlands: Enschede.

Caplan, S. R. & Essig, A. (1986). *Bioenergetics and Linear Nonequilibrium Thermodynamics: The Steady State*. Cambridge, MA: Harvard University Press.

Chandrashekar, M. & Wong, F. C. (1982). Thermodynamic systems analysis. I. A graph theoretic approach. *Energy*, **7**, 539–66.

Curtis, W. D. & Miller, F. R. (1985). *Differential Manifolds and Theoretical Physics*. New York: Academic Press.

Dame, D. F. & Patten, B. C. (1981). Analysis of energy flows in an intertidal oyster ref. *Mar. Ecol. Prog. Ser.*, **5**, 115–24.

Fidelman, M. L. & Mikulecky, D. C. (1986). Network thermodynamic modeling of hormone regulation of active Na^+ transport in cultured renal epithelium. *Am. J. Physiol.*, **250**, C928–91.

Finkelstein, A. & Mauro, A. (1963). Equivalent circuits as related to ionic system. *Biophys. J.*, **3**, 214–37.

Gebben, V. B. (1979). Bond graph bibliography. *J. Franklin Inst.*, **308**, 361–9.

Harrary, F. (ed.) (1967). *Graph Theory and Theoretical Physics*. New York: Academic Press.

Hill, T. L. (1977). *Free Energy Transduction in Biology: The Steady State and Kinetic Formalism*. New York: Academic Press.

Huf, E. G. & Mikulecky, D. C. (1986). Role of topology in bioenergetics of sodium transport in complex epithelia. *Am. J. Physiol.*, **250**, F1107–8.

King, E. L. & Altman, C. (1956). A schematic method of deriving the rate laws for enzyme catalyzed reactions. *J. Phys. Chem.*, **60**, 1375–8.

Kirchhoff, G. R. (1976). On the solution of the equations obtained from the investigation of the linear distribution of galvanic currents. In *Graph Theory, 1736–1936*, ed. N. L. Biggs, F. K. Lloyd & R. J. Wilson, pp. 131–6. Oxford: Oxford University Press.

Lewis, E. R. (1977). *Network Models in Population Biology*. New York: Springer-Verlag.

MacDonald, N. (1983). *Trees and Networks in Biological Models*. New York: Wiley.

Mason, S. J. & Zimmerman, H. J. (1960). *Electronic Circuits, Signals, and System*. New York: Wiley.

Maloney, C. L., Bergh, M. O., Field, J. G. & Newell, R. C. (1986). The effect of sedimentation and microbial nitrogen regeneration in a plankton community: a simulation investigation. *J. Plankton Res.*, **8**, 427–45.

May, J. & Mikulecky, D. C. (1983). Glucose utilization in rat adipocytes: the interaction of transport and metabolism as affected by insulin. *J. Biol. Chem.*, **258**, 4771–7.

Mikulecky, D. C. (1984). Network thermodynamics: a simulation and modeling method based on the extension of thermodynamic thinking into the realm of highly organized systems. *Math. Biosci.*, **72**, 157–79.

Mikulecky, D. C. (1985). Network thermodynamics in biology and ecology: an introduction. In *Ecosystem Theory for Biological Oceanography*, ed. R. E. Ulanowicz & T. Platt, pp. 163–75. *Can. Bull. of Fisheries and Aq. Sci.*, **213**.

Mikulecky, D. C. (1987*a*). Geometric vs. topological information in biological networks: an approach to organization and complexity. *Fed. Proc.*, **46**, 673.

Mikulecky, D. C. (1987*b*). Topological contributions to the chemistry of living systems. In *Applications of Graph Theory and Topology in Chemistry*, ed. R. B. King & D. Rouvray. Studies in Physical and Theoretical Chemistry 51. pp. 115–23. Amsterdam: Elsevier.

Mikulecky, D. C. (1988). Network thermodynamics and complex systems theory: an approach to understanding the relationship between structure and function in biological systems. In *Network Thermodynamics, Heat and Mass Transfer in Biotechnology*, ed. K. R. Diller, pp. 1–7. New York: American Society Mechanical Engineers.

Mikulecky, D. C. & Peusner, L. (1985). Network thermodynamics in bioenergetics: some useful new results and their implications. *Biophys. J.*, **47**, 417a.

Mikulecky, D. C., Sauer, F. A. & Peusner, L. (1986). The choice of coordinate system for the description of dynamical living systems. In *Biophys. of Membrane Transport VIII*, ed. J. Kucera & X. Przestalski, pp. 217–39. Wroclaw: Agricultural Academy.

Mikulecky, D. C. & Thomas, S. R. (1979). Some network thermodynamic models of coupled, dynamic physiological systems. *J. Franklin Inst.*, **308**, 309–26.

Nahikian, H. M. (1964). *A Modern Algebra for Biologists*. Chicago: The University of Chicago Press.

Nash, C. & Jen, S. (1983). *Topology and Geometry for Physicists*. New York: Academic Press.

Odum, H. T. (1972). An energy circuit language for ecological and social systems: its physical basis. In *Systems Analysis and Simulations in Ecology, II*, ed. B. C. Patten. pp. 139–211. New York: Academic Press.

Odum, H. T. & Pinkerton, R. C. (1955). Time's speed regulator: the optimum efficiency for maximum power output in physical and biological systems. *Am. Sci.*, **43**, 331–43.

Oken, D. E., Thomas, S. R. & Mikulecky, D. C. (1981). A network thermodynamic model of glomerular dynamics: application in the rat. *Kidney Int.*, **19**, 359–73.

Onsager, L. (1931*a*). Reciprocal relations in irreversible processes – I. *Phys. Rev.* Ser. 2, **37**, 405–26.

Onsager, L. (1931*b*). Reciprocal relations in irreversible processes – II. *Phys. Rev.* Ser. 2, **38**, 2265–79.

Oster, G. F., Perelson, A. S. & Katchalsky, A. (1973). Network thermodynamics: dynamic modeling of biophysics systems. *Q. Rev. Biophys.*, **6**, 1–134.

Patten, B. C., Bosserman, R. W., Finn, J. T. & Cale, Wm. G. (1976). Propagation of Cause in Ecosystems. In *Systems Analysis and Simulation in Ecology IV*, ed. B. C. Patten. pp. 457–579. New York: Academic Press.

Peusner, L. (1970). *The Principles of Network Thermodynamics and Biophysical Applications*, Ph.D. Thesis. Cambridge, MA: Harvard University Press.

Peusner, L. (1982). Global reaction-diffusion coupling and reciprocity in linear asymmetric networks. *J. Chem. Phys.*, **77**, 5500–7.

Peusner, L. (1983*b*). Electrical network representation of n-dimensional chemical manifolds. In *Chemical Applications of Topology and Graphy Theory*, ed. R. B. King, pp. 379–91, Studies in Physical and Theoretical Chemistry 28. Amsterdam: Elsevier.

Peusner, L. (1983*b*). Hierarchies of irreversible energy conversion systems. I. Linear steady state without storage. *J. Theor. Biol.*, **102**, 7–39.

Peusner, L. (1985*a*). Network thermostatics. *J. Chem. Phys.*, **83**, 1276–91.

Peusner, L. (1985*b*). Premetric thermodynamics: A topological graphical model. *J. Chem. Soc., Faraday Trans.*, **8**, 1151–61.

Peusner, L. (1985*c*). Hierarchies of irreversible energy conversion systems. II. Network derivation of linear transport equations. *J. Theor. Biol.*, **115**, 319–35.

Peusner, L. (1986*a*). *Studies in Network Thermodynamics*. Studies in Modern Thermodynamics No. 5. New York: Elsevier.

Peusner, L. (1986*b*). Hierarchies of irreversible energy conversion systems. III. Why are Onsager equations reciprocal? The Euclidean geometry of fluctuation-dissipation space. *J. Theor. Biol.*, **122**, 125–55.

Peusner, L., Mikulecky, D. C., Caplan, S. R. & Bunow, B. (1985). Unifying graphical approaches to dynamic systems: network thermodynamics, Hill and King–Altman diagrams in reaction–diffusion kinetics. *J. Chem. Phys.*, **83**, 5559–66.

Plant, E. E. & Horowitz, J. M. (1979). Energy conversion in biological systems: I Chemical reactions and ion transport. *J. Franklin Inst.*, **308**, 269–80.

Platt, T., Mann, K. H. & Ulanowicz, R. E. (eds.) (1981). *Mathematical Models in Biological Oceanography*, pp. 59–62. Paris: The Unesco Press.

Rapundalo, S. & Feher, J. J. (1981). Computer simulation of a proposed model of calcium uptake in sarcoplasmic reticulum. *Virginia J. Sci.*, **32**, 136.

Rosen, R. (1985*a*). Information and complexity, In *Ecosystem Theory for Biological Oceanography*, ed. R. Ulanowicz & T. Platt, pp. 221–33. Can. Bull. Fisheries and Aquat. Sci., **213**.

Rosen, R. (1985*b*). Organisms as casual systems which are not mechanisms: an essay into the nature of complexity. In *Theoretical Biology and Complexity*, ed. R. Rosen, pp. 165–203. New York: Academic Press.

Rosen, R. (1985*c*). *Anticipatory Systems*. London: Pergamon.

Rosen, R. (1986). On the scope of syntactics in mathematics and science: the machine metaphor. *Proc. of Brain Research, Artificial Intelligence and Cognitive Science*. At the Systems Interface Workshop, Abisko, Sweden, May 12–16.

Rosen, R. (1990). Biology and physics: an essay in natural physiology. *Physica D* (in press).

Tellegen, B. D. H. (1952). A general network theorem, with applications. *Philips Res. Rep.*, **7**, 259–69.

Thoma, J. U. (1975). *Introduction to Bond Graphs and Their Applications*. New York: Pergamon.

Thomas, S. R. & Mikulecky, D. C. (1978). Network thermodynamic modeling of transepithelial salt and water flow: the kidney proximal tubule. *Am. J. Physiol.*, **235**, F638–48.

Walz, D. & Caplan, S. R. (1986). The thermostatics and thermodynamics of cotransport revisited: a restatement of the Zeroth Law. *Biochem. Biophys. Acta*, **859**, 151–64.

White, J. C. & Mikulecky, D. C. (1986). Application of network thermodynamics to the computer modeling of the pharmacology of anticancer agents: a network model for methotrexate action as a comprehensive example. In *Membrane Transport of Antineoplastic Agents*, ed. I. D. Goldman, pp. 73–111. Section 118 International Encyclopedia of Pharmacology and Therapeutics. New York: Pergamon.

4

Improving predictability in networks: system specification through hierarchy theory

T. F. H. ALLEN AND R. V. O'NEILL

4.1 Introduction

Most ecological research papers fall into one of two categories; those whose authors appear clearly to be empiricists and those written by theoreticians. There are some ecologists who cannot be classified as one type or the other, but the majority will be either an empiricist or theorist, and would be happy to admit the fact. This division is not a bad thing, but what is disturbing is the apparent lack of communication between theory and practice. For example, very few insights from network theory are used or cited in the work of the empiricists who make up the body of the ecological scientific community. We see hierarchy theory as having a special role to play in remedying the situation. The focus on prediction by hierarchy theory is an important characteristic which facilitates communication between theory and empiricism.

Despite the efforts of an ever larger cadre of ecologists, the progress of the discipline seems incremental. One is entitled to ask what is new beyond the description of new details linked to the particulars of the study site or organism. A fair defense might be that we do not have commonplace ecological systems sufficiently described, and so the continuing growth of the database is necessary before moving on to a more synthetic stage. Even so, we are uneasy and find ourselves reminded of the astronomers who collected ordered images with the world's most powerful telescope for some extended time before they thought to calculate how long it would be before they had an image for the whole night sky. They found that the time was unworkably long; the whole sky could not be mapped. The time is also too short to apply the prevailing ideas and data collecting protocols to any sizable portion of the world's ecological systems at the level of detail that characterizes the modern literature. There are too many species interacting in too many combinations for us to describe even a significant minority of critical simple paired relationships.

If the field biologists and experimentalists have a problem, we might look to theory to offer focus. Unfortunately, ecological theorists appear not to be helping at this juncture. Once again it is not a matter of lack of talent or effort. Ecological theorists are very busy, but the problem is that they are working in isolation. Theorists talk to each other, but theory is too often dislocated from practice.

Ecology would benefit considerably if it were organized like physics. In the 1890s in Germany, in the 1920s in Britain, and in the 1940s in the United States, physics reorganized itself into two complementary camps. Today, most physicists are even more clear in their minds than are ecologists as to whether they are experimentalists or theoreticians. Only very rarely does the empiricist do an experiment in physics to see, in some vague way, what will happen; almost always the experiment is to distinguish between the predictions of two explicit competing theories offered by their theoretical counterparts. Would that this were so in ecology. Unfortunately, empirical aspects of ecology are generally not based in theory, or rather are theoretically unfocused. This is because experimentalists in ecology generally do not use the work of theorists, perhaps because they do not have access to it. Theorists and experimentalists in physics work closely in tandem. The theoreticians work to break out of the intellectual frame of the prevailing view. The experimentalists devise ways to press upon the assumptions of the accepted paradigm. In ecology, a similar working relationship needs to be fashioned, and we submit that theory couched in terms of prediction is the way to build the bridge.

We identify a need for theory in ecology that has tangible consequences for field biologists and experimentalists. It is our hope that hierarchy theory can play that role. When we say hierarchy theory, we mean something distinct from network theory. Network theory is deeply hierarchical, and many of its principles translate straight across into hierarchy theory as we narrowly define it. There is, nevertheless, some profit in making a distinction between hierarchy theory, as we apply the term, and network theory. With the distinction drawn, the differences in their strengths can be recognized and each can make its own contribution. Let us explain the division that we see.

4.2 Network algebra and hierarchical sets

Network theory starts with a square matrix of interactions. The columns and rows are often a plant or animal taxon, the interactions represent animals eating each other or plants being eaten. It is not necessary that the columns and rows be so defined, for any ecological entity, for example a nutrient pool, can be represented as a column and a row. From that matrix, network theory proceeds along a path that is substantially algebraic. Many counter-intuitive patterns have emerged from that algebra. Also the formality of the approach has allowed explicit formulation of ideas about ecological interactions that were theretofore taken only on faith. The approach is powerful, and its contributions are undeniable.

We see hierarchy theory as closely related to network theory, but it conducts its business with a difference in style, even when it addresses the same questions. Hierarchy theory makes much of the way that our experience of nature is significantly, although not completely, discrete. Simon (1962) calls this aspect of our world, 'near decomposability'. The 'decomposability' part indicates the discreteness of firm sets, while the 'near' prefix recognizes that some sets are fuzzy (Zadeh, 1965). Hierarchy theory is, therefore, an emphatically set theoretic approach to ecology. Network theory by no means ignores set theory, but it does not use it as a standard opener as it does algebra.

Hierarchy theory is based formally in set theory (e.g. Mesarovic *et al.*, 1970). Often hierarchists use word models, appearing not to be mathematical at all (e.g. Allen & Wileyto, 1983). Any impression of lack of rigor is false, for just under the surface of those word models are formal set theoretic relationships. As an explicit example, Roberts (1987) presents a word model for environment/vegetation relations, having already published a formal set theoretic approach to vegetation analysis (Roberts, 1986).

Field ecologists are not necessarily strong algebraists. This state of affairs often isolates network theory from biologists who focus on fieldwork. The special advantage of hierarchy theory is its intuitive appeal and consequent accessibility. Nouns amount to sets, so hierarchies are readily translated into word based models, which are much more meaningful to most ecologists than the same models expressed overtly in mathematical terms. Because hierarchies and networks often share avenues of investigation, hierarchy theory can help network theory gain the acceptance that it deserves in recognition for its achievements. Network theory might then receive application, which will be of benefit to ecology as a whole. With a language that is comfortable for both network theorists and experimentalists, it may well be possible to use hierarchical approaches to translate the theoretical predictions of network theory into action on the part of field ecologists. The predictions of hierarchists can, in turn, be readily sharpened by the precision tools of their network colleagues. These valuable network theory contributions can be translated into experiments with greater incisiveness because of the language that hierarchy theory uses.

Because hierarchy theory is employed before the algebraic operations on networks, and because of the intuitive appeal of hierarchical word models, there might be a temptation to dismiss the categorization and level identification stages in hierarchy theory as undeserving of much attention. Despite the intuitive appeal of the hierarchical end-product, scaled specification of ecological systems is non-trivial; if it were easy, it would have been done before by thoughtful ecologists of prior decades.

In many new areas of study, the first ground is broken with a formal approach to entification. If one cannot identify what are the appropriate entities, then one has nothing on which to operate algebraically. Set theory, therefore, belongs in the vanguard. Algebra compared to set theory has a very different set of strengths and applications. The formal proofs that algebraic approaches offer are invaluable, and one is hard pressed to imagine how we might otherwise gain the insights that these proofs achieve. Nevertheless, a focus on algebraic analyses belongs later in the succession of ideas. We feel that there is much to be done in defining what the columns and rows of the interaction matrix should be.

4.3 The special role of hierarchy theory

The network theorist works with a defined matrix for the proof at hand. The hierarchy theorist, who could easily be the same person, has a very different role to play. The hierarchist works at defining the columns and rows of the interaction matrix. If the definition of the entities in the system were obvious, then there would be no need for a formal hierarchy theory stage in the modeling process. Definitions are necessary, and every effort must be made to find good ones.

Having found good definitions, the network theorist still has to pay a price. Definitions are always heuristic; they are not right or wrong in any general sense. This is so despite the wrangling about things like the 'true' definition of niche. The danger of a powerful compartmentalization of the world is that it may have too much appeal and result in an overcommitment to what is, after all, a subjective decision. So even if the network theorist defines entities with great heuristic value, he still needs to consider alternative configurations. This is where hierarchy theory can help. We present examples of system respecifications for published models, but first we outline the advantages, in general terms, of system respecification.

The value of alternative entifications is twofold. First the new definitions may work even better. For a given question, the first satisfactory definition is often the one that prevails. Simon (1957) identifies that businessmen do not maximize and use the best decision; they satisfice, that is they muddle through using the first satisfactory decision. It would be foolhardy for ecologists to imagine that they usually end up using the best definitions; ecologists satisfice too. It is good policy to extend the selection of system configurations beyond the first choice. The point here is not that other interesting questions can be addressed by alternative system specifications. Rather the point is that, even for a single question, it is presumptuous to assume that any good definition is the most powerful.

The second value of using alternative definitions for the columns and rows of a network is probably more important. The alternative configuration leads to robustness. Consider the simple case of a system specified as a feedback loop. The loop in question is a two compartment system that amounts to regulation of a population carrying capacity. One compartment is population size and the other is, say, food. In the vicinity of the carrying capacity, should the population numbers increase, then the food will be consumed and its scarcity will bring the population back down again. It is, of course, possible to respecify the compartment called 'food' such that its role is played by a new compartment called 'famine'. All the relationships to famine will be the opposite for food, so the respecification will look very different. The switch changes a plus–minus loop into a minus–plus loop. Nevertheless, the original feedback loop was negative, and the respecified loop, despite its very different formulation, is also negative. The connections in the system may be different, but the property of feedback stability is robust.

The robustness of simple feedback loops to respecification is trivial, but the robustness of other system attributes discovered through network analysis cannot be taken for granted. The network theorist may prove startling system properties, but without additional tests on different but equivalent specifications, he cannot know how much of the counter-intuitive property comes trivially from the particulars of the original specification. A system property that is local to only one system specification should be accepted as significant only with caution.

4.4 Examples of system respecification

There is resistance to specifying systems in alternative ways. W. Scott Overton (personal communication) proposed to use his hierarchical FLEX REFLEX modeling program (White & Overton, 1977) to perform independent specifications for the models for each US biome of the International Biological Program (IBP) in the 1970s. His preliminary effort uncovered errors in some of the models that he was able to reprogram. The proposal remains unfunded, so we do not know what new insights might have been achieved above and beyond increased confidence in the original IBP models had the FLEX REFLEX versions turned up the same patterns.

Although the term 'network theory' was not in the vernacular at the time, Shugart *et al.* (1976) performed a network analysis on a simulated Liriodendron forest in East Tennessee. They were interested in finding if there was much difference in the ability to control ecosystem calcium across changes in the points used to monitor system calcium. They used frequency response analysis and Nyquist analysis on calcium cycling. The model had 29 compartments conceptualized as four subsystems: canopy and soil subsystem, subcanopy subsystem, shrub subsystem and herb subsystem. They variously pulsed the system with calcium from the outside and monitored the compartments for the levels of calcium. There being 29 compartments, the analysis was of the stability of 29 feedback loops from each compartment in turn and from the external source. Control was implemented by using the measurements at the various places in the system to adjust the calcium input from the source. They identified the points in the system that gave the best control of the calcium levels. Each feedback loop between the system compartment and the source is a different specification of the forest. The results indicate that there are significantly different consequences from specifying the different loops as the calcium control system. Shugart *et al.* (1976) speak to the practical implications for management by applied ecologists: 'The results proscribe arbitrary assignment of components as ecosystem monitoring points.' Theory can be brought to bear on practical problems of empirical ecologists.

Taking the Shugart *et al.* (1976) 29×29 network as a point of departure, Chambers *et al.* (1980) undertook major system respecifications. They aggregated the columns and rows according to four separate criteria producing four much smaller models of five, nine, ten and ten compartments. The plant physiologist–soil geologist five compartment model aggregated all organic material into non-metabolizing and fast metabolizing (the physiologists' world view) and three soil water components defined by frequency of movement (soil geologist). The forester's nine compartment version of the Shugart *et al.* (1976) model aggregated all underground material, living and dead organic material and inorganic soil components all in one compartment. The above-ground material was stratified into herb layer and shrub layer. The canopy itself, the most interesting part of the forest for a forester, was left as five separate parts: bole, branch, leaf, fruit, and dead. The third model was stratified by layer but accented compartments with high leaching. The last model was not logically conceptual according to a type of empiricist, but rather was designed to test the theoretical consideration of lumping compartments with commensurate turnover times.

Each aggregation had distinctive properties that were not derivable from the other system specifications. The properties of the new networks were investigated with a Monte Carlo simulation method. For each aggregation, the Monte Carlo method randomly assigned transfer coefficients so that multiple deterministic steady states were calculated. The original model could be considered as a surrogate for the real forest with one critical advantage; it is defined and known and so is a useful reference. The models were variously close to the 'ground truth' of the calcium budget in the original 29 compartment model. All models are representations of external reality, and accordingly compromise aspects of the world that could be captured in some other specification. The four alternative specifications departed from the original model from which they were directly descended, but should be seen not so much as less than the original, but rather as embodiments of alternative world views of Liriodendron forests.

Before either network theory or hierarchy theory were explicitly applied to ecology, Webb *et al.* (1967) investigated the properties of community respecification in a tropical forest. Their purpose was to see which aggregations of species would give the best match with the 'true' arrangement of stands, as defined by a cluster analysis of the full list numbering 818 species. They found that all 269 big tree species gave a perfect reproduction of the original total set. More remarkable was that a mere 65 big tree species, those represented by tall standing individuals, gave an almost perfect reproduction of the original classification. The reason is that the vegetation sits in the context of the dominant individuals. Also, to be a big tree, an individual must have passed through all strata below. At some time in the past big trees must have experienced the environment of the rest of the species. The same is true for the lianes, a small group of species that made a very creditable showing behind big trees.

This result is interesting enough, but of more significance to the present discussion are the groups that failed to characterize the whole community. Webb *et al.* (1967) used some randomly chosen groups as benchmarks against which they could measure success of groups defined by biologically rational criteria. Herbs and epiphytes were spectacular in their failure to represent the total list of species. They were much worse than the random groups with the same number of species, and produced classifications that seemed very much at odds with the community at large. It emerged that herbs and epiphytes were actively demonstrating an ecology that was independent of the ecology of the other species. They appeared to be ordered on the degree of openness of the canopy, a factor that is incidental to most of the other species in the system. In this case, a respecification of the data matrix showed a whole new ecology which was swamped in the community matrix.

4.5 Finding where predictions can be made

Concepts have emerged in parallel in network and hierarchy theory; the principles of incorporation and constraint (O'Neill *et al.*, 1986) have also emerged in network analysis. The advantage of a hierarchical formulation of this broad class of ideas is that they can, indeed, be translated into explicit tests.

Hierarchy theory has a vocabulary distinct from that of network theory. There are some powerful concepts that as yet have no equivalent in network theory. One such conception is the middle number system (Allen & Starr, 1982; Weinberg, 1975). Let us show how the idea of middle number systems along with the concept of constraint can be used to guide experimental design and change conventional investigative strategy.

The phrase 'middle number system' was first used by Weinberg (1975) to mean systems that are unpredictable because they have an unworkable number of parts. Unworkable here does not just mean too many, it also means too few parts. There are two types of systems that fall into the class of predictable systems. These systems have either many fewer or many more parts than the obstinate middle number systems. Small number systems, like planetary systems, are predictable: one gravitational equation involving mass and velocity, an assumption that the sun is the center, substitute various planetary masses and velocities and one has a predictive model of a solar system. In small number systems each part gets its own equation. Large number systems are also predictable. In large number systems there are enough parts to subsume all the astronomical number of parts under the behavior of a few average parts. There are enough parts that their individuality does not matter when it comes to prediction. The gas laws are large number system specifications. A very large number of gas particles is represented by a single mean, the perfect particle, which is the average particle. Small and large number systems are actually the same, but are considered from different perspectives. In a small number system, the significant entities, like the planets, have astronomical numbers of parts. In a sense, the planet Earth is an average. There are so many parts that we do not even bother to consider them. We forget them all and reify the aggregate, the planet. This is because any one part is quite unable to influence significantly the critical average, namely the center of the gravity for the purposes of building a solar system. In middle number systems, the parts are too many to have one equation for each, but too few to ignore their individuality by using an average value.

It is not easy to see how network theory could invoke a tool like the concept of middle number systems. This is because the difference between a small, middle or large number system arises before the network is even specified. The 'number' in middle number systems does not refer to the number of columns and rows in the network matrix. The quality of the columns and rows makes the difference, not the number *per se*. The number in middle number systems refers more to the number of parts that make up the entity which is the column or row. Are there enough microscopic parts to make the entity represented by the column well behaved? The difference between the types of systems is a matter of system specification, not of network algebra once the system is specified.

It may well be that the network theorists can work on middle number systems from the other end. Rather than first identifying a system as middle number and then asserting that it is unpredictable, network theory may be able to start by investigating the patterns of unpredictability and identifying from there that the system is middle number. However, all this is speculation at this point. For the time being, it is not known how the concept of middle number systems relates to the algebra of networks. This is unfortunate, particularly given the unhappy separation of theory and empiricism in ecology to which we referred earlier. Predictions and predictability are common currency for theory and practice. Application of notions of small, middle and large number systems allow hierarchy theory a means of bridging the gap. The notion of predictions applied to small, middle and large number systems is only available to network theory, at the moment, via its association with hierarchy theory.

The reason why states of a middle number system cannot be reliably predicted can be explained in strictly hierarchical terms using the notion of constraint. It is because the upper level, whose behavior is the object of prediction, does not have unambiguous constraint over all the parts. A control system is implied in any assertion as to what is a whole and what are its parts. In a middle number system, this pattern of control does not operate consistently enough to allow reliable prediction. Individual parts in a middle number system can, all by themselves, make a difference to the outcome of what is predicted for the whole system.

Kenneth Boulding is given to one line remarks that startle his audiences. One of these is that it is easy to predict in physics because nothing happens in the systems that physicists study. The prediction is, in fact, that nothing will change. Biology and the social sciences study situations, he says, where all is change, and that is why prediction in biology is so difficult. In a way we agree, but suggest that prediction is possible in biology when, as in physics, nothing happens. All is not change in biology, and the times when there is no change are times when the scientist is working in the vicinity of a constraint. The prediction in biological and ecological systems is that the constraint will persist and continue to prevent the lower levels in the system from achieving certain states. Being impressed with the overtly dynamical nature of their material, ecologists often couch the predictions in terms of non-random change. The non-random change is a reflection of the fact that the constraint has not done anything. When the constraint shifts, then as Boulding asserts, all is change. When the constraint shifts, we have a middle number system. Hierarchy theory with its focus on middle number systems and constraint, has the power of flexibility which allows system respecification so that prediction can be achieved.

Hierarchy theory may provide a key to deciding which specific mechanisms must be understood to predict system behavior. Network theory works on ecosystem interaction or community matrices, the values of which are predicated upon a particular mechanism (e.g. predation). However, that mechanism may not be constraining and so may not be predictive. Since network theory does not often use constraint as an explicit criterion, it does not ask the questions that might lead to a distinction between, on the one hand, constraining mechanisms that would give prediction and, on the other hand, other mechanisms that are valid but not predictive. Even if network theory does derive power from intelligently choosing a constraining mechanism as the basis of the terms in the matrix, it does not address the question of new predictions from new constraining mechanisms. Even when the particular network analysis is powerful, the algebraic methods of network theory do not usually lead from a successfully completed network analysis to a new network based on specifications pertaining to fresh constraining mechanisms. By contrast, successful application of hierarchy theory does lead to new system specifications based on alternative constraining mechanisms. Hierarchy theory is more than a preliminary to network operations.

Most, perhaps all, of the tangible physical systems that are known to be unpredictable are middle number ways of specifying the world. Remember that there are too many parts to enter each as a contingency, but too few to ignore their individuality. The middle number specification of a glass of champagne asks where the individual streams of bubbles will arise. The answer is that we never know, because any one of too-many-to-investigate minor imperfections in the glass could form a nucleus and change the outcome. The predictable large number specification of the same situation would ask a question that is posed at a different level of organization. 'What will happen when we pour champagne into a glass?' So long as it is a good bottle, the prediction will be, 'It will fizz'. By measuring a small set of physical variables that are surrogates for 'good bottle', we can predict if it will fizz or not. That specification is large number because it depends on vapor pressure of particles in excess of Avogadro's number. Note that a network analysis of the glass of champagne would come too late in the decision making process to offer a solution to the problem of prediction. The critical difference comes in specifying the streams of bubbles as the columns of the matrix as opposed to a small set of physical parameters that define a 'good bottle'.

The problem of middle number unpredictability cannot be overcome by more complete specification of initial conditions, or by the collection of more detailed data. The reductionist agenda, which might go down to algal physiology when species as a whole fail to give prediction, is mistaken if the problem is in fact that the specification is middle number. Even full information of states, dynamics and interactions at the level of the whole does not help. It is not a problem of the way the world is ontologically constructed, it is a problem of system description and the consequences of that specification for system measurement. Note the actual glass of champagne can be either a middle number or large number system, depending on the specification that is implied in the question. The middle number system is one of the manifestations of the observation problem well known in quantum mechanics, but much more widely applicable. Hierarchy theory does not so much help us with middle number systems themselves, rather it tells us when we have one and indicates how to respecify in a new way that avoids the pathological unpredictability. Hierarchy theory takes it as a challenge to find in the entification stage of system description a more workable set of constraints implied by the columns and rows of the network matrix.

4.6 Conclusions

Lawton (1987) identifies some promise in hierarchy theory, but complains, fairly enough, that heretofore hierarchy theory has offered few 'clear and new predictions'. We concur, but we have indicated where we think we can remedy that situation. Lawton also expressed disappointment that, 'The ideas [of hierarchy theory] are more qualitative than quantitative'. We agree with his characterization of hierarchy theory, but disagree that this is disappointing. To equate qualitative analysis with lack of rigor is a common misconception. Topologists and logicians would probably be amused to hear that qualitative analysis is denigrated in ecology as being less rigorous. Without any algebra or arithmetic, Allen *et al.* (1984) developed a completely formal method for moving between levels.

In ecology at large, our hope is for a firm partnership in ecological observation between those with a predilection for quantification and those with a greater facility for qualitative description. In the realm of theory, we look forward to teamwork between the algebraists and those who find that they can contribute using set theoretic approaches. We have proposed the contrast between hierarchy theory and network theory not so that the two types of theoreticians shall ignore or compete with the other, for together we can gain more ground. Our purpose in focussing on the differences is to allow each group to retain its identity as we all pull together. That way the practitioners of each type of theory can hold on to their respective points of purchase and so apply leverage from their respective best directions. There is more strength in an equal partnership. Together we can make network theory, or at least its insights, part of the tool kit of empirical investigations performed by the main stream of ecology.

4.7 Summary

A working relationship needs to be fashioned between empiricists and theoreticians in ecology. Hierarchy theory's focus on prediction is an important characteristic which can facilitate communication.

There is a distinction between hierarchy theory and network theory. Network theory is quantitative and algebraic, whereas hierarchy theory is qualitative and a formally set theoretic approach to entification. Its use of word based models gives it intuitive appeal and accessibility to empiricists. Because they share avenues of investigation, hierarchy theory can help network theory gain acceptance and application.

The hierarchist works at defining the columns and rows of the interaction matrix with which the network theorists work. Scaled specification of the ecological system should come before algebraic analyses. Considering alternative configurations leads to robust conclusions and new questions about constraining mechanisms. We outline the advantages of system respecification and present examples from the literature.

There are powerful concepts that as yet have no equivalent in network theory, for example, middle number systems. Hierarchy theory with its focus on unpredictable middle number systems and the constraining mechanisms on system behavior, provides a formal basis for respecification as large or small number systems so that prediction can be achieved. Prediction and predictability are common currency for theory and practice.

References

Allen, T. F. H., O'Neill, R. V. & Hoekstra, T. W. (1984). Interlevel relations in ecological research and management: some working principles from hierarchy theory. USDA Forest Service General Technical Report RM-110, 11pp. Fort Collins: Rocky Mountain Forest and Range Experiment Station.

Allen, T. F. H. & Starr, T. B. (1982). *Hierarchy: Perspectives for Ecological Complexity*, 310pp. Chicago: University of Chicago Press.

Allen, T. F. H. & Wileyto, E. P. (1983). A hierarchical model for the complexity of plant communities. *J. Theor. Biol.*, **101**, 529–40.

Chambers, B. D., O'Neill, R. V. & Gardner, R. H. (1980). *Analysis of uncertainty in ecological models where paradigm is problematic.* Internal document of Environmental Sciences Division of Oak Ridge National Laboratory, Oak Ridge, TN.

Lawton, J. (1987). Problems of scale in ecology. *Nature*, **325**, 206.

Mesarovic, M. D., Macko, D. S. & Takahara, Y. (1970). *Theory of Hierarchical, Multilevel Systems.* New York: Academic Press.

O'Neill, R. V., DeAngelis, D. L., Waide, J. B. & Allen, T. F. H. (1986). *A Hierarchical Concept of Ecosystems.* Princeton: Princeton University Press.

Roberts, D. W. (1986). Ordination on the basis of fuzzy set theory. *Vegetatio*, **66**, 123–31.

Roberts, D. W. (1987). A dynamical systems perspective on vegetation theory. *Vegetatio*, **69**, 27–33.

Shugart, H. H. Jr., Reichle, D. E., Edwards, N. T. & Kercher, J. R. (1976). A model of calcium-cycling in an East Tennessee Liriodendron forest: model structure, parameters and frequency response analysis. *Ecology*, **57**, 99–109.

Simon, H. A. (1957). *Models of Man.* New York & London: Wiley.

Simon, H. A. (1962). The architecture of complexity. *Proc. Amer. Phil. Soc.*, **106**, 467–82.

Webb, L. J., Tracey, J. G., Williams, W. T. & Lance, G. N. (1967). Studies in the numerical analysis of complex rain-forest communities. II. The problem of species sampling. *J. Ecol.*, **55**, 525–38.

Weinberg, G. M. (1975). *An Introduction to General Systems Thinking*. New York: Wiley.

White, C. & Overton, W. S. (1977). *Users manual for the FLEX2 and FLEX3 model processors for the FLEX modelling paradigm.* Corvallis: Forest Research Laboratory, OSU.

Zadeh, L. A. (1965). Fuzzy sets. *Inf. Control*, **8**, 338–53.

Part II

Network approaches to problems in ecosystems ecology

5

Network trophic dynamics: an emerging paradigm in ecosystems ecology

M. HIGASHI, B. C. PATTEN AND T. P. BURNS

5.1 Classical trophic dynamics

5.1.1 *The Lindeman–Hutchinson paradigm*

Lindeman's (1942) paper, 'The trophic-dynamic aspect of ecology', opened a new approach to understanding, in terms of their trophic interactions, how different life-forms co-exist. He took an ecosystem approach, considering not only living but also non-living entities that exchange energy and matter. The use of energy as a common measurable currency allows a general specification of the entire trophic-interaction structure of an ecosystem, thus making it possible to compare ecosystems on a holistic basis. Comparison of the same ecosystem at different times reveals successional change. Biogeographic comparison between ecosystems can reveal the divergence of a single species and the convergence of distinct species in different communities. Lindeman applied Hutchinson's (1941) notion of progressive efficiency of energy transfer between trophic levels as an index of ecosystem function with which to make such comparisons. This whole attempt was certainly a benchmark for a holistic and quantitative, albeit phenomenological (Ulanowicz, 1986), approach to ecology.

5.1.2 *The dilemma*

Lindeman (1941 *a*, *b*, 1942) approximated energy flows in the Cedar Bog Lake 'food-cycle', confirming that trophic interactions in nature fit a cyclical network model with detritivory and non-predatory mortality connecting the grazing and detrital food chains, and omnivory further confounding the identity of trophic levels. To apply the notions of trophic level and progressive efficiency (Hutchinson, 1941), Lindeman

(1942) reduced the ecosystem to a single simple chain of trophic levels connected in series by a one-way flow of energy (Fig. 5.1).

Imposing the single food-chain model on the real ecosystem led to a logical dilemma between two tendencies that he identified in his data:

(i) increasing progressive efficiencies

$$\frac{\lambda_2}{\lambda_1} < \frac{\lambda_3}{\lambda_2} < \ldots \tag{1}$$

and, (ii) increasing respiration-loss ratios

$$\frac{R_1}{\lambda_1} < \frac{R_2}{\lambda_2} < \ldots \tag{2}$$

where λ_k and R_k, respectively, represent the gross production rate and respiration rate of the kth trophic level. These two factual relations are a logical contradiction in the single food-chain model, which neglects feedbacks through detrital-decomposer compartments and jump-forward flows, thereby proving the significance of those neglected flows and the invalidity of the simple food chain as a model of ecosystems, where in general compartments cannot be assigned unambiguously to trophic levels (Burns, 1989).

Evidence for the significance of feedbacks (e.g. detritivory) and jump-forward flows, often due to omnivory, are accumulating for diverse ecosystems, including marine (e.g. Pomeroy, 1974, 1985), freshwater (e.g. Kerfoot & DeMott, 1984) and soil ecosystems (e.g. Swift *et al.*, 1979; Coleman, 1985). Nevertheless, the single food-chain model and its

Fig. 5.1. The single chain model used by Lindeman (1942) as the theoretical basis for ecosystem trophic dynamics, where compartments are assigned wholly to trophic levels and the latter are connected in series by a one-way flow of energy. The symbols λ_k, NP_k, R_k and Λ_k, respectively, represent the summed gross production rates, net production rates, respiration rates and standing stocks of the compartments constituting the kth trophic level.

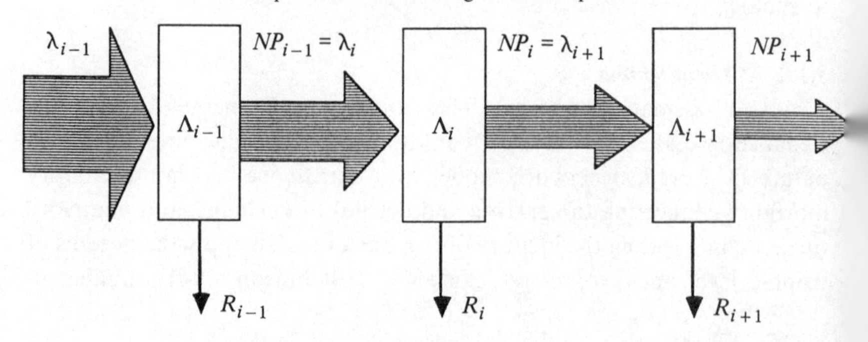

variants have been employed for years in theoretical studies of trophic dynamics, and have led naturally to the conclusion that low efficiencies of energy transfer cause a rapid decrease in available energy with increasing trophic level, and ultimately, a shortage of energy by the fourth or fifth step (e.g. Patten, 1959; Odum, 1968; Gates, 1985).

In the following sections, we will discuss two ways to expand the Lindeman–Hutchinson paradigm of trophic dynamics from the unrealistic single food-chain model to the most general case of any 'food network' with jump-forward flows and feedbacks (cycles). The first approach (Patten *et al.*, 1990) generalizes the notion of progressive efficiency for the general network case so as to make it useful in quantifying the trophic relations (i.e. the energy-availing relations) among species (or more aggregated units) within a natural ecosystem. The second approach (Higashi *et al.*, 1989.) focuses on the idea of trophic levels, and derives a transformation of any given food network such that its organization along the 'trophic-level axis' becomes explicit.

5.2 Energy transfer and utilization coefficients in food networks

In this section, we briefly discuss two different generalizations for the notion of progressive efficiency.

First, progressive efficiencies may be considered *transfer coefficients* for single food-chain models. That is, in a single food-chain model, what fraction of the total inflow to (i.e. gross production of) compartment j is ultimately transferred to compartment $j+1$ is given by λ_{j+1}/λ_j, the one-step progressive efficiency. And, the fraction of the total inflow to j that is transferred to i ($i > j$) is given by the progressive efficiency from j to i, which is the product of the one-step progressive efficiencies:

$$\frac{\lambda_i}{\lambda_j} = \frac{\lambda_{j+1}}{\lambda_j}\frac{\lambda_{j+2}}{\lambda_{j+1}}\cdots\frac{\lambda_i}{\lambda_{i-1}}. \tag{3}$$

Therefore, we can generalize the notion of progressive efficiency, which is defined only for the unnatural single food-chain model, by characterizing it as a transfer coefficient, which can be defined and evaluated for general food networks, as we now show in the following.

To evaluate correctly the transfer coefficients for general food networks, we must account for all paths from one compartment to another that *do not contain cycles including the latter compartment*. Consider the general case of a food network consisting of n compartments with stationary flows and standing stocks. Let x_i denote the standing stock in compartment i, f_{ij} the flow from compartment j to i, z_i the inflow from the out-of-system environment to compartment i, and y_i the outflow from compartment i to the environment. The constancy of the standing stock in compartment i implies the balance between the total flow into i and the total flow out of i; let T_i denote this balanced total flow through compartment i, i.e. $T_i \equiv \sum_{j=1}^{n} f_{ij} + z_i = \sum_{j=1}^{n} f_{ji} + y_i$. For any path ψ connecting any two compartments j and i, let $g(\psi)$ denote the product of *donor normalized flows*, $g_{kh} \equiv f_{kh}/T_h$, associated with path ψ; for instance, if path ψ is represented by $(j \to k \to h \to i)$, then $g(\psi) = g_{ih} \cdot g_{hk} \cdot g_{kj} = (f_{ih}/T_h) \cdot (f_{hk}/T_k) \cdot (f_{kj}/T_j)$. Then, for any path ψ connecting j to i, $g(\psi)$ clearly represents the fraction of T_j that will follow path ψ to i. Thus, the transfer coefficient from j to i (the fraction of T_j that is *transferred* to i) is given by $\sum g(\psi)$, where the summation is taken over all the *parallel paths* ψ from j to i, that is, the paths that contain no cycles including i.

Consider 'particles' that travel in the network and assume that they transfer from compartment j to i by probability g_{ij} and leave the network from compartment i to the environment by probability y_i/T_i. Then, clearly $g(\psi)$ represents the probability that a particle initially in compartment j will follow path ψ to compartment i. Thus, the above summation $\sum g(\psi)$ also represents the probability that a particle initially in compartment j will ever reach i before going out of the network (termed *reaching probability*), and therefore, must lie between 0 and 1. This probability of reaching from j to i, which equals the transfer coefficient from j to i, can be calculated from the *normalized flow* (*transfer probability*) *matrix* $G = (g_{ij}), i, j = 1, 2, \ldots, n$, as follows (Appendix A5.3):

$$[\text{transfer coefficient from } j \text{ to } i] = \frac{n_{ij}}{n_{ii}} \tag{4}$$

where n_{ij} is the (i,j)th element of *fundamental matrix*, $N \equiv (I-G)^{-1}$, of the stochastic process that governs the particles.

Lindeman's progressive efficiency can be extended thus for the general food network case. As a transfer coefficient, a measure easily derivable from normalized flow matrix G, it represents the total resource-consumer relationship in ecosystems determined collectively by the direct and indirect energy (matter)-supply pathways.

The total portion of T_j that is ultimately experienced by consumer i, as opposed to transferred to it, can be evaluated as the summation $\sum T_j g(\psi)$ over all paths ψ from j to i, *including recycling paths around* i (i.e. paths containing cycles including i). This *utilization coefficient* of consumer i with respect to resource T_j is then given by the summation $\sum g(\psi)$ over all paths ψ. For networks without cycles, utilization coefficients equal the corresponding transfer coefficients, because where there are no recycling paths there can be no energy reutilization (by this we mean reincorporation, not retransformation of the energy). Thus, for the single food-chain model, the simplest case of a cycle-free network, utilization coefficients equal transfer coefficients, and therefore, the notion of utilization coefficient is another extension of the concept of progressive efficiency.

The utilization coefficient of i with respect to T_j can be calculated simply as n_{ij}, because $n_{ij} = \sum_{k=0}^{\infty} g_{ij}^{(k)}$ (Appendix A5.2) and $g_{ij}^{(k)}$, the (i,j)th element of the kth power G^k of matrix G, equals the summation $\sum g(\psi)$ taken over all paths of length k connecting j to i (Appendix A5.1). The quantity n_{ij} can be interpreted also as the average number of visits to compartment i that a particle initially in j will make before leaving the network (Appendix A5.3).

While transfer coefficients, being probabilities, lie between 0 and 1, utilization coefficients, if feedback flows via cycles in the network are great enough, can exceed one due to reutilization of the energy recycling around closed paths again and again until all of it dissipates, or is lost from the system (Patten, 1985). This is clear also from their interpretation as average number of visits.

Not only can utilization coefficients exceed one, but they may also increase with trophic levels along a food chain in the network. The portion of T_i, the energy actually experienced by consumer i at each unit moment, that originates in primary production z_j is given as z_j multiplied by n_{ij}. Thus, given primary production z_j a consumer 'further away' from the producer, but with a greater utilization coefficient, could actually utilize in its production more z_j-originating energy than a consumer 'closer' to the input. Increasing utilization coefficients, n_{ij}, implies an increase of the productivity of (i.e. ultimate inflow into) j, which denies the classic dogma that, due to the second law of thermodynamics, energy input (available energy) to a compartment necessarily decreases with trophic level (*sensu* Lindeman, 1942), causing energy shortages at higher trophic levels and, consequently, limits on the length of food chains.

5.3 Unfolding food networks: implementation of the trophic partitioning idea

5.3.1 *Trophic partitioning*

The difficulties in determining the trophic level to which any given organism uniquely belongs motivates the idea that an organism may belong to more than one trophic level (Riley, 1966; Cummins *et al.*, 1966;

Odum, 1968). The essence of this idea is an abstract partitioning of each compartment of a food network into portions belonging to different trophic levels. A more recent implementation of the idea partitions the energy flow through each compartment into several classes according to how many steps of intercompartmental transfer (i.e. how many trophic levels) it experienced before it arrived at the compartment (Ulanowicz & Kemp, 1979; Levine, 1980; Ulanowicz, 1990). In the next section, we present a general method for 'unfolding' any given food network along the direction of trophic process, from the bottom (source components or primary producers) to higher trophic levels, by partitioning all network components, both standing stocks (compartments) and flows, according to the past history of the energy in the system (Higashi *et al.*, 1989). This *food network unfolding* will serve in the subsequent sections as the basis for a second expansion of the Lindeman–Hutchinson paradigm of trophic dynamics.

5.3.2 *Food network unfolding*

To illustrate the unfolding of food networks, let us first consider two simple examples:

Fig. 5.2. (*a*) A hypothetical simple food web with four compartments, discussed in Example 1. (*b*) The partitioning, along the trophic-level axis, of each compartment's standing stock and each intercompartmental flow in the food web depicted in (*a*).

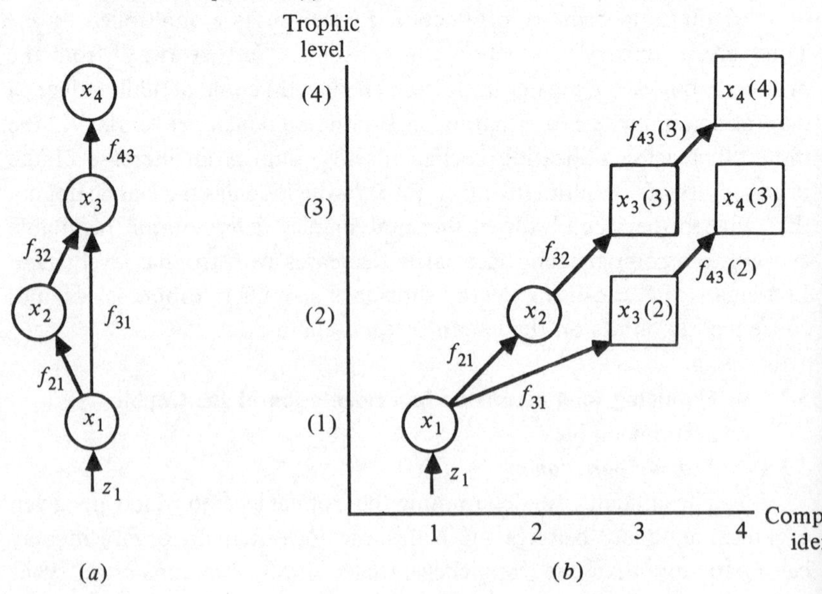

Example 1. In the simple food web depicted in Fig. 5.2(*a*), compartment 3 feeds on both compartments 1 and 2, thus it may be considered to, in part, belong to the second trophic level (2) and also, in part, to trophic level 3. Fig. 5.2(*b*) explicitly shows this partitioning of biomass (standing stock) of compartment 3 along the trophic-level axis. If we only know the flows and standing stocks, then it is reasonable to make this partitioning in proportion to flow ratios. Thus, the standing stock x_3 of compartment 3 has component $x_3 f_{31}/(f_{31}+f_{32})$ on trophic level 2, and component $x_3 f_{32}/(f_{31}+f_{32})$ on level 3. Compartment 4, which feeds on compartment 3, should be partitioned into two parts, in proportion to the partitioning of x_3: the part feeding on the trophic level 2 component of compartment 3 belongs to level 3, and that feeding on the level 3 part of 3 belongs to trophic level 4. Further, intercompartmental flow f_{43} can be partitioned into two parts: one connecting the components of compartment 3 and 4 at lower levels, and the rest connecting the second components on higher levels.

Example 2. Next, consider the food network with cycle depicted in Fig. 5.3(*a*). If the current biomass x_4 of compartment 4 is traced backward (against the arrows) in time to the point where it entered the system via the inflow z_1, a portion of it will lead to z_1 in four steps, $4 \to 3 \to 2 \to 1 \to z_1$; this portion may be considered to belong to trophic level 4. But the rest of x_4, branching at compartment 2 in the time-backward tracing, will lead back to 4, through path $4 \to 3 \to 2 \to 4$. A portion of this remainder will, with further tracing backward, lead to z_1 through the same path the first portion took, $4 \to 3 \to 2 \to 1 \to z_1$, now seven steps from the start; this portion may be viewed as being on trophic level 7. The portion of x_4 on level 4 is evaluated as $x_4 f_{21}/(f_{21}+f_{24})$, while that on level 7 is given as $x_4[f_{24}/(f_{21}+f_{24})][f_{21}/(f_{21}+f_{24})]$. Continuing the same procedure, we may partition x_4 into those portions belonging to levels $(4+3k), k = 0, 1, \ldots,$ evaluated as $x_4[f_{24}/(f_{21}+f_{24})][f_{21}/(f_{21}+f_{24})]^k$. In a similar way, x_2 and x_3 also may be decomposed along the trophic-level axis, and each of the intercompartmental flows may be partitioned in accordance with the partitioning of its donor compartment's biomass [Fig. 5.3(*b*)].

We now extend to the general case of any food network the ideas of partitioning flows and standing stocks, and unfolding networks along the trophic-level axis. Consider a food network consisting of n compartments with stationary standing stocks and flows. We make a theoretical assumption: for any $i, j (i, j = 1, 2, \ldots, n$ and $i \neq j)$, the fraction of the standing stock in compartment i that has last (in the previous step) visited compartment j is given by the flow ratio $g_{ij}' \equiv f_{ij}/T_i$. Under this assumption, as we show below, we can derive the *trophic-level partitioning* for standing stocks and flows, i.e. the partitioning into portions each of which belongs to one of the trophic levels, where the portion belonging to the kth trophic level is defined as the portion that has experienced $(k-1)$ steps of intercompartmental transfer since arriving in the system (see Section 5.7 for a comment on the relationship between transfer- and assimilation-based trophic levels in food networks).

First, we derive the trophic-level partitioning of standing stock x_j, $j = 1, 2, \ldots, n$. This follows from a more fundamental idea, depicted in Fig. 5.4, of partitioning x_j into portions each of which corresponds uniquely to one of the paths connecting a source input z_h to compartment j. In this

Fig. 5.3. (*a*) A hypothetical cyclic food web, discussed in Example 2. (*b*) The partitioning, along the trophic-level axis, of each compartment's standing stock and each intercompartmental flow in the food web depicted in (*a*).

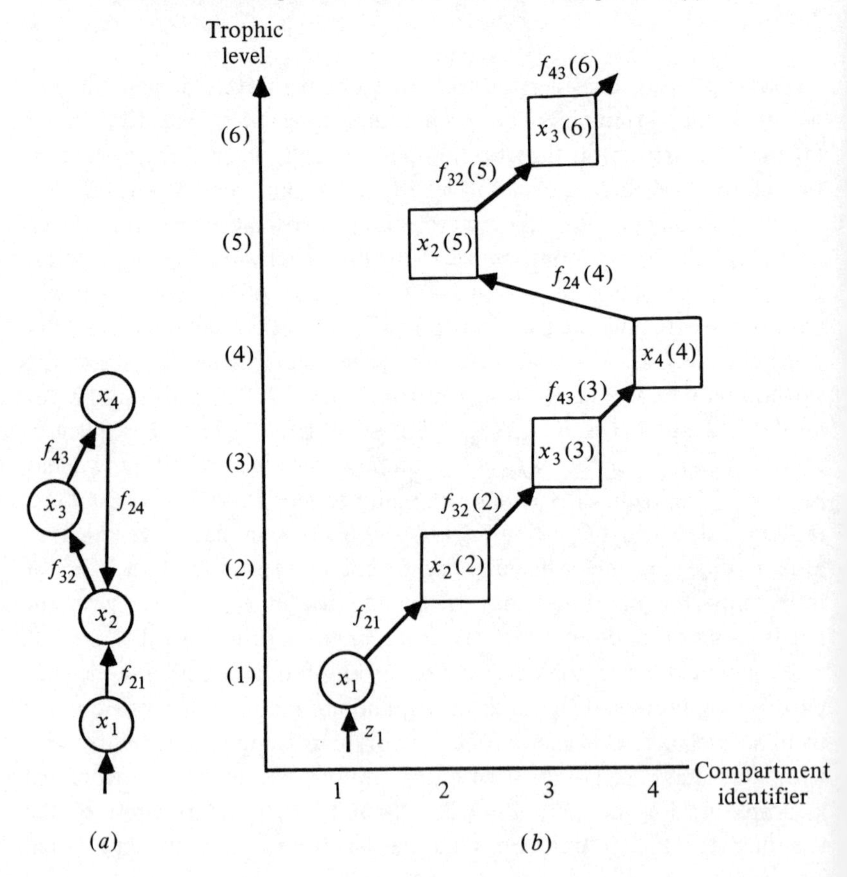

correspondence, the portion of x_j associated with a path ψ connecting h to j is defined as the portion of x_j that, when its history is traced back, goes back all the way along path ψ to the source inflow z_h by which it came into the system. These portions, therefore, do not overlap each other, and cover the entire x_j; they constitute a partition of x_j.

In this *path-partitioning* of x_j, by applying repeatedly the theoretical assumption that the fraction of x_i that originates in x_j is given by the ratio $g_{ij}' = f_{ij}/T_i$, it follows that the *fraction* of x_j associated with a path ψ connecting h to j is given by $g'(\psi)(z_h/T_h)$. Then, the kth trophic-level fraction of x_j, which is the fraction associated with the paths of length $(k-1)$ connecting h $(h = 1, 2, \ldots, n)$ to j, equals

$$\sum_{h=1}^{n} \sum_{\psi} g'(\psi)(z_h/T_h), \tag{5}$$

where the latter summation is taken over all paths ψ of length $(k-1)$ connecting h to j. The summation $\sum_{\psi} g'(\psi)$ in (5) can be calculated as $g'^{(k-1)}_{jh}$, the (j,h)th element of the $(k-1)$th power of matrix $G' \equiv (g_{ij}')$, $i,j = 1, 2, \ldots, n$, of *recipient normalized flows* $g_{ij}' = f_{ij}/T_i$. [This follows from applying to *recipient normalized flow matrix* $G' = (g_{ij}')$ the same argument

Fig. 5.4. Path-partitioning of standing stock in compartment j, x_j. In this partitioning, each portion of x_j is associated with a path ψ leading to j from the source inflow z_h by which it came into the system.

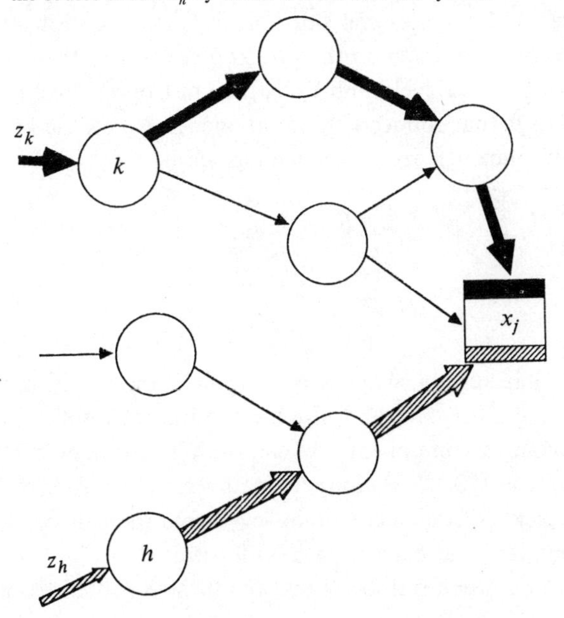

as developed for donor normalized flow matrix G in Appendix A5.1.] Therefore, the kth trophic-level fraction of x_j is evaluated as

$$\sum_{h=1}^{n} g'_{jh}{}^{(k-1)}(z_h/T_h). \tag{6}$$

We thus have the following trophic-level partitioning of standing stock x_j:

$$x_j = x_j(1) + x_j(2) + \ldots, x_j(k) + \ldots, \tag{7}$$

where $x_j(k), k = 1, 2, \ldots,$ denotes the *kth trophic-level portion of* x_j, and can be evaluated as

$$x_j(k) = x_j \sum_{h=1}^{n} g'_{jh}{}^{(k-1)}(z_h/T_h). \tag{8}$$

The trophic-level partitionings of flows f_{ij} and y_j $(i, j = 1, 2, \ldots, n; i \neq j)$ are clearly proportional to that of the standing stock of their donor compartment, x_j, that is, the kth trophic-level fractions of f_{ij} and y_j equal that of x_j:

$$f_{ij} = f_{ij}(1) + f_{ij}(2) + \ldots + f_{ij}(k) + \ldots \tag{9}$$

$$y_j = y_j(1) + y_j(2) + \ldots + y_j(k) + \ldots \tag{10}$$

where $f_{ij}(k)$, the kth trophic-level portion of f_{ij} (the portion of f_{ij} that connects the component $x_j(k)$ on the kth trophic level to the component $x_i(k+1)$ on the $(k+1)$th level), and $y_j(k)$, the kth trophic-level portion of y_j (the portion of y_j that connects the component $x_j(k)$ on the kth trophic level to the environment), are respectively given by

$$f_{ij}(k) = f_{ij} \sum_{h=1}^{n} g'_{jh}{}^{(k-1)}(z_h/T_h) \tag{11}$$

$$y_j(k) = y_j \sum_{h=1}^{n} g'_{jh}{}^{(k-1)}(z_h/T_h). \tag{12}$$

Now that we have established, by eqs. (7)–(12), the partitioning of all elements (standing stocks, x_j, and flows, f_{ij} and y_j) of the food network into their trophic-level components, we can 'unfold' the network from the source inputs z_j $(j = 1, 2, \ldots, n)$ along the pathways of transfer by 'cutting out' the components of the next trophic level from the network elements as they are encountered along any path, as illustrated in Fig. 5.5, where it is assumed for the sake of simplicity that the turnover rate is common to all compartments so that the size of standing stock is proportional to that of throughflow.

5.3.3 *Characteristics of unfolded food networks*

The network resulting from unfolding any food network does not involve cycles, because all paths run toward higher trophic levels (Fig. 5.5).

Fig. 5.5. Illustration of trophic network unfolding. The figure on the right shows the unfolding of the hypothetical food network with three compartments pictured on the left, where it is assumed for simplicity that the turnover rate is equal for all compartments so that the size of standing stock is proportional to that of throughflow.

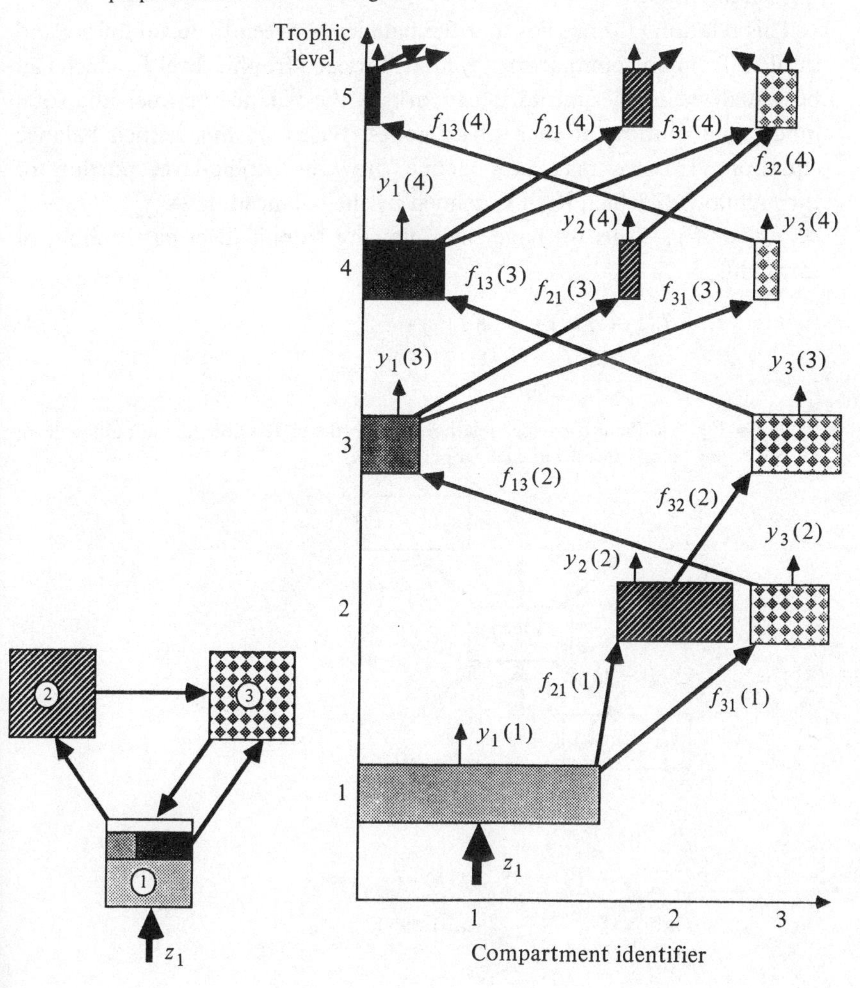

Note that, for $k = 1, 2, \ldots$, we have the following relation (Fig. 5.6).

$$\sum_{m=1}^{n} f_{jm}(k-1) = \sum_{i=1}^{n} f_{ij}(k) + y_j(k) = T_j \sum_{h=1}^{n} g'_{jh}{}^{(k-1)}(z_h/T_h) \tag{13}$$

where for $k = 1$ the left-hand side of (13) is defined as z_j, because (i) the left-hand side and the middle of (13) are, respectively, the kth trophic-level portions of the total inflow to j ($\sum_{m=1}^{n} f_{jm} + z_j$) and the total outflow from j ($\sum_{i=1}^{n} f_{ij} + y_j$); (ii) the total inflow to j and outflow from j both equal T_j; and (iii) the k-th trophic-level fractions of the total inflow to j and outflow from j both equal that of the standing stock in j, which is $\sum_{h=1}^{n} g'_{jh}{}^{(k-1)}$ (z_h/T_h). (For a more formal proof along this heuristic line of reasoning, see Appendix A5.5.)

This relation (13) implies that the balance between the total inflow and outflow from any compartment j holds for each trophic level k, which can be visualized in the unfolded network as the balance between the total inflow and outflow at each of its nodes. Based on this refined balance equation (13), we may now define the kth trophic-level portion of throughflow T_j, which itself is defined by the balanced flows $\sum_{m=1}^{n} f_{jm} + z_j = \sum_{i=1}^{n} f_{ij} + y_j$. Thus we have the following trophic-level partitioning of throughflow T_j

$$T_j = T_j(1) + T_j(2) + \ldots + T_j(k) + \ldots, \tag{14}$$

Fig. 5.6. Refined balance relation between the total inflow to and outflow from any compartment j at each trophic level k.

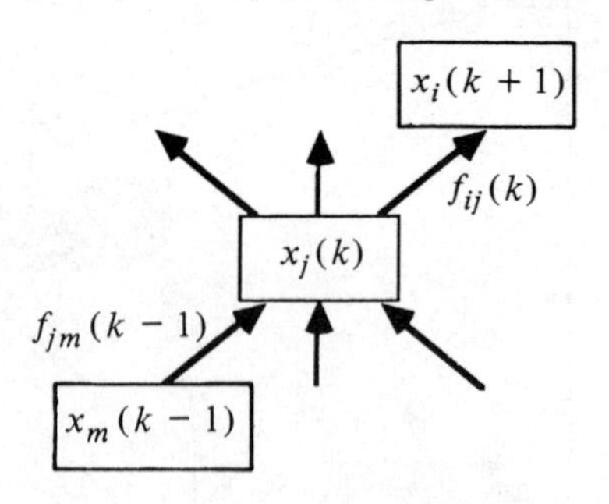

$$\sum_{m=1}^{n} f_{jm}(k-1) = \sum_{i=1}^{n} f_{ij}(k) + y_j(k)$$

(inflows) (outflows)

where $T_j(k)$, the kth trophic-level portion of throughflow T_j, is given by

$$T_j(k) = T_j \sum_{h=1}^{n} g'_{jh}{}^{(k-1)} (z_h/T_h). \tag{15}$$

Equation (15) indicates that the partitioning of throughflow T_j is also proportional to that of x_j; $T_j(k)/T_j = x_j(k)/x_j$.

In deriving this trophic network unfolding, we employed the time-backward perspective, whereby we traced energy (matter) being consumed by compartment j back to its origins thereby determining its trophic status. However, the network resulting from unfolding a food network permits the time-forward tracing of flows consistent with the corresponding tracing in the original network. That is, if we trace the future fate of each flow (i.e. evaluate what fraction follows which path) in an unfolded network, it would be consistent with the corresponding tracing in the original network before unfolding. For instance, for any path $\psi = (h \to a \ldots \to b \to j)$ of length $k-1$ connecting source inflow z_h to j, the fraction of inflow z_h that follows path ψ from h to j is evaluated as

$$\frac{f_{jb}(k-1)}{T_b(k-1)}, \ldots, \frac{f_{ah}(1)}{T_h(1)}, \tag{16}$$

in the unfolded network, which coincides with the same quantity evaluated in the original network:

$$\frac{f_{jb}}{T_b}, \ldots, \frac{f_{ah}}{T_h}. \tag{17}$$

This is because the forward donor normalized flow (the fraction of flow transferred in one step) from compartment j to i remains the same after unfolding a food network, i.e. for any $i, j = 1, 2, \ldots, n$ $(i \neq j)$, and $k = 1, 2, \ldots,$

$$\frac{f_{ij}(k)}{T_j(k)} = \frac{f_{ij}}{T_j}, \tag{18}$$

which follows from (11) and (15). Therefore, unfolding a food network and the resultant acyclic network (Fig. 5.5) may be viewed as a visual tracing of the fate of source inflows z_h, $h = 1, 2, \ldots, n$, along the trophic process (levels) until it dissipates from one of the compartments. This elucidates how each inflow z_j will branch and be distributed at each trophic transfer.

The notion of trophic level (*sensu* Burns 1989) is unambiguous in the unfolded representation even for food networks with many jump-forward and cyclic flows. This fact establishes a foundation for the second approach to expanding the Lindeman–Hutchinson paradigm of trophic dynamics from the single food-chain model to the most general case of food networks.

5.4 A generalization of Lindeman–Hutchinson trophic dynamics

Any given food network, however complicated, can be unfolded by the foregoing method, and the classical notions of trophic dynamics from Lindeman (1942) and Hutchinson (1941) find their place in this new domain. Indeed, those notions are naturally extended to the unfolded representation.

5.4.1 *The macroscopic trophic chain*

As illustrated in Fig. 5.7, we can construct a single 'macro' trophic chain from any unfolded trophic network by combining those components (flows or standing stocks) on the same trophic level. Formally, to a given unfolded network consisting of the trophic-level components, $f_{ij}(k), y_j(k), T_j(k), x_j(k)$, for $k = 1, 2, \ldots$, there corresponds a macro chain comprising gross production rates λ_k, respiration rates R_k, and standing stocks Λ_k on trophic levels $k = 1, 2, \ldots$, which are defined as

$$\lambda_k = \sum_{j=1}^{n} T_j(k) = \sum_{j=1}^{n} \left[\sum_{i=1}^{n} f_{ij}(k) + y_j(k) \right] \tag{19a}$$

$$R_k = \sum_{j=1}^{n} y_j(k) \tag{19b}$$

$$\Lambda_k = \sum_{j=1}^{n} x_j(k), \tag{19c}$$

for $k = 1, 2, \ldots$ And, in particular,

$$\lambda_1 = \sum_{j=1}^{n} z_j. \tag{20}$$

The macro chain is the vertical structure of the corresponding unfolded network, because it preserves the information regarding interlevel movement of energy or matter in the network while neglecting the information regarding intercompartmental movement, i.e. for $f_{ij}(k)$, k is preserved while i and j are neglected.

Fig. 5.7. 'Macro' trophic chain for the unfolded trophic network in Fig. 5.5, obtained by combining those components (flows or standing stocks) on the same trophic level, to form the gross production rates λ_k, respiration rates R_k, and standing stocks Λ_k of that trophic level $k = 1, 2, \ldots$.

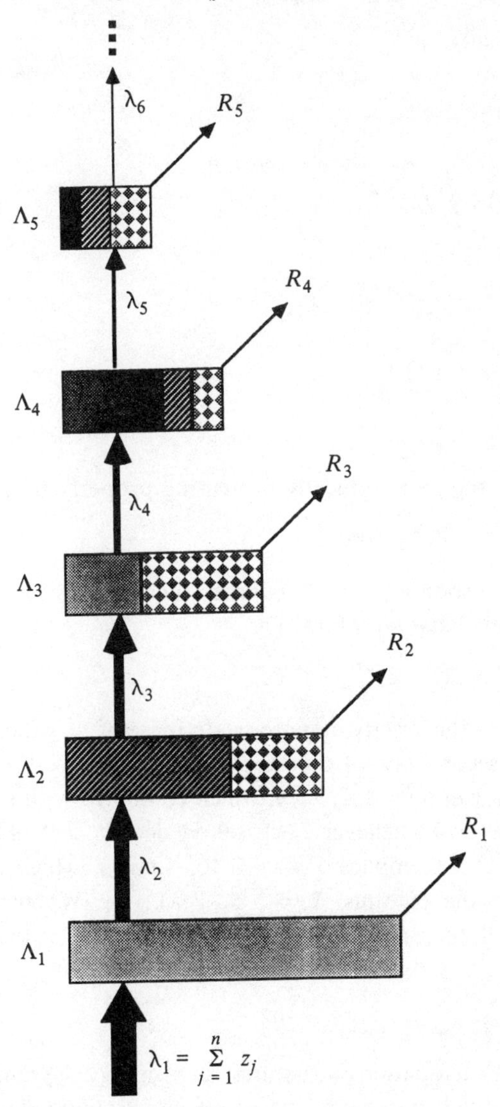

5.4.2 *Energy pyramid*

By definition

$$\lambda_k = R_k + \lambda_{k+1} \tag{21}$$

for $k = 1, 2, \ldots$ Formally,

$$\begin{aligned}
\lambda_k &= \sum_{j=1}^{n} T_j(k) \\
&= \sum_{j=1}^{n} [y_j(k) + \sum_{i=1}^{n} f_{ij}(k)] \\
&= \sum_{j=1}^{n} y_j(k) + \sum_{i=1}^{n} \sum_{j=1}^{n} f_{ij}(k) \\
&= R_k + \sum_{i=1}^{n} T_i(k+1) \\
&= R_k + \lambda_{k+1}
\end{aligned} \tag{22}$$

This equality implies the monotonically decreasing property of λ_k, i.e.

$$\lambda_1 \geqslant \lambda_2 \geqslant \ldots. \tag{23}$$

Thus, once inflow to a specific level k^* is zeroed (i.e. $\lambda_{k^*} = 0$) for the first time, subsequent levels have no inflow, i.e.

$$\lambda_1 \geqslant \lambda_2 \geqslant \ldots > \lambda_{k^*} = \lambda_{k^*+1} = \ldots 0. \tag{24}$$

We can further show the strictly monotonic decrease of λ_k, which stems ultimately from the second law of thermodynamics. Due to the second law, we have $y_j > 0$ for any $j = 1, 2, \ldots, n$, which combined with (11) and (15), implies that $y_j(k) > 0$ whenever $T_j(k) > 0$. Hence, $R_k > 0$ as long as $\lambda_k > 0$, [$\lambda_k = \sum_{j=1}^{n} T_j(k) > 0$ implies $T_j(k) > 0$ for some j, which in turn implies $y_j(k) > 0$ for that j, thus $R_k = \sum_{j=1}^{n} y_j(k) > 0$]. We may thus conclude that if $\lambda_k > 0$, then $R_k = \lambda_k - \lambda_{k+1} > 0$, thus $\lambda_k > \lambda_{k+1}$; therefore, we have *always*

$$\lambda_1 > \lambda_2 > \ldots > \lambda_{k^*} = \lambda_{k^*+1} = \ldots 0. \tag{25}$$

This is the energy dissipation structure, or 'energy pyramid', a generalization of the Lindeman–Hutchinson energy pyramid (Lindeman, 1942) now proven to necessarily hold for their simple chain model. In other words, network unfolding is a transformation of trophic networks that reveals the energy dissipation (pyramidal) structure of a given network. Other efforts to achieve this transformation are not general, because they require acyclic networks (trees), or exclude cycled energy (e.g. Ulanowicz & Kemp, 1979; Ulanowicz, 1990).

We note that the sequence of standing stocks, $\Lambda_1, \Lambda_2, \ldots$, does not necessarily decrease monotonically. To see this, we first note the relationship derived from (8) and (15) together with (19*c*):

$$\Lambda_k = \sum_{i=1}^{n} x_i(k) = \sum_{i=1}^{n} (x_i/T_i)\, T_i(k). \tag{26}$$

Comparing this with the definition of λ_k, eq. (22), we note that Λ_k is a weighted sum of $T_i(k)$ with weighting factors (x_i/T_i), the turnover time at compartment i, whereas λ_k is the simple sum of $T_i(k), i = 1, 2, \ldots, n$. It is easy to see the possibility that $\Lambda_k < \Lambda_{k+1}$ despite $\lambda_k > \lambda_{k+1}$, because of uneven weighting factors, i.e. differences in turnover time among compartments. Therefore, standing stocks Λ_k do *not* always appear as pyramids, even though flows (gross production rates) λ_k must do so in the trophic-level unfolding.

5.4.3 *Progressive efficiency*

Because of the energy pyramid for interlevel flows λ_k, we can now define 'progressive efficiency' as the ratio

$$0 \leqslant \lambda_{k+1}/\lambda_k < 1 \tag{27}$$

for any level k for which inflow exists, $\lambda_k > 0$. This is an extension of Lindeman's progressive efficiency which was defined only for his single chain model. The network unfolding transformation makes possible a definite measure for progressive (transfer) efficiency of energy (matter) processing along trophic levels in any given food network.

5.5 Distribution pattern of flows connecting successive trophic levels

In the previous section, we showed how food network unfolding provides a characterization of the 'vertical' trophic structure of ecosystems with the macro chain, the energy dissipation pyramid and progressive efficiency. We now derive measures of 'horizontal' trophic structure made possible by network unfolding.

5.5.1 *Some ideas from information theory for characterizing flow distribution pattern*

MacArthur (1955) presented the idea that the diversity of alternative paths over which energy passes from primary producers to consumers should have a positive correlation with stability of a community (food web). Recently, Ulanowicz (1980, 1986) proposed a measure, called *ascendency*, that is the product of the total flow through a system and the mutual information measure, which represents the degree of articulation (specificity) in flow from one compartment to another in the network. He considered the total flow as a measure of growth and the mutual information measure as representing the degree of development, and hypothesized that the ascendency of a flow network increases as the network grows and develops.

Applying these information (communication) theory ideas in the context of unfolded trophic networks, we derive below a measure that represents the degree of articulation or specification in energy (matter) transfer from the components on the kth trophic level to those on the $(k+1)$th level, and a measure that represents the degree of generalization or specialization of a component on the $(k+1)$th trophic level feeding on the kth level.

5.5.2 *Some measures to characterize the distribution of flows between successive trophic levels*

We now introduce several measures to characterize the configuration of flows between successive trophic levels and their ecological interpretation. For relevant knowledge about Shannon information theory, see Appendix AI.2 from the Introduction to this book.

Focusing on the flows transferred from trophic level $k-1$ to k, that is $f_{ij}(k-1), i,j = 1,2,\ldots,n$ (Fig. 5.8) and ignoring the flows from level $k-1$

Fig. 5.8. The configuration of flows transferred from trophic level $k-1$ to k, $f_{ij}(k-1)$, $i,j = 1,2,\ldots,n$, ignoring the flows from level $k-1$ that are not transferred to the next level, but instead leave the system, $y_j(k-1)$, $j = 1,2,\ldots,n$.

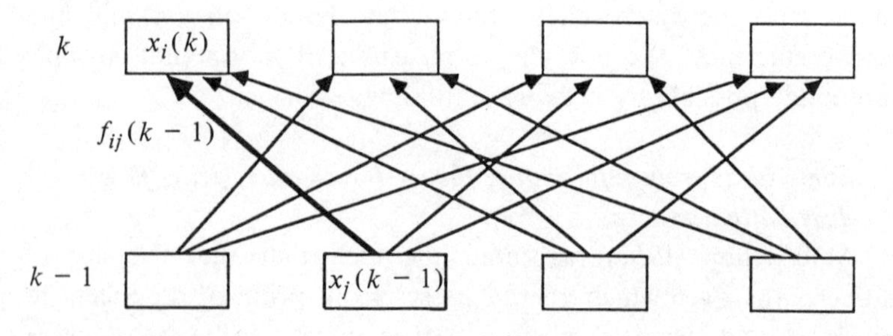

that are not transferred to the next level, but instead leave the network system, $y_j(k-1), j = 1, 2, \ldots, n$, we note the following relations:

$$\lambda_k = \sum_{i=1}^{n} \sum_{j=1}^{n} f_{ij}(k-1) \tag{28a}$$

$$T_i(k) = \sum_{j=1}^{n} f_{ij}(k-1) \tag{28b}$$

$$T_j(k-1) - y_j(k-1) = \sum_{i=1}^{n} f_{ij}(k-1). \tag{28c}$$

In these equations, λ_k represents the total gross production at trophic level k, $T_i(k)$ the gross production by the kth trophic-level component of compartment i, and $T_j(k-1) - y_j(k-1)$, the net production by the $(k-1)$th level component of compartment j (the gross production by it that is transferred to and utilized by the next level components for their production).

Now imagine that we can label an energy 'particle' and trace its movement within an ecosystem, we observe that it has left trophic level $k-1$ and arrives at level k. We ask where (which component on level $k-1$) it has come from. With only knowledge of the flow pattern between levels $k-1$ and k ($f_{ij}(k-1), i, j = 1, 2, \ldots, n$), it is reasonable to answer that the

Fig. 5.9. The diversity of the resources utilized by the kth trophic level as a whole which feeds on the $(k-1)$th level components in accordance with the flow distribution, $[T_j(k-1) - y_j(k-1)]$, $j = 1, 2, \ldots, n$.

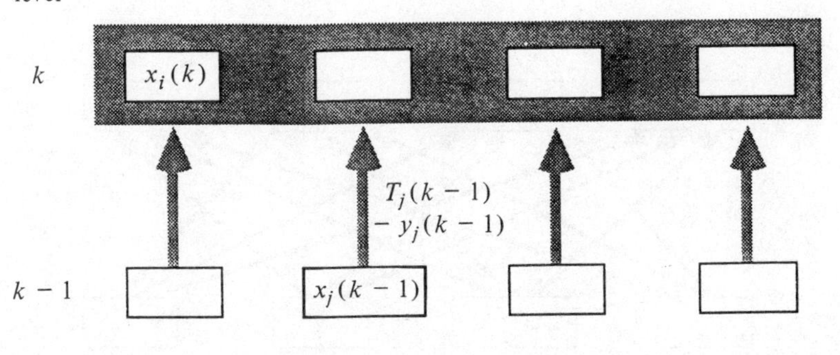

$$\lambda_k = \sum_{j=1}^{n} [T_j(k-1) - y_j(k-1)]$$

particle has come from the level $(k-1)$th component of compartment j with probability proportional to the flow ratio:

$$p_j(k) \equiv [T_j(k-1) - y_j(k-1)]/\lambda_k \tag{29}$$

(Fig. 5.9). The uncertainty about which compartment on level $k-1$ a particle arriving at level k has come from is measured by the Shannon entropy of the probability distribution, $(p_1(k), p_2(k), \ldots, p_n(k))$, i.e.

$$D(k) \equiv H(p_1(k), p_2(k), \ldots, p_n(k))$$

$$= -\sum_{j=1}^{n} p_j(k) \log p_j(k). \tag{30}$$

Note that $D(k)$ may be considered a measure of the *diversity of resources utilized by trophic level k as a whole*, feeding on the $(k-1)$th level components.

The uncertainty about the departure place on level $k-1$ of a particle arriving at level k, provided that it arrives at compartment i on level k, is expressed as the entropy

$$C_i(k) \equiv H(c_{i1}(k), c_{i2}(k), \ldots, c_{in}(k))$$

$$= -\sum_{j=1}^{n} c_{ij}(k) \log c_{ij}(k) \tag{31}$$

where $c_{ij}(k)$ denotes the conditional probability that a particle has come

Fig. 5.10. The diversity of the resources utilized by the kth trophic level component of compartment i which feeds on the $(k-1)$th level components in accordance with the flow distribution $(f_{i1}(k-1), f_{i2}(k-1), \ldots, f_{in}(k-1))$.

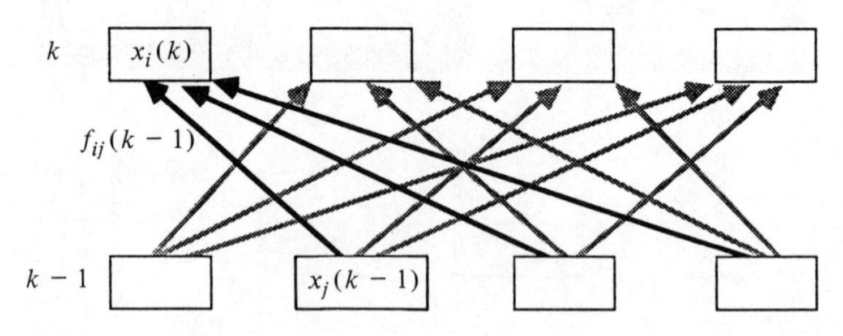

from compartment j on level $k-1$, given that it arrives at compartment i on level k, viz.

$$c_{ij}(k) \equiv f_{ij}(k-1)/T_i(k). \tag{32}$$

We note here that $C_i(k)$ may be interpreted as the *diversity of the resources utilized by the kth trophic-level component of compartment i*, feeding on the $(k-1)$ level components (Fig. 5.10); thus, it may be considered as a measure for the *degree of trophic generality* of that component.

It is possible that $C_i(k)$, the uncertainty about the departure place on level $k-1$ of a particle arriving at level k, under the condition that it arrives at compartment i on level k, is greater than $D(k)$, the uncertainty about the same without any condition, that is, without knowledge of where it arrives. Fig. 5.11 gives an example; the diversity of compartment 1 on level k, $C_1(k)$, attains its maximum ($\log_2 2 = 1$), while the diversity of level k as a whole, $D(k)$, equals nearly zero, because the flow from compartment 1 is super-dominant.

If we assume that a particle coming from level $k-1$ arrives at compartment i on level k with probability proportional to the flow ratio,

$$p'_i(k) \equiv T_i(k)/\lambda_k \tag{33}$$

Fig. 5.11. An illustrative example for the case where $C_i(k)$, the uncertainty about the departure place on level $k-1$ of a particle arriving at level k, given that it arrives at the compartment i on level k, is greater than $D(k)$, the uncertainty about the same without any condition, i.e. without any knowledge of where it arrives.

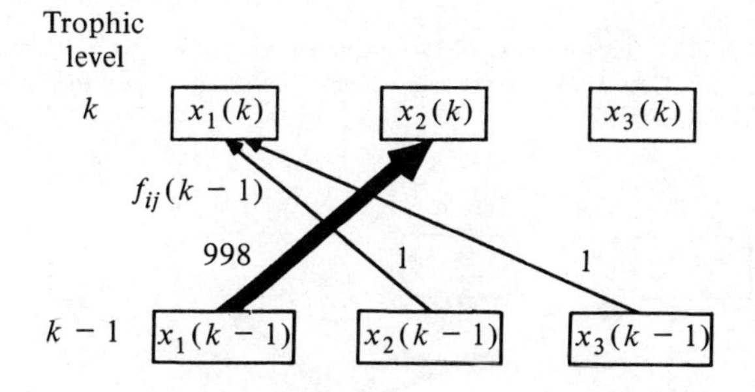

$$D(k) = -\left(\frac{1}{1000}\log_2\frac{1}{1000} + \frac{1}{1000}\log_2\frac{1}{1000} + \frac{998}{1000}\log_2\frac{998}{1000}\right) \approx 0$$

$$C_1(k) = -\left(\tfrac{1}{2}\log_2\tfrac{1}{2} + \tfrac{1}{2}\log_2\tfrac{1}{2}\right) = \log_2 2 = 1$$

$$\therefore D(k) < C_1(k)$$

(Fig. 5.12), then the average uncertainty about the departure place on level $k-1$ of a particle arriving at level k, given the knowledge of its arrival place at level k, is evaluated as the average of entropies $C_i(k)$ weighted by these probabilities $p'_i(k), i = 1, 2, \dots, n$,

$$C(k) \equiv \sum_{i=1}^{n} p'_i(k)\, C_i(k). \tag{34}$$

Note that this conditional entropy $C(k)$ can be interpreted as the *average diversity of the resources utilized by the kth trophic-level components*, feeding on the $(k-1)$th level components; thus, it may be considered as a measure for the *average degree of trophic generality of the kth trophic-level components*.

Also note that

$$D(k) \geqslant C(k) \tag{35}$$

always holds (Appendix AI.2 of the Introduction to this book). That is, the diversity of resources utilized by trophic level k, as a whole, is greater than that of its individual components on average. This relation (35) may be interpreted as follows: When we know which compartment on level k a particle arrives at after leaving level $k-1$, we can better infer which component on level $k-1$ the particle came from. The uncertainty about the place of the particle's departure is reduced from $D(k)$ to $C(k)$ due to knowledge about the place of its arrival. This reduction of uncertainty, $D(k) - C(k)$, represents the amount of information about the place of the particle's departure gained by knowing the place of its arrival.

Fig. 5.12. The diversity of the consumers on the kth trophic level feeding on the $(k-1)$th trophic level as a whole in accordance with the flow distribution $T_i(k)$, $i = 1, 2, \dots, n$.

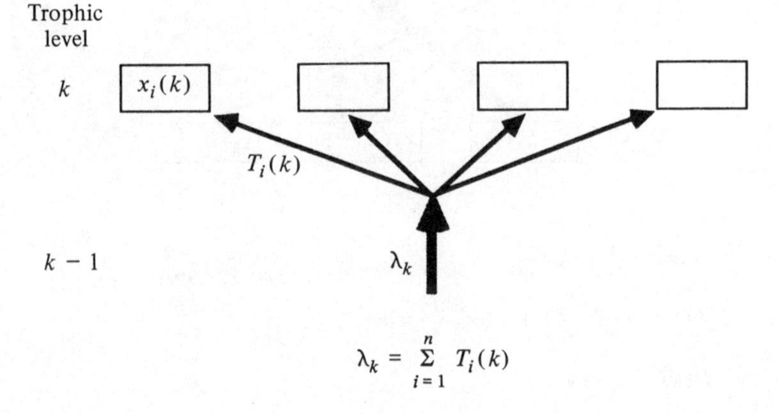

The quantities presented above are derived from the time-backward perspective, or the viewpoint of resource utilization by consumers. Alternatively, we can derive the time-forward counterparts of these quantities, based on the viewpoint of the energy (matter) supply by resources.

First, let $D'(k)$ be the entropy of probability distribution $(p'_1(k), p'_2(k), \dots, p'_n(k))$ given by (33) (Fig. 5.12):

$$D'(k) \equiv H(p'_1(k), p'_2(k), \dots, p'_n(k))$$

$$= -\sum_{i=1}^{n} p'_i(k) \log p'_i(k). \tag{36}$$

This $D'(k)$ represents the *diversity of the consumers on trophic level* k *feeding on trophic level* $k-1$ *as a whole.*

The uncertainty about the arrival place on level k of a particle coming from level $k-1$, provided that it departs from compartment j on level $k-1$, is expressed as the entropy

$$C'_j(k) \equiv H(c'_{1j}(k), c'_{2j}(k), \dots, c'_{nj}(k))$$

$$= -\sum_{i=1}^{n} c'_{ij}(k) \log c'_{ij}(k), \tag{37}$$

where $c'_{ij}(k)$ denotes the conditional probability that a particle will go to compartment i on level k, given that it departs from compartment j on level $k-1$:

$$c'_{ij}(k) = f_{ij}(k-1)/[T_j(k-1) - y_j(k-1)]. \tag{38}$$

$C'_j(k)$ represents the *diversity of the consumers on trophic level* k *feeding on the* $(k-1)$*th level component of compartment* j (Fig. 5.13).

The average uncertainty about the arrival place at trophic level k of a particle coming from level $k-1$, given the knowledge of its departure place at level $k-1$, is evaluated as the average of entropies $C'_i(k)$ weighted by $(p_1(k), p_2(k), \dots, p_n(k))$, the probabilities about the departure place on trophic level $k-1$ of a particle that arrives in trophic level k:

$$C'(k) \equiv \sum_{i=1}^{n} p_i(k)\, C'_i(k) \tag{39}$$

This conditional entropy $C'(k)$ can be interpreted as the *average diversity of the consumers on trophic level* k *feeding on the* $(k-1)$*th level components.*

Note that, like (35), the relation

$$D'(k) \geqslant C'(k) \tag{40}$$

always holds (Appendix AI.2 of the Introduction of this book). That is, the diversity of consumers of trophic level $k-1$ as a whole is greater than that of its individual components, on average. This relation (40) may be interpreted as follows: When we know which component on level $k-1$ a particle departed from, we can better infer which component on level k the particle will go to. The uncertainty about the place of the particle's arrival is reduced from $D'(k)$ to $C'(k)$ due to the knowledge about the place of its departure. This reduction of uncertainty, $D'(k)-C'(k)$, represents the amount of information about the place of the particle's arrival gained by knowing the place of its departure.

We now note the following relation that relates the time-backward and forward perspectives:

$$\begin{aligned} M(k) &\equiv D(k)-C(k) \\ &= D'(k)+D(k)-J(k) \\ &= D'(k)-C'(k) \end{aligned} \tag{41}$$

where $J(k)$ denotes the entropy of the distribution $(p_{ij}(k)\,|\,i,j=1,2,\ldots,n)$ of joint probabilities that a particle has come from compartment j on level $k-1$ and is arriving at compartment i on level k, i.e.

$$\begin{aligned} J(k) &\equiv H(p_{ij}(k)\,|\,i,j=1,2,\ldots,n) \\ &= -\sum_{i=1}^{n}\sum_{j=1}^{n} p_{ij}(k)\log p_{ij}(k), \end{aligned} \tag{42}$$

Fig. 5.13. The diversity of the consumers on the kth trophic level that feed on the $(k-1)$th trophic level component of compartment j in accordance with the flow distribution $(f_{i1}(k-1), f_{2j}(k-1), \ldots, f_{nj}(k-1))$.

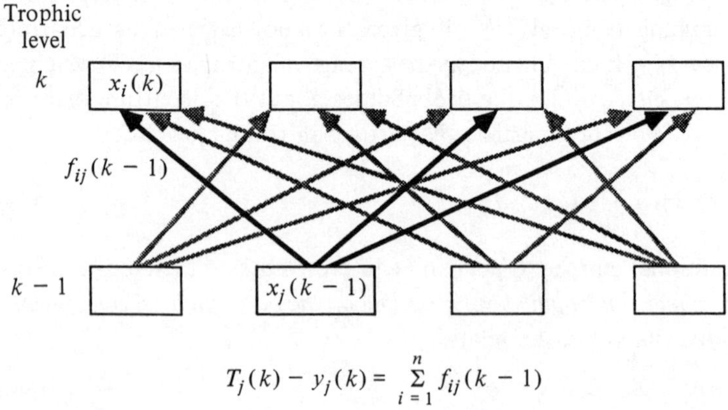

Trophic level

k $x_i(k)$

$f_{ij}(k-1)$

$k-1$ $x_j(k-1)$

$$T_j(k) - y_j(k) = \sum_{i=1}^{n} f_{ij}(k-1)$$

with

$$p_{ij}(k) = p'_i(k)\, c_{ij}(k) = c'_{ij}(k)\, p_j(k) = f_{ij}(k-1)/\lambda_k. \tag{43}$$

This $J(k)$ represents the *diversity of the overall flow configuration of energy-matter transfer from trophic levels* $k-1$ *to* k.

Relation (41) indicates that for a particle that has left level $k-1$ and arrives at level k, the amount of information about the particle's departure place gained by the knowledge of its arrival place equals the amount of

Fig. 5.14. Hypothetical situations to illustrate the characteristic properties and develop an interpretation of 'mutual information', $M(k) = D(k) - C(k)$.

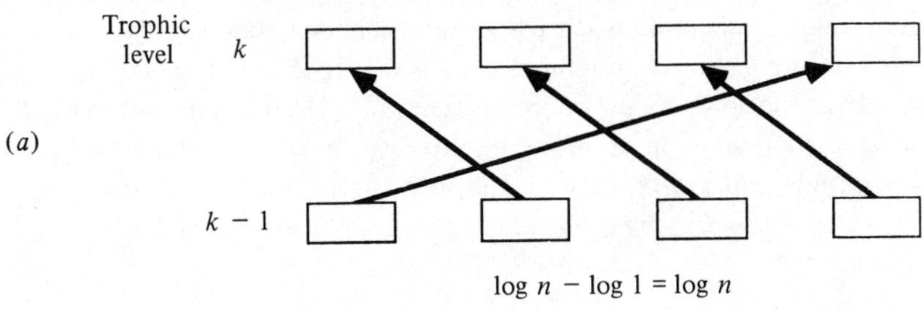

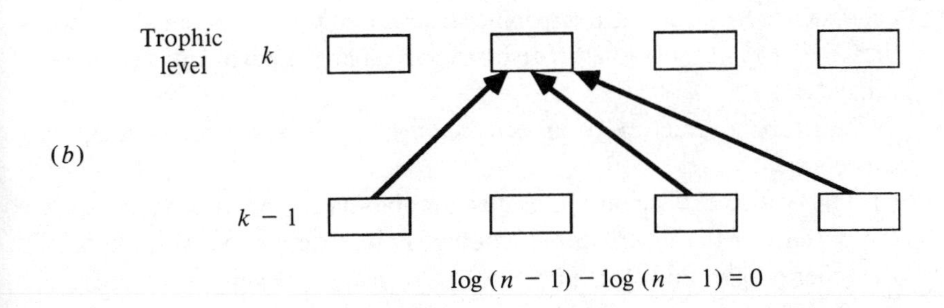

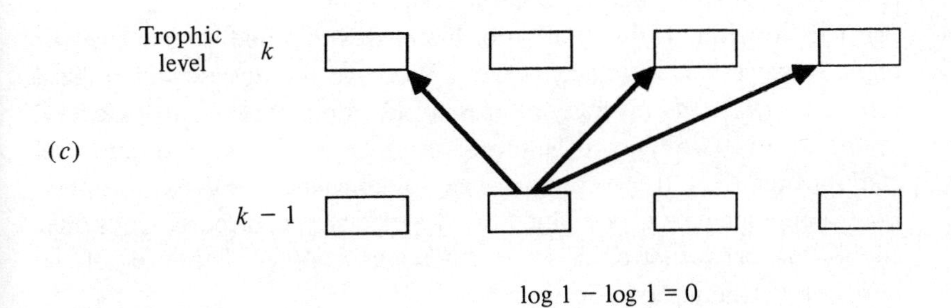

information about the particle's arrival place gained by the knowledge of its departure place. Based on this relationship, let $M(k)$ be called the mutual information between the departure and arrival place of a particle transferred from trophic level $k-1$ to k. The middle expression of $M(k)$ in (41) suggests that $M(k)$ represents a degree of correlation (interdependence) between the two probability distributions about the departure and arrival places; the sum, $D'(k)+D(k)$, represents the uncertainty about the departure and arrival place if they are independent of each other, and $J(k)$ represents the actual uncertainty, which is always less due to the interdependence of the two events involved.

Note that $M(k)$ increases as $D(k)$ increases or $C(k)$ decreases, and that it attains its maximum when $D(k)$ attains its maximum, $\log n$, at the same time $C(k)$ goes to its minimum, $\log 1 = 0$ (Fig. 5.14). That is, $M(k)$ gets larger as the components of trophic level $(k-1)$ are connected to the kth level components in a more one-to-one fashion by flows of equal magnitude. Thus, $M(k)$ may be interpreted as a *measure of specificity in the resource–consumer connection* (*correspondence*) *between trophic levels* $k-1$ *and* k.

5.6 Trophic niche and ecosystem structure

Unfolding a food network reveals two complementary views of ecosystem organization: the trophic structure of the ecosystem as a whole and the trophic position or diversity of each compartment within the ecosystem.

The trophic structure of an ecosystem comprises two complementary aspects:

(i) The 'vertical' distribution of flow and biomass over trophic levels, i.e. the pattern of the macroscopic trophic chain. The dissipative pattern of the energy pyramid for flows between trophic levels, the series of progressive efficiencies, λ_{k+1}/λ_k, defined for the macro chain and the energy content of trophic levels, Λ_k, are quantitative expressions (indices) of the vertical structure (see Section 5.4).

(ii) The 'horizontal' distribution of biomass within one trophic level, or of flow between two adjacent levels. We have the diversity of utilized resources, $D(k)$, and diversity of gross production, $D'(k)$, of trophic level k, the average diversity of utilized resources of the kth trophic-level components, $C(k)$, the average diversity of consumers of the kth trophic-level components, $C'(k)$, and the specificity in resource–consumer connection between trophic levels $k-1$ and k, $M(k)$, as indices of the horizontal structure (see Section 5.5).

A *general* and operational method for characterizing the trophic structure of ecosystems, in particular one with *quantitative* indices, makes it possible to compare different ecosystems and detect trends in ecosystem change, such as succession.

Quantitative measures of the trophic position or diversity of compartments constituting a food network are useful when studying, on an objective basis, the functional divergence of a single group (e.g. species population, trophic guild) and the convergence of distinct groups under the influence of the ecosystems in which they live. If a food network is constructed at such a resolution that each compartment of the network corresponds to one species population, then the unfolding of the network explicitly captures the 'trophic niche' of each member species along the two complementary axes:

(i) The trophic niche of species i is represented by the distribution, $T_i(1)/T_i, T_i(2)/T_i, \ldots$, of species i over all trophic levels, i.e. the 'vertical' axis. $T_i(k)/T_i$ is the fraction of species i belonging to trophic level k; it indicates the extent to which species i serves as resource for trophic level $k+1$ and at the same time feeds on trophic level $k-1$. From this latter aspect the quantity

$$D_i = -\sum_{k=1}^{\infty} \frac{T_i(k)}{T_i} \log_2 \frac{T_i(k)}{T_i} \tag{44}$$

may be considered to represent the trophic-level diversity of species i; that is, D_i represents *the degree to which species i is a generalist–specialist on the trophic-level axis*. Since species i belongs to trophic level k with probability $T_i(k)/T_i$, the *average trophic level of species i* is given by

$$[\text{average trophic level of species } i] = \sum_{k=1}^{\infty} k \frac{T_i(k)}{T_i}, \tag{45}$$

which can be calculated from elements of matrix $N = (I-G')^{-1}$ by the formula

$$\sum_{j=1}^{n} n'_{ij} \tag{46}$$

(Appendix A5.6). This average, together with the diversity, D_i, of species i's trophic-level distribution, characterizes the vertical trophic position of the species.

(ii) The position or niche of species i on the 'horizontal' axis of the kth trophic level is represented by the resource utilization distribution of i as a *consumer*, $c_{ij}(k) = f_{ij}(k)/T_i(k)$, $j = 1, 2, \ldots, n$ and the energy (matter)

Table 5.1. *The values of* Λ_k *(standing stock in kcal/m*^{2}*),* λ_k *(gross production rate in kcal/m*$^2\cdot$*day),* λ_{k+1}/λ_k*,* $D(k)$*,* $C(k)$ *and* $M(k)$ *for trophic levels* $k = 1, 2, \ldots, 9$ *of the oyster-reef ecosystem model.*

k	Λ_k	λ_k	λ_{k+1}/λ_k	$D(k)$	$C(k)$	$M(k)$
1	2000	41.47	0.393	0	0	0
2	761	16.31	0.714	0.147	0	0.147
3	30	11.66	0.450	1.428	0.596	0.832
4	145	5.25	0.713	1.464	0.466	0.998
5	77	3.74	0.561	1.795	0.748	1.047
6	34	2.10	0.584	1.915	0.816	1.099
7	27	1.23	0.614	1.801	0.741	1.060
8	14	0.75	0.578	1.887	0.799	1.088
9	9	0.44	0.597	1.864	0.780	1.084

supply distribution from i as a *resource*, $c'_{mi}(k) = f_{mi}(k)/[T_i(k) - y_i(k)]$, $m = 1, 2, \ldots, n$. $C_i(k)$ represents species i's degree of generality–specificity in exploiting trophic level $k-1$, whereas $C'_i(k)$ represents the extent to which species i is a popular (common) resource shared and utilized by the components on trophic level (k) (see Section 5.5).

Similar measures exist for compartments in models with coarser resolution (e.g. trophic guilds), but the interpretation and importance of these for ecological theory has yet to be explored.

5.7 An example

To conclude, we apply the network unfolding analysis to a real food network, the oyster-reef ecosystem model which was presented in the Introduction of this book as a common reference model.

Many published food or energy flow networks include non-living or mixed compartments, for example 'ooze', 'detritus and decomposers' and 'particulate organic carbon', which combine both living and non-living entities; the oyster-reef model includes a non-living 'deposited detritus' compartment with transport losses in lieu of respiration. Models with such compartments cause discrepancies between a transfer based definition of trophic level and one based on assimilation events. Only in networks where all compartments comprise living, respiring organisms and where energy transfers include only assimilated energy will the transfer based analysis presented here coincide with the assimilation definition of trophic level. This is a problem for modeling and model analysis, but does not reduce the significance of the foregoing general development, which presumes an equivalence between transfer and assimilation events. Models can be constructed according to the above requirements or analysis programs written to distinguish and correct for non-living compartments and egestive flows. We have done neither for this example, and as a result it is meant only as a demonstration of network trophic dynamics.

The 'macro' trophic chain for this detritus based food network with cycles is shown in Fig. 5.15. Table 5.1 summarizes several indices characterizing the trophic structure of the system: the 'vertical'

Fig. 5.15. The 'macro' trophic chain for the food network of the oyster-reef ecosystem. Values in boxes are kcal/m^2 and those associated with arrows are kcal/m$^2\cdot$day.

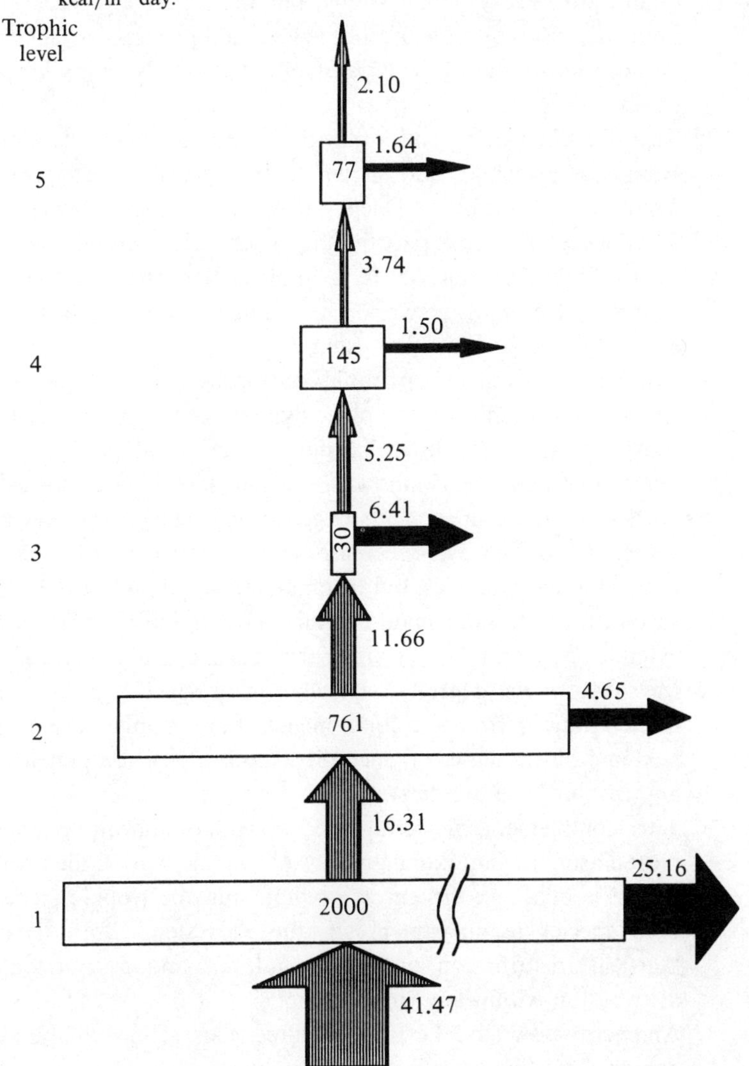

distributions of standing stock, Λ_k, and gross production, λ_k, over trophic levels, and the progressive efficiency between successive trophic levels, λ_{k+1}/λ_k; the 'horizontal' diversities in utilized resources of the trophic level as a whole, $D(k)$, and of its individual components on average, $C(k)$; and the specificity in resource–consumer connection between successive trophic levels, $M(k)$, for trophic levels $k = 1, 2, \ldots, 9$.

5.8 Summary

1 Lindeman and Hutchinson set the benchmark for a holistic and quantitative ecosystems ecology, but were unable to accurately implement the concepts of trophic level and progressive efficiency because they reduced the ecosystem to the overly simple trophic chain.

2 Transfer efficiencies and utilization efficiencies are developed from first principles and shown to be extensions and generalizations of progressive efficiency to food (trophic) networks.

3 The concept of trophic partitioning of compartments (e.g. species) into different trophic levels is implemented through the 'unfolding' of trophic networks according to the past history of energy (matter) within the system.

4 Trophic level, energy pyramid and progressive efficiency of Lindeman–Hutchinson trophic dynamics are generalized to ecosystem networks by collapsing the unfolded network to a macroscopic trophic chain, where trophic levels are composed of parts of many compartments (e.g. species). The energy flow into successive trophic levels decreases in accordance with the second law of thermodynamics, but trophic-level standing stocks are not so constrained. Lindeman–Hutchinson progressive efficiency is extended to trophic levels along the 'macro' chain.

5 Measures of the degree of articulation or specificity in energy-matter transfer from the components of any trophic level to the next and of the degree of generality–specificity of components on any trophic level are derived.

6 Two complementary views of ecosystem organization are explored using the unfolding analysis of food networks: the trophic structure of the ecosystem as a whole and the trophic niche of each species (compartment) in the ecosystem. Both have a 'vertical' distribution over trophic levels and a 'horizontal' distribution within any trophic level.

7 An energy-flow model of an oyster-reef ecosystem is analyzed to demonstrate this new trophic analysis.

Notations

The following is a list of notations used in this chapter, considering a trophic network system of n compartments with stationary standing stocks and flows.

λ_k: the gross production rate of compartment k in the simplified single chain model, or trophic level k in the unfolded network representation of any food network

Λ_k: the standing stock of compartment k (or trophic level k)

R_k: the respiration rate of compartment k (or trophic level k)

x_i: the standing stock of energy or matter in compartment i

f_{ij}: the flow of energy or matter from compartment j to i

z_i: the inflow from the out-of-system environment to compartment i

y_i: the outflow from compartment i to the environment

T_i: the total flow through (i.e. into or out of) compartment i, i.e. $T_i \equiv \sum_{j=1}^{n} f_{ij} + z_i = \sum_{j=1}^{n} f_{ji} + y_i$

g_{ij}: the donor normalized flow from j to i, i.e. $g_{ij} \equiv f_{ij}/T_j$

G: the donor normalized flow matrix, i.e. $G \equiv (g_{ij}), i, j, = 1, 2, \ldots, n$

G^k: the kth power of matrix G

$g_{ij}^{(k)}$: the (i,j)th element of matrix G^k

N: the matrix defined as $N \equiv (I-G)^{-1} = I+G+G^2+\ldots$ (for the latter equality, see Appendix A5.2)

n_{ij}: the (i,j)th element of matrix $N = (I-G)^{-1} = I+G+G^2+\ldots$

$g(\psi)$: the product of the series of donor normalized flows ($g_{kh} \equiv f_{kh}/T_h$) associated with path ψ [for example, if path ψ is represented by $(j \to k \to h \to i)$, then $g(\psi) = g_{ih} \cdot g_{hk} \cdot g_{kj} = (f_{ih}/T_h) \cdot (f_{hk}/T_k) \cdot (f_{kj}/T_j)$]

g'_{ij}: the recipient normalized flow from j to i, i.e. $g'_{ij} \equiv f_{ij}/T_i$

G': recipient normalized flow matrix, i.e. $G' \equiv (g'_{ij}), i, j, = 1, 2, \ldots, n$

G'^k: the kth power of matrix G'

$g'^{(k)}_{ij}$: the (i,j)th element of matrix G'^k

$g'(\psi)$: the product of the series of recipient normalized flows ($g'_{kh} \equiv f_{kh}/T_k$) associated with path ψ [for example, if path ψ is represented by $(j \to k \to h \to i)$, then $g'(\psi) = g'_{ih} \cdot g'_{hk} \cdot g'_{kj} = (f_{ih}/T_i) \cdot (f_{hk}/T_h) \cdot (f_{kj}/T_k)$]

N': the matrix defined as $N' \equiv (I-G')^{-1} = I+G'+G'^2+\ldots$ (the latter equality holds because the argument in Appendix A5.2 is applicable also to matrix G')

n'_{ij}: the (i,j)th element of matrix $N' = (I-G')^{-1} = I+G'+G'^2+\ldots$

$x_i(k)$: the portion of x_i that belongs to the kth trophic level (i.e. that has experienced $k-1$ steps of intercompartmental transfer since arriving in the system)

$f_{ij}(k)$: the portion of f_{ij} that belongs to the kth trophic level

$y_i(k)$: the portion of y_i that belongs to the kth trophic level

$T_i(k)$: the portion of T_i that belongs to the kth trophic level

$$p_j(k) \equiv [T_j(k-1) - y_j(k-1)]/\lambda_k$$

$$D(k) \equiv H(p_1(k), p_2(k), \ldots, p_n(k)) = -\sum_{j=1}^{n} p_j(k) \log p_j(k)$$

$$p'_i(k) \equiv T_i(k)/\lambda_k$$

$$D'(k) \equiv H(p'_1(k), p'_2(k), \ldots, p'_n(k)) = -\sum_{i=1}^{n} p'_i(k) \log p'_i(k)$$

$$c_{ij}(k) \equiv f_{ij}(k-1)/T_i(k)$$

$$C_i(k) \equiv H(c_{i1}(k), c_{i2}(k), \ldots, c_{in}(k)) = -\sum_{i=1}^{n} c_{ij}(k) \log c_{ij}(k)$$

$$C(k) \equiv \sum_{i=1}^{n} p'_i(k)\, C_i(k)$$

$$c'_{ij}(k) = f_{ij}(k-1)/[T_j(k) - y_j(k)]$$

$$C'_i(k) \equiv H(c'_{i1}(k), c'_{i2}(k), \ldots, c'_{in}(k)) = -\sum_{i=1}^{n} c'_{ij}(k) \log c'_{ij}(k)$$

$$C'(k) \equiv \sum_{i=1}^{n} p_i(k)\, C'_i(k)$$

$$p_{ij}(k) \equiv f_{ij}(k-1)/\lambda_k$$

$$J(k) \equiv H(p_{ij}(k) \mid i,j = 1, 2, \ldots, n) = -\sum_{i=1}^{n} \sum_{j=1}^{n} p_{ij}(k) \log p_{ij}(k)$$

$$M(k) \equiv D(k) - C(k) = D'(k) + D(k) - J(k) = D'(k) - C'(k)$$

Appendix

A5.1 The meaning of $g_{ij}^{(k)}$

We prove the following:

For any k ($k = 1, 2, \ldots,$), $g_{ij}^{(k)}$, the (i,j)th element of the kth power G^k of matrix G, equals the summation $\sum g(\psi)$ taken over all paths ψ of length k from j to i.

(i) For any i, j, the (i,j)th element $g_{ij}^{(2)}$ of the second power G^2 of matrix G, by definition, equals the summation $\sum_{h=1}^{n} g_{ih} g_{hj} = \sum g(\psi)$, where the latter summation is taken over all paths of length 2 connecting j to i.

(ii) Suppose for any i, j, the (i,j)th element $g_{ij}^{(k-1)}$ of the $(k-1)$th power G^{k-1} equals the summation $\sum g(\psi)$ taken over all paths of length $k-1$ connecting j to i. Then, for any i, j, because

$$g_{ij}^{(k)} = \sum_{h=1}^{n} g_{ih} g_{hj}^{(k-1)}, \tag{A1}$$

$g_{ij}^{(k)}$ equals the summation $\sum g(\psi)$ taken over all paths of length k connecting j to i.

From (i) and (ii), by mathematical induction with respect to k, we conclude the proposition.

A5.2 The expansion of n_{ij}

We here note a proof for the following basic theorem:

If $G^k \to 0$ as $k \to \infty$, then

$$\sum_{k=0}^{\infty} G^k = (I-G)^{-1} \tag{A2a}$$

i.e.

$$\sum_{k=0}^{\infty} g_{ij}^{(k)} = n_{ij}, \quad (i,j = 1, 2, \ldots, n) \tag{A2b}$$

where n_{ij} is the (i,j)th element of matrix $N = (I-G)^{-1}$.

First, for any integer $k \geqslant 0$, we have

$$(I-G)\cdot(I+G+G^2+\ldots+G^k) = I-G^{k+1}. \tag{A3}$$

Taking the determinants of the matrices on both sides of this equation, we get

$$|I-G|\cdot|I+G+G^2+\ldots+G^k| = |I-G^{k+1}|. \tag{A4}$$

Assume that $G^k \to 0$ as $k \to \infty$. Then, as $(k+1) \to \infty$, $(I - G^{k+1}) \to I$, and thus $|I - G^{k+1}| \to |I| \neq 0$, due to the continuity of the determinant operation. Hence, for a large k, $|I - G^{k+1}| \neq 0$, and from (A4), $|I - G| \neq 0$. Therefore, the inverse $(I-G)^{-1}$ exists. Multiplying this inverse $(I-G)^{-1}$ from the left on both sides of (A3), and making $k \to \infty$, we get (A2*a*).

A5.3 **The meaning of n_{ij}**

Imagine 'particles' that travel within the network under consideration, transferring from compartment j to i by probability $g_{ij} = f_{ij}/T_j$, and leaving the network system from compartment i to the environment by probability y_i/T_i.

We first show the following:

►► n_{ij}, the (i,j)th element of matrix $N = (I-G)^{-1}$, equals the average number of visits to compartment i that a particle in compartment j will make before leaving the network system.

(i) Because clearly $g(\psi)$, the product of the series of one-step transfer probabilities ($g_{kh} = f_{kh}/T_h$) associated with path ψ from j to i, represents the probability that a particle now in compartment j will follow path ψ up to i; $g_{ij}^{(k)}$, which equals the summation $\sum g(\psi)$ taken over all paths of length k connecting j to i (Appendix A5.1), represents the probability that a particle initially in j will be in i after k steps.

(ii) Now let X_k be a random variable defined as

$$X_k = \begin{cases} 1 \text{ (if a particle in compartment} \\ \quad j \text{ will be in } i \text{ after } k \text{ steps)} \\ 0 \text{ (otherwise)} \end{cases} \qquad \text{(A5)}$$

Then, clearly the probability that $X_k = 1$ equals $g_{ij}^{(k)}$, i.e.

$$\text{Prob.}[X_k = 1] = g_{ij}^{(k)} \qquad \text{(A6)}$$

It is also clear that the summation $\sum_{k=0}^{\infty} X_k$ represents the total number of visits to compartment j that the particle will make before leaving the network system. Taking the average (expectation, E) of this sum, we have

$$E\left[\sum_{k=0}^{\infty} X_k\right] = \sum_{k=0}^{\infty} E[X_k], \qquad \text{(A7)}$$

where from (A6),

$$E[X_k] = 1 \cdot \text{Prob.}[X_k = 1] + 0 \cdot (1 - \text{Prob.}[X_k = 1])$$
$$= \text{Prob.}[X_k = 1]$$
$$= g_{ij}{}^{(k)}. \quad \text{(A8)}$$

Thus,

$$E\left[\sum_{k=0}^{\infty} X_k\right] = \sum_{k=0}^{\infty} g_{ij}{}^{(k)} = n_{ij}. \quad \text{(A9)}$$

This means that n_{ij} equals the average number of visits to compartment i that a particle in compartment j will make before leaving the network system.

Next, we present a simple proof for the following proposition:

---

n_{ij}/n_{ii} represents the reaching probability from j to i, i.e. the probability that a particle in j will ever reach i before going out of the network to the environment.

Let r_{ij} denote this reaching probability from j to i. Then, we have two different expressions for the expected (average) number of visits that a particle in j will make to i before leaving the network system:

$$n_{ij} = r_{ij} n_{ii}. \quad \text{(A10)}$$

The former expression (the left-hand side of (A10)) is proved in the first part of this appendix. The latter one (the right-hand side of (A10)) is derived from the following reasoning: For the case that the particle in j never reaches i, whose probability is $1 - r_{ij}$, the average number of its visits to i is zero. On the other hand, for the case that it reaches i, with probability r_{ij}, the average number of its visits to i equals n_{ii}, the total number of visits that a particle that has arrived in i, including this arrival, will make to i before leaving the network system. This is because the behavior of a particle after its arrival in i is independent of that before the arrival. Equation (A10) yields

$$r_{ij} = \frac{n_{ij}}{n_{ii}}, \quad \text{(A11)}$$

which means the proposition.

A5.4 The meaning of n'_{ij}

Imagine the situation where we trace back in time the movement of particles that have been travelling within the network under consideration, assuming that if we trace back a particle now in i by one step we lead to compartment j by probability $g'_{ij} = f_{ij}/T_i$, and that tracing back by one step a particle now in i would lead back to the system environment by probability z_i/T_i. Then, noting that the results in Appendices A5.1 and A5.2 hold also for matrix G', we get the following propositions from essentially the same line of logic as used above in A5.3:

►► n'_{ij}, the (i,j)th element of matrix $N' = (I-G')^{-1}$, equals the average number of visits to compartment j that a particle now in compartment i has made after entering the network system.

►► n'_{ij}/n'_{jj} represents the reaching back probability from i to j, i.e. the probability that a particle now in i has ever been in j after entering the network from the environment.

A5.5 Proof of relation (13).

►►
$$\sum_{m=1}^{n} f_{jm}(k-1) = \sum_{i=1}^{n} f_{ij}(k) + y_j(k) = T_j \sum_{h=1}^{n} g'_{jh}{}^{(k-1)}(Z_h/T_h). \tag{13}$$

The left-hand side of (13), by use of (11) in the text, may be written as

$$\sum_{m=1}^{n} f_{jm} \sum_{h=1}^{n} g'_{mh}{}^{(k-2)}(z_h/T_h), \tag{A12}$$

which may be further rewritten as

$$\sum_{m=1}^{n} T_j(f_{jm}/T_j) \sum_{h=1}^{n} g'_{mh}{}^{(k-2)}(z_h/T_h) = T_j\left[\sum_{m=1}^{n} g'_{jm} \sum_{h=1}^{n} g'_{mh}{}^{(k-2)}(z_h/T_h)\right]$$
$$= T_j\left[\sum_{h=1}^{n}\left\{\sum_{m=1}^{n} g'_{jm} g'_{mh}{}^{(k-2)}\right\}(z_h/T_h)\right]. \tag{A13}$$

It is clear, from the definition of matrix multiplication, that (A13) equals the right-hand side of (13).

The middle of (13), rewritten by use of (11) in the text, becomes

$$\left(\sum_{i=1}^{n} f_{ij} \sum_{h=1}^{n} g'_{jh}{}^{(k-1)} (Z_h/T_h)\right) = \left[\sum_{h=1}^{n} g'_{jh}{}^{(k-1)} (z_h/T_h)\right]\left(\sum_{i=1}^{n} f_{ij} + y_j\right) \quad \text{(A14)}$$

which equals the right-hand side of (13), because $T_j = \sum_{i=0}^{n} f_{ij}$

A5.6 Proof of identity of equations (45) and (46)

$$\sum_{k=1}^{\infty} k \frac{T_i(k)}{T_i} = \sum_{j=1}^{n} n'_{ij}. \qquad \text{(A15)}$$

From (15) in the text, we have

$$\frac{T_i(k)}{T_i} = \sum_{j=1}^{n} g'_{ij}{}^{(k-1)} \frac{z_j}{T_j}; \qquad \text{(A16)}$$

the right-hand side of this equation represents the probability that a particle located now in i has experienced $k-1$ steps since it entered the system at one of the compartments (as it is clear from the result for the reverse process in Appendix A5.1). Thus, the left-hand side of (A15) can be interpreted as the average of the number of steps (i.e. the number of visits to the system compartments) that a particle now located in i made before arriving at i and after entering the system (including the entrance as a step), which is in turn given by the right-hand side of (A15) (Appendix A5.4). Therefore, we have (A15).

References

Burns, T. P. (1989). Lindeman's contradiction and the trophic structure of ecosystems. Ecology **77**, 1355–62.

Coleman, D. C. (1985). Through a ped darkly: an ecological asssessment of root-soil-microbial-faunal interactions. In *Ecological Interactions in Soil*, ed. A. H. Fitter *et al.*, pp. 1–21. Oxford: Blackwell.

Cummins, K. W., Coffman, W. P. & Roff, P. A. (1966). Trophic relationships in a small woodland stream. *Verh. Int. Ver. Limnol.*, **16**, 627–38.

Gates, D. M. (1985). *Energy and Ecology*. Sunderland: Sinauer.

Higashi, M., Burns, T. P. & Patten, B. C. (1989). Food network unfolding: an extension of trophic dynamics for application to natural ecosystems. *J. Theor. Biol.*, **140**, 243–61.

Hutchinson, G. E. (1941). *Lecture notes on limnology*. Distributed upon request from the Osborn Zool. Lab., Yale University, New Haven, Connecticut, (copyright author).

Kerfoot, W. C. & DeMott, W. R. (1984). Food web dynamics: dependent chains and vaulting. In *Trophic Interactions within Aquatic Ecosystems*, ed. D. G. Meyers & J. R. Strickler, pp. 347–82. AAAS Selected Sympos. 85, Boulder: Westview.

Levine, S. (1980). Several measures of trophic structure applicable to complex food webs. *J. Theor. Biol.*, **83**, 195–207.

Lindeman, R. L. (1941 *a*). *Ecological Dynamics in a Senescent Lake*. Dissertation. University of Minnesota, Minneapolis, Minnesota.

Lindeman, R. L. (1941 *b*). Seasonal food-cycle dynamics in a senescent lake. *Am. Midl. Nat.*, **26**, 636–73.

Lindeman, R. L. (1942). The trophic-dynamic aspect of ecology. *Ecology*, **23**, 399–418.

MacArthur, R. H. (1955). Fluctuations of animal populations, and a measure of community stability. *Ecology*, **36**, 533–6.

Odum, E. P. (1968). Energy flow in ecosystems: a historical review. *Am. Zool.*, **8**, 11–18.

Odum, E. P. (1983). *Basic Ecology*. Philadelphia: Saunders.

Patten, B. C. (1959). An introduction to the cybernetics of the ecosystem: the trophic-dynamic aspect. *Ecology*, **40**, 221–31.

Patten, B. C. (1985). Energy cycling in the ecosystem. *Ecol. Mod.*, **28**, 1–71.

Patten, B. C., Bosserman, R. W., Finn, J. T., Higashi, M. & Hill, J. (1990). *Ecological Networks: Theory, Analysis, Applications*. New York: Prentice-Hall. (In press.)

Patten, B. C., Higashi, M. & Burns, T. P. (1990). Trophic dynamics in ecosystem networks: significance of cycles and storage. *Ecol. Mod.* **51**, 1–28.

Pomeroy, L. R. (1974). The ocean's food web: a changing paradigm. *BioScience*, **24**, 499–504.

Pomeroy, L. R. (1985). The microbial food web of the southeastern U.S. continental shelf. In *Oceanography of the Southeastern U.S. Continental Shelf*, ed. L. P. Atkinson, D. W. Menzel & K. A. Bush, pp. 118–29. Washington D.C.: Am. Geophys. Union.

Riley, G. A. (1966). Theory of food-chain relations in the ocean. In *The Sea.*, vol. II, ed. M. N. Hill, pp. 438–63. London: Interscience.

Swift, M. J., Heal, O. W. & Anderson, J. M. (1979). *Decomposition in Terrestrial Ecosystems*. Oxford: Blackwell.

Ulanowicz, R. E. (1980). An hypothesis on the development of natural communities. *J. Theor. Biol.*, **85**, 223–45.

Ulanowicz, R. E. (1986). *Growth and Development: Ecosystems Phenomenology*. New York: Springer-Verlag.

Ulanowicz, R. E. (1990). Ecosystem trophic foundations: Lindeman exonerata. In *Complex Ecology: The Part-Whole Relation in Ecosystems*. ed. B. C. Patten & S. E. Jørgensen. The Hague: SPB Academic Publishing. (In press.)

Ulanowicz, R. E. & Kemp, W. M. (1979). Toward canonical trophic aggregations. *Am. Nat.*, **114**, 871–83.

6

Positive feedback and ecosystem organization

D. L. DEANGELIS AND W. M. POST

6.1 Introduction

Feedback occurs in a system when the output of one component affects one or more of the other components, which exert an influence back on the first component. The behavior of the system derives in large part from the network of feedbacks among its components.

People of the twentieth century are so used to thinking in terms of the feedback relationships in natural and cultural systems that they would find it hard to describe these systems without the concept of feedback. Yet the concept is relatively modern, first taking hold in eighteenth-century England (Mayr, 1986). Mayr argues that the idea of feedback acting in a self-regulating way in economic systems entered the thinking of such theorists as Adam Smith at about the same time that self-regulating feedback mechanisms like the Watt steam governor were being developed at the beginning of England's Industrial Revolution. This trend in thinking was less rapid in becoming established in continental Europe, where the use of older imagery of clockwork mechanisms for describing economic and natural systems continued somewhat longer. The clockwork mechanism embodied the idea of a central or authoritarian management of a system, rather than of self-regulation arising from feedback interactions among freely behaving individuals.

As a general mode of understanding the complex physical and biological systems of the world, the concept of feedback has now replaced that of the clockwork mechanism. The newer concept has been applied more consciously and systematically in ecology than in many of the other sciences. Various sub-fields of ecology that differ in other respects often converge in their recognition of the importance of feedback. For example, systems ecologists have emphasized feedbacks of energy, matter, and information flows between ecosystem components (e.g. E. P. Odum, 1971; Van Dyne, 1966; Patten, 1965; Gutierrez & Fey, 1980; Ulanowicz, 1986), while population and community ecologists have described the dynamics of populations in terms of feedbacks between species (e.g. Levins, 1975; Pimm, 1982; Puccia & Levins, 1985).

Most of the theory developed by ecologists in relation to feedback has stressed its role as a stabilizing or homeostatic mechanism that returns a system to its original state following a deviation from that state. This is exemplified in theories as early as that of Thomas Malthus at the end of the eighteenth century. According to Malthus, human populations tend to grow faster than their resources. Competition, leading to war or starvation (negative feedbacks), eventually follows, reducing the population to a level that can be maintained, at least temporarily. This same reasoning has been applied to biological species in general (e.g. Lotka, 1925; Elton, 1966; Tanner, 1966), though there has been some debate as to how important biotic feedbacks are compared with physical environmental factors in regulating populations (Andrewartha & Birch, 1984).

This emphasis on negative feedback and homeostasis is understandable, since it is a conspicuous occurrence that ecological systems tend to return to their former state following fires, floods, hurricanes, and other disturbances. However, in our view this emphasis has led to relative neglect of an equally important aspect of feedback, its role in organizing and altering ecological systems. This role usually involves positive feedback, which can be defined to occur when a deviation in one component of a system causes deviations in other components that feed back to the first in a way that reinforces the original deviation. A rapid amplification of the original deviation results. When such a process gets started in machinery, the end result is usually destructive, which may explain why positive feedback has not received as much attention as negative feedback, except as something to be eliminated.

There has been a slow trend, however, toward assigning a more constructive role to positive feedback in ecological systems. This trend was evident as early as Maruyama's (1963) essay on positive feedback, and is evident in recent books on ecology and evolutionary theory (e.g. Berryman, 1981, 1986; Dawkins, 1986).

Our own interest in positive feedback began with the study of mutualistic systems (Travis & Post, 1979). Such systems are characterized by positive feedback, which should be mathematically destabilizing (May, 1973*a*). In fact, countless mutualisms exist relatively stably within all ecosystems. They can persist because the tendencies toward unstable amplification are eventually constrained by negative feedbacks of various kinds. Our research disclosed many stable ecological systems that have heterogeneous arrays of positive and negative feedbacks, as well as systems that are unstable for short or long periods of time because dominant positive feedback is driving them to new states (DeAngelis *et al.*, 1986).

In this paper we will outline some important ways in which positive feedback serves as a factor in the development and organization of ecosystems. First, we review some basic ideas concerning feedback.

The term 'network' can be used in a large number of ways. We define a 'feedback network' as a set of system variables linked by negative or positive effects to each other. A hypothetical example is shown in Fig. 6.1 (*a*). For concreteness, we could say this system represents a food web composed of three autotrophs, 1, 2, and 3, a consumer, 4, and a detritivore, 5. The state variables, X_i, stand for either population numbers or biomasses. Autotrophs 1 and 2 compete with each other, but 2 and 3 are mutualistic. The detritivore obtains food from two of the other species, but does not exert a negative influence on their populations.

The simple feedback network in Fig. 6.1 (*a*) can be decomposed into six distinct feedback cycles (Figs. 6.1 *b*, *c*, *d*, *e*, *f*, *g*) ranging from unit length to a five-link cycle. Four of these feedback cycles (*d*, *e*, *f*, *g*) are positive; that is, the product of the signs around the cycle is positive in each case. Such cycles are called 'deviation-amplifying' cycles. For example, if cycle (*f*) were the only one influencing components 1, 2, 4, and 5, then a deviation of X_1 away from equilibrium would be reinforced by feedback from X_4. Two of the feedback cycles (*b*, *c*) are negative, and are called 'deviation-counteracting'. For example, a deviation to X_1 in cycle (*b*) would produce a feedback that tended to counteract the deviation. Readers interested in the mathematical representations of feedback systems are referred to Appendix A6.1.

It is important to realize that a given system that includes positive feedbacks can be quite stable if there are negative feedbacks that balance the positive feedbacks at a particular value of the state variable vector,

$$X^* = (x_1^*, x_2^*, \ldots, x_n^*). \quad (1)$$

Nonetheless, most of the interest in positive feedback has focused on non-equilibrium situations, where the positive feedback acts to push the system from one state to another. An early explicit diagram of a positive feedback cycle was presented by Lotka (1925), who described the coupling of economic and physiological debilitation by the following loop: economic stress → malnutrition → lowered resistance to infection → tuberculosis → loss of appetite → incapacitation → more economic stress → ... Diakonov (1969) described the early growth of irrigation-based states by positive feedback cycles of the following type: irrigation → increased wealth →

Fig. 6.1. (*a*) A hypothetical food web with five species. The individual feedback loops are shown in (*b*), (*c*), (*d*), (*e*), (*f*), and (*g*).

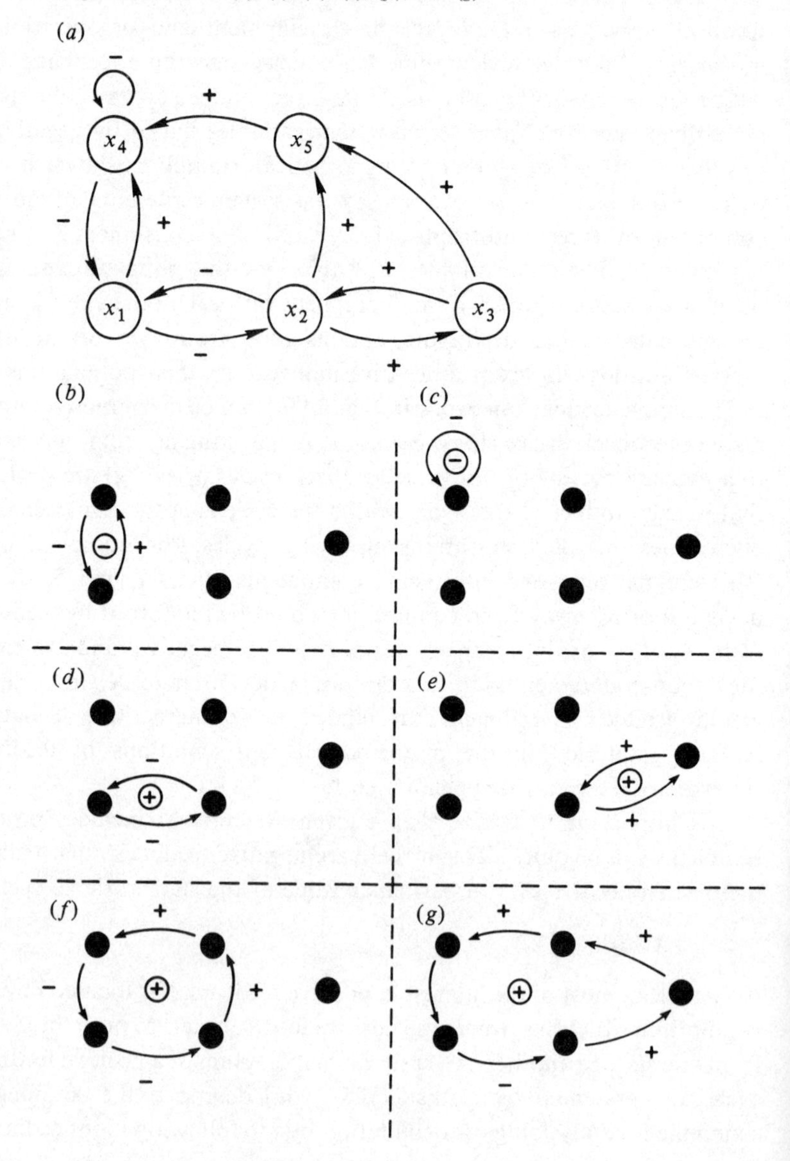

differentiation into rich and poor → class conflict → increase in state power → increased irrigation → .... A variety of geophysical phenomena have been interpreted as positive feedback cycles. For example, John (1979) described a scenario that could feasibly lead to an ice age: abnormally high winter snowfall → extensive snow cover on land surface → increased albedo → lowering of summer temperature → summer survival of snow cover → continued cooling → abnormally high winter snowfall → .... Positive feedback cycles have been identified as possible causes of ecological evolutionary trends, such as the evolution of *k*-selection (Horn, 1978): large body size → damping of environmental fluctuations → steady and crowded population → few local opportunities for young → large young and parental care → large size → ....

We will describe the role of positive feedback in ecological systems in much greater detail below.

6.2 Ecosystem organization

We wish to examine the role played by positive feedback in various attributes of ecosystem organization. We will consider three aspects of ecosystem organization. The first is the pattern of interactions between components (species populations or physical variables such as nutrient reserves) in the ecosystem and how the pattern stabilizes a system or results in directional change. The second aspect encompasses the temporal patterns the system variables undergo during change. The third aspect covers the spatial patterns that develop from ecological interactions extended over space. Each aspect will be discussed below and illustrated by examples.

6.2.1 *Patterns of interactions between ecological entities*

Ecological communities are structured by a combination of positive and negative feedbacks. The 'up-down' or consumer–resource relationships between different trophic levels in the food web involve negative feedback; a consumer on one trophic level is affected positively by the resources it feeds on at lower trophic levels, and it affects them negatively by removing their biomass. On the other hand, the 'width' of food webs is often shaped by positive feedbacks. Competitive (–, –) interactions limit the number of species that can occupy similar ecological niches. Mutualistic (+, +) interactions link numerous species in a network of dependencies.

Among the most prevalent mutualisms are those involving pollination of angiosperm plants and the dispersal of plant seeds by animals. The animals gain some reward, such as fruit or nectar, in return for their assistance in plant reproduction and dispersal. Often the animals perform services for several different plant species. In this case, they have been called 'mobile links' by Gilbert (1980). Gilbert also defined 'keystone mutualists' as plants that provide critical support to large complexes of mobile links. A keystone mutualist, therefore, indirectly helps support a large number of other plants. Loss of such a species would cause a drastic narrowing of the food web. Tropical food webs frequently contain keystone mutualists.

Fig. 6.2. A hypothetical network of species connected by positive feedbacks (competition and mutualism). In (*a*) species 6 and 7 are excluded by competition but are present in small numbers due to constant low immigration rates. (*b*) Configuration of species following removal of species 3.

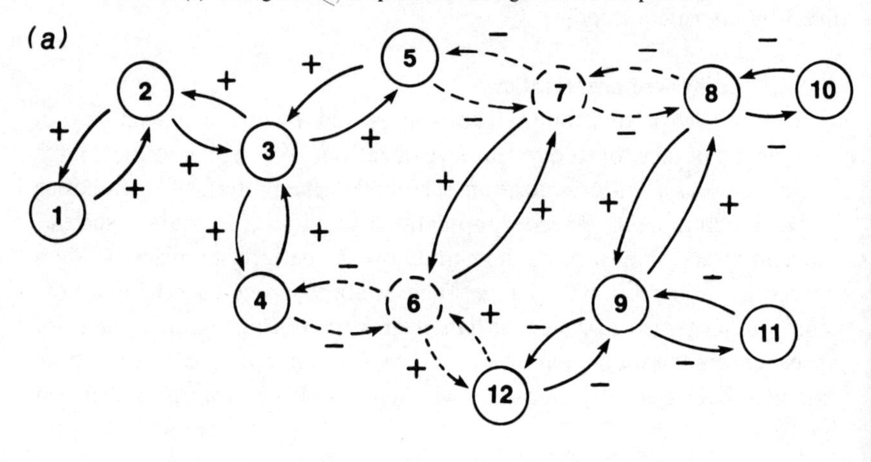

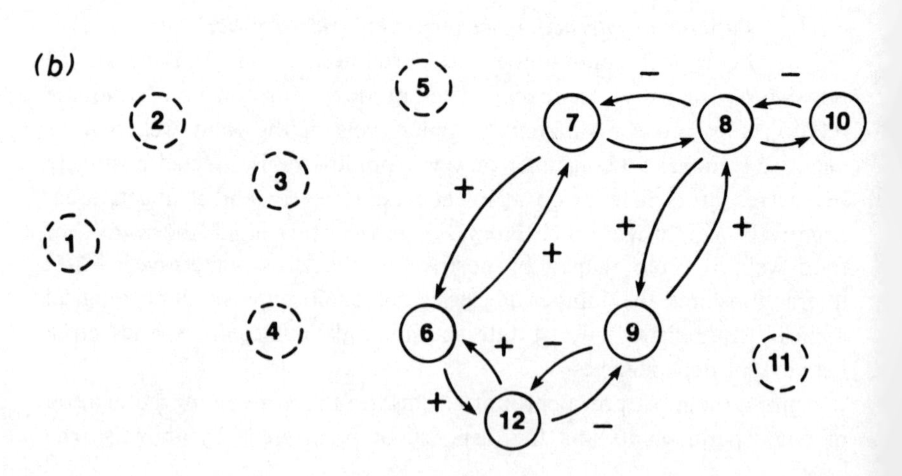

As an example of the propagation of effects through a system connected by positive feedbacks, both competition and mutualism, consider the hypothetical system shown in Fig. 6.2(*a*), in which 12 species are linked together. Note that species 6 and 7 are missing from the initial network, although a very small immigration of members of these species is allowed, providing 'seeds' for growth if conditions change. Table 6.1 is the set of

Table 6.1. *Equations for the system pictured in Fig. 6.2. The functions* $h_i X_i^3/(1+h_i X_i^3)$ *for several species represent an Allee-type effect, a decline in the intrinsic growth rate of population size as population size becomes small.*

$$\frac{dX_1}{dt} = (r_1 - g_1 X_1 + g_2 X_2) X_1 + h_1 X_1^3/(1+h_1 X_1^3)$$

$$\frac{dX_2}{dt} = (r_2 - g_3 X_2 + g_4 X_1 + g_3 X_3) X_2 + h_2 X_2^3/(1+h_2 X_2^3)$$

$$\frac{dX_3}{dt} = (r_3 - g_6 X_3 + g_7 X_2 + g_8 X_4 + g_9 X_5) X_3 + h_3 X_3^3/(1+h_3 X_3^3)$$

$$\frac{dX_4}{dt} = (r_4 - g_{10} X_4 + g_{11} X_3 - g_{12} X_6) X_4 + h_4 X_4^3/(1+h_4 X_4^3)$$

$$\frac{dX_5}{dt} = (r_5 - g_{13} X_5 + g_{14} X_3 - g_{15} X_7) X_5 + h_5 X_5^3/(1+h_5 X_5^3)$$

$$\frac{dX_6}{dt} = (r_6 - g_{16} X_6 - g_{17} X_4 + g_{18} X_7 + g_{19} X_{12}) X_6$$

$$\frac{dX_7}{dt} = (r_7 - g_{20} X_7 - g_{21} X_5 + g_{22} X_6 - g_{23} X_8) X_7$$

$$\frac{dX_8}{dt} = (r_8 - g_{24} X_8 - g_{25} X_7 + g_{26} X_9 + g_{27} X_{10}) X_8 + h_8 X_8^3/(1+h_8 X_8^3)$$

$$\frac{dX_9}{dt} = (r_9 - g_{28} X_9 + g_{29} X_8 - g_{30} X_{12} - g_{31} X_{11}) X_9$$

$$\frac{dX_{10}}{dt} = (r_{10} - g_{32} X_{10} - g_{33} X_8) X_{10}$$

$$\frac{dX_{11}}{dt} = (r_{11} - g_{35} X_{11} - g_{36} X_9) X_{11}$$

$$\frac{dX_{12}}{dt} = (r_{12} - g_{37} X_{12} - g_{34} X_9 + g_{38} X_6) X_{12} + h_{12} X_{12}^3/(1+h_{12} X_{12}^3)$$

equations we have used to describe the system. In Fig. 6.3 the equilibrium values of each of the species is shown at time $t = 0$. At a later time, $t = t_1$, species 3 is removed from the system, simulating a possible natural or anthropogenic extinction. The effect on the system is profound. The mutualistic activities of species 3 had been essential to the survival of species 2, 4, and 5. When these species declined, species 1 also declined. Species 6, 7, and 12, suppressed previously by species 4 and 5, increased in size, causing decreases in species 8, 9, and 11 through competition. The changes in community structure are shown in Fig. 6.2(*b*). This example is purely hypothetical and contrived, but it is probably not an exaggeration of the large and unexpected consequences that can sometimes follow the removal of even a single species from an ecosystem.

Fig. 6.3. Dynamics of the populations in the system pictured in Fig. 6.2(*a*) following the removal of species 3.

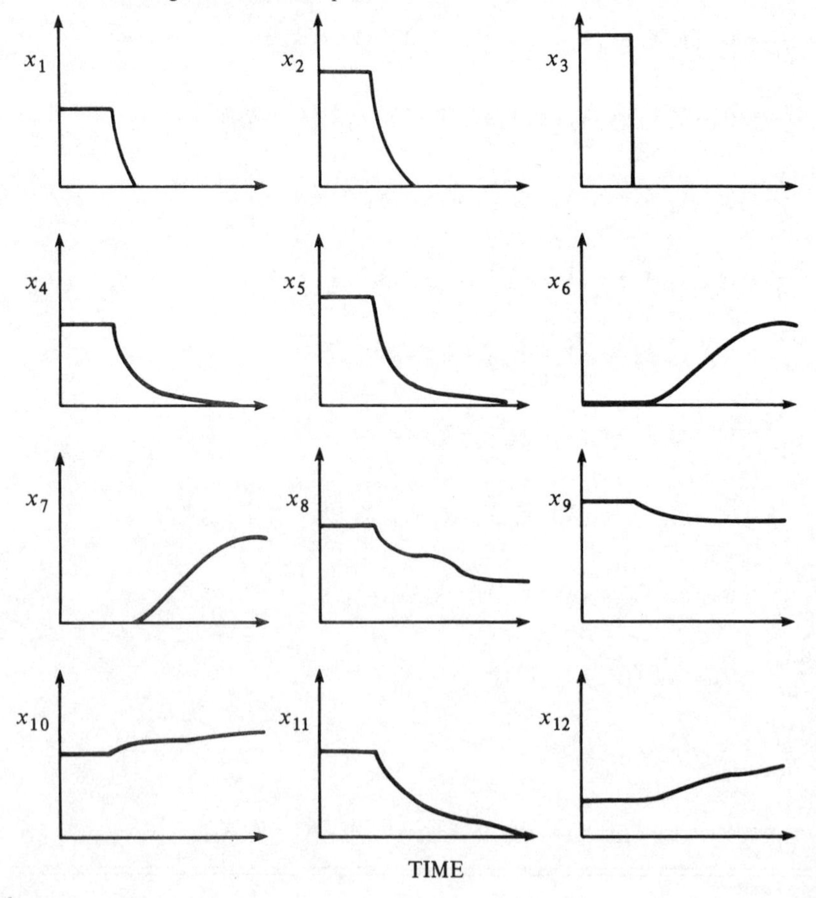

The maintenance of the static food web structures discussed here is not the only way in which positive feedbacks enter into ecosystem organization, and species populations are not the only variables that can be included in feedback networks. Positive feedback also acts in a general sense to drive systems from old to new states, especially from states of lesser to states of greater complexity. Connell & Orias (1964) and Whittaker (1969) proposed that over evolutionary time scales, species diversity in an ecosystem might increase, because each additional species that becomes established could provide opportunities for new species to exploit in some way. Positive feedback should promote the initial acceleration of biotic diversity in a new environment, because the more species that become established, the greater the number of possible niches there will be; that is, each new species will open up new possibilities for parasites, predators, and mutualists. Eventually negative feedbacks, in the form of various limitations, will diminish the possibility of new species entering the system. Gutierrez & Fey (1980) examined the positive feedback cycles connecting such ecosystem components or properties as primary production, community biomass, and inorganic nutrients. In this example, mutual reinforcements of production, nutrient cycling, and diversity through feedback can be thought of as driving succession over ecological time. The precise ways in which the temporal changes take place in ecological systems takes us into the second aspect of ecosystem organization, and this will now be discussed.

6.2.2 *Temporal patterns resulting from positive feedbacks*

If one were asked to identify the three most characteristic types of temporal change that variables in ecological systems (or perhaps any systems) undergo, one might reasonably answer 'exponential growth, logistic growth, and oscillatory or periodic behavior'. Positive feedbacks are often important (or even essential) factors in these types of behaviors.

Malthusian exponential growth of a population (before the negative feedbacks of war and starvation set in) is an obvious example of positive feedback. The number of new individuals added to a population per unit time is affected positively by and affects positively the current population level. This population phenomenon is directly analogous to many autocatalytic chemical reactions and to chain reactions in nuclear fission. The same basic equations can be used to describe such various different variables as population size, new chemical species formed in chemical reactions, and heat and neutrons released in nuclear fusion.

Positive feedback mechanisms can work in the opposite direction of population growth, causing catastrophic exponential decline. Depensatory mortality is one example. Consider predator and prey populations where the rate of consumption of prey individuals is described by a Holling Type II interaction; that is,

$$\text{rate of consumption of prey} = f_1 N_1 N_2/(b_1 + N_1), \tag{2}$$

where N_1 is the number of prey individuals in the population, N_2 is the number of predators, and f_1 and b_1 are constants. The probability of a particular prey individual being consumed at a given time (so that N_1 becomes $N_1 - 1$), is

$$\text{Prob}\,(N_1 - 1; N_1; \Delta t) = f_1 N_2/(b_1 + N_1). \tag{3}$$

Suppose predator individuals can survive for a long time even if the supply of prey, N_1, is decreasing (perhaps the predators feed on other prey as well), so that N_2 stays relatively constant over periods of time. Then, as N_1 decreases, $\text{Prob}\,(N_1 - 1; N_1; \Delta t)$ increases, at least as long as N_2 remains constant. The lower N_1 is, the higher the probability of mortality per prey individual becomes. The positive feedback structure of this effect is shown in Fig. 6.4. Computer simulations of the predator–prey interaction (with N_2 unchanging) shows a roughly linear decrease of N_1 toward zero.

The Allee effect is another depensatory mechanism, but it usually acts to decrease per capita reproduction as the population decreases, rather than to increase per capita mortality (Allee, 1938). The Allee effect can occur in social animals that require a critical population size to be able to reproduce efficiently.

Exponential growth (or decline) is usually only a short-term phenom-

Fig. 6.4. Positive feedback relation involved in depensatory mortality.

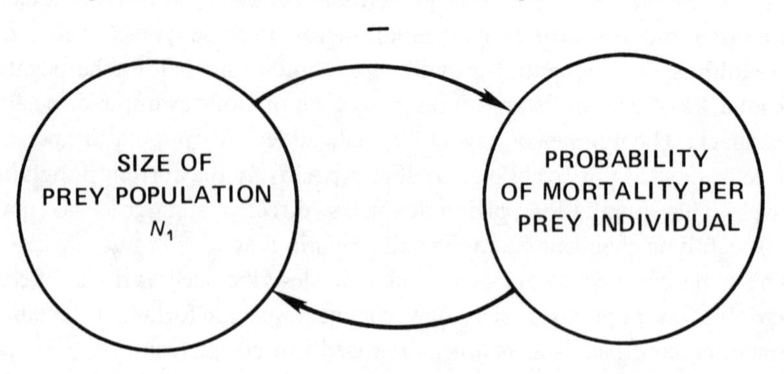

enon for populations. Over longer time periods, behavior is more likely to be logistic. For example, the settlement of a new area by colonizing species is likely to proceed exponentially at first, and then to level off as space fills up. These dynamics can be represented by the well-known Pearl–Verhulst equation, but it is interesting to see if logistic-like behavior can be obtained from individual-by-individual models of populations that do not explicitly assume this form of behavior.

To illustrate this we have used a simple model of settlement of an initially bare intertidal area by sessile organisms such as barnacles. We assumed that the propagules settled randomly and attachment was greatly enhanced if a propagule landed within a distance P_1 of another established organism. This is a positive feedback effect that has been noted in studies of barnacle recruitment, perhaps as a result of reducing the likelihood of being 'bulldozed' by limpets (Underwood *et al.*, 1983). The attachment probability on top of another barnacle was assumed to be zero (Fig. 6.5*a*). Fig. 6.5(*b*) shows the number of barnacles settling in an area as a function of time in a typical simulation. The exponential phase was due to feedback effects of barnacles already present, while the leveling off was due to crowding.

The final mode of temporal behavior we will discuss is oscillatory behavior. Discussion of oscillatory behavior may be surprising, as this behavior is usually thought of as the result of underdamped negative feedback. Examples of such negative feedback oscillations include predator–prey cycles. These cycles can result when the predator is able to consume the prey down to very low levels, which is followed by a crash of the predator population, a subsequent recovery of the prey, and then a repeat of the predator devastation of the prey.

Positive feedbacks can also be implicated, however, as a partial cause of certain oscillatory phenomena. A physical example will illustrate this point. Consider a rod suspended by springs (Fig. 6.6) across which a steady current of air is blown. Any slight fluctuation of the rod in one direction or the other is self-amplifying because the Bernoulli effect increases 'lift' in that direction. Thus the rod moves away from its equilibrium position either up or down. The force from the springs eventually slows the rod, which causes the Bernoulli lift to cease, and the stored potential energy in the spring forces the rod back in the opposite direction, overshooting the equilibrium position. The Bernoulli effect now reinforces movement in the new direction. Hence, continuous oscillations ensue, thanks to the existence of the positive feedback.

It is not difficult to find analogous ecological situations. One example

Fig. 6.5. Barnacle settlement on an empty surface (*a*) Previous settlement by barnacles makes new successful settlement nearby (denoted by dotted circle) more likely. (*b*) Number of barnacles on initially clean surface as a function of time in computer simulation.

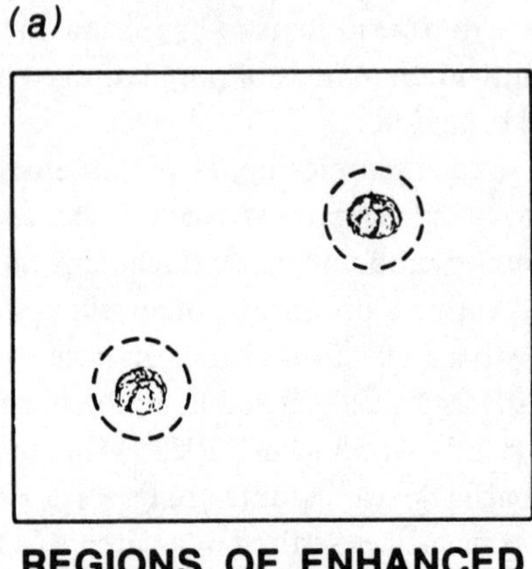

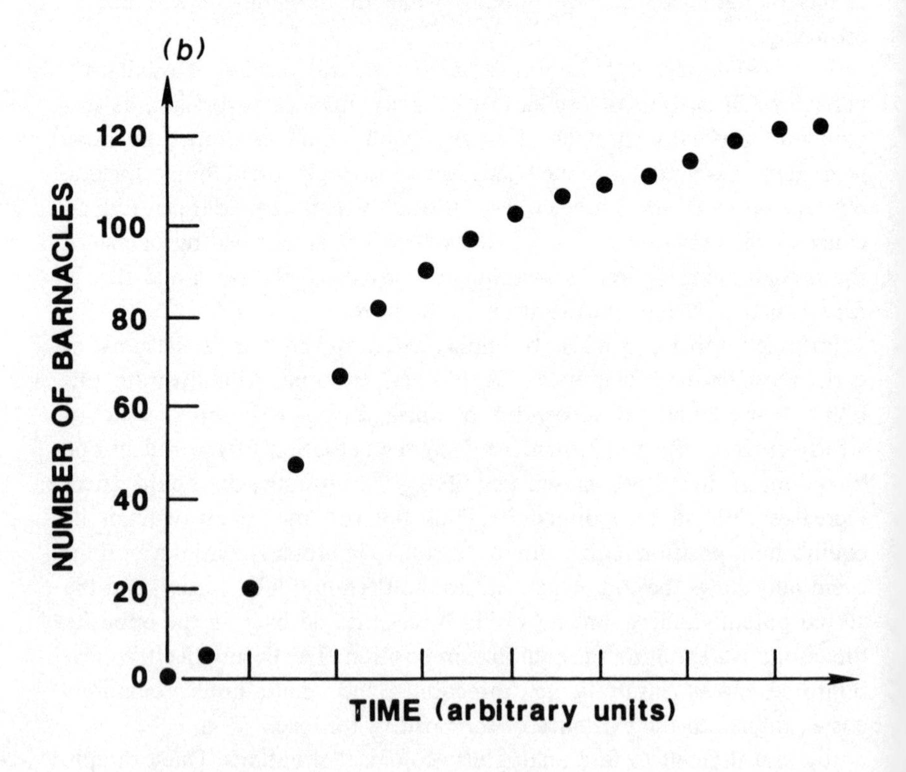

is based on the competitive behavior of two sympatric populations. Let the competing populations be described by the Lotka–Volterra equations:

$$\frac{dX}{dt} = r_x(1 - X/K_x - C_{yx}\,Y)X - G(X) \qquad (4a)$$

$$\frac{dY}{dt} = r_y(1 - Y/K_y - C_{xy}X)\,Y - G(Y). \qquad (4b)$$

If competition is very strong ($C_{xy}\,C_{yx} > 1/(K_x\,K_y)$), then, depending on the initial conditions, positive feedback will drive the population to either of the stable equilibrium points $X^* = K_x$, $Y^* = 0$, or $X^* = 0$, $Y^* = K_y$. However, suppose for each species there is a high probability of

Fig. 6.6. Physical oscillations stimulated by positive feedback. A slight displacement of the rod suspended by springs will be reinforced by the Bernoulli effect due to the change in air flow pattern seen from the rod's frame of reference. This leads to oscillations, as explained in the text.

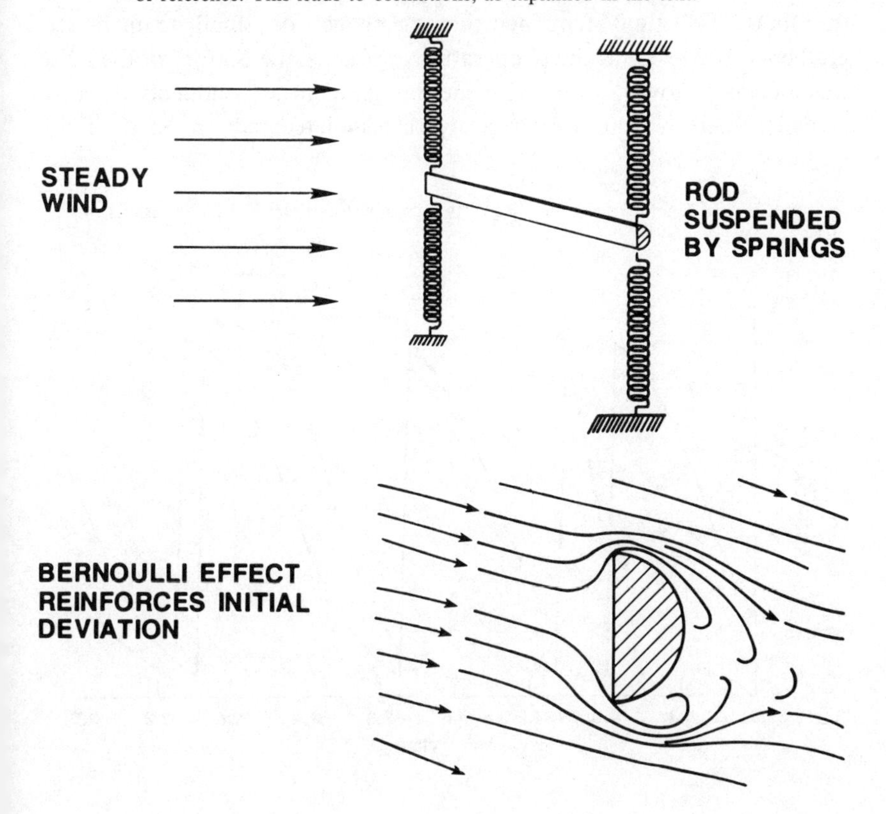

catastrophic decline, $G(X)$ and $G(Y)$ (say, a viral infection) when a high population level is reached. Then neither of these equilibrium points will be reached and the population will oscillate in a quasi-periodic manner (Fig. 6.7). Hence, we see that positive feedback can be a driving force in oscillatory phenomena, but strong negative feedbacks are needed as well.

6.2.3 *Spatial patterns developing from positive feedbacks*

Interactions within systems create patterns in space as well as in time, and positive feedback is often the driving force in the creation of these spatial patterns.

As in the case of temporal behavior, it is useful here to first describe a relatively well-understood example from physics to show the generality of positive feedback as a mechanism. The development of Bénard convective cells is an illustrative case. The nineteenth-century French physicist Bénard conducted experiments in which he heated a thin layer of fluid (spermaceti oil) from below, while leaving its upper surface relatively cool by contact with air. This allowed a temperature gradient to develop across the fluid. The fluid remained homogeneous for small temperature gradients. However, as the temperature applied at the bottom of the fluid was increased slowly, a threshold thermal gradient was suddenly reached at which there was an appearance of a tessellated pattern on the fluid

Fig. 6.7. Simulated dynamics of two competitors described by equations (4*a*, *b*).

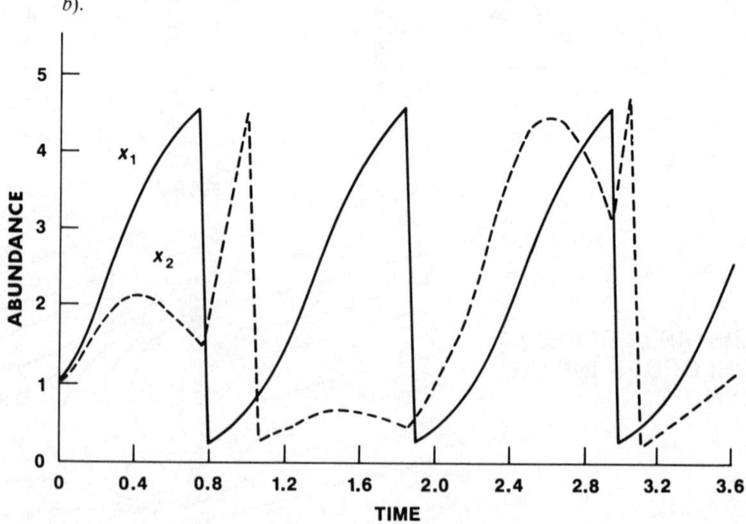

surface: usually a mosaic of hexagons, indicating a convective circulation of the fluid. The explanation for the onset of this behavior can be understood by examining a cross section of the fluid (Fig. 6.8*a*). The temperature gradient creates a gradient in density, with denser, heavier fluid on top. If a parcel of fluid somewhere in the column is given a slight upward motion due to a random fluctuation, it moves into fluid that is relatively denser. Hence, the movement is reinforced by increased buoyancy, and if the force of buoyancy is enough to overcome the fluid viscosity, the column of fluid moves upwards, balanced by downward movement of other fluid columns. The result is the creation of a pattern of convective cells (Fig. 6.8*b*).

Can one find ecological analogs to this sort of spatial pattern formation? There are at least two common types of ecological behavior that illustrate positive feedback pattern formation; the aggregation of colonial organisms and the formation of sharp boundaries between competing species.

Grasse (1959) described the early development of a termite nest. At first, building material is distributed somewhat randomly by the termites. But when by chance a deposit at one location is large enough, termites will preferentially deposit more material there. Deneubourg (1977) hypothesized that chemical attractants accounted for the aggregation and formulated a simple mathematical model. Let C be the concentration of insects depositing material and P the concentration of building material. A chemical attractant is mixed with the building material and has

Fig. 6.8. (*a*) Diagram of Bénard's experiment. A slight upward displacement of a fluid parcel in the temperature and density gradient in the oil is reinforced by positive feedback, because its buoyancy is less than that of the surrounding fluid. This positive feedback instability can lead to circulation cells shown in (*b*).

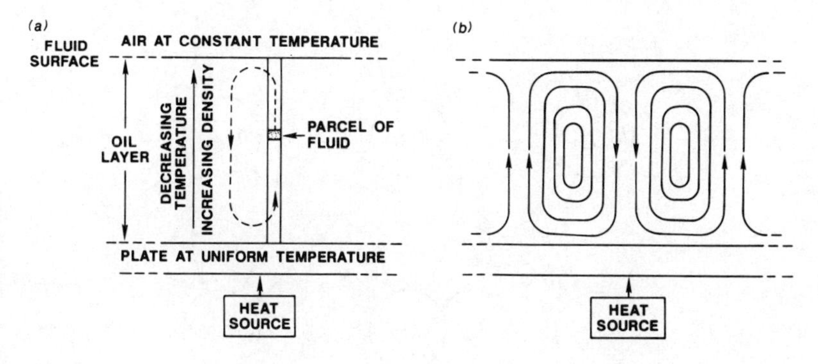

concentration H, which decays through time. Deneubourg's equations for the dynamics of this system are:

$$\frac{\partial P}{\partial t} = k_1 C - k_2 P \tag{5a}$$

$$\frac{\partial H}{\partial t} = k_1 P - k_2 H + D_H \nabla^2 H \tag{5b}$$

$$\frac{\partial C}{\partial t} = F^c - k_1 C + D\nabla^2 C + \gamma\nabla(C\nabla H) \tag{5c}$$

where $D_H\nabla^2 H$ and $D\nabla^2 C$ are terms representing diffusion of attractant and insects and $\gamma\nabla(C\nabla H)$ represents chemotactic force on the insects.

DeAngelis *et al.* (1986) showed this system could be represented by a feedback diagram (Fig. 6.9) relating p, h, and c at a series of discrete spatial points. Some of the cycles in this feedback network are positive. They tend to reinforce any local aggregation of insects and building material at the expense of other nearby areas. If the critical parameter of chemotactic force, γ, is large enough, the feedbacks leading to development of local aggregation will dominate and an initially empty homogeneous plain invaded by termites will develop regularly-spaced colonies. At an abstract level, the parameter γ resembles the temperature of the bottom of the fluid in Bénard's experiments.

Fig. 6.9. Feedback cycles associated with the spatially discretized version of equations (5*a*, *b*, *c*). The quantities s_{i-1}, s_i, and s_{i+1} are points along a spatial dimension.

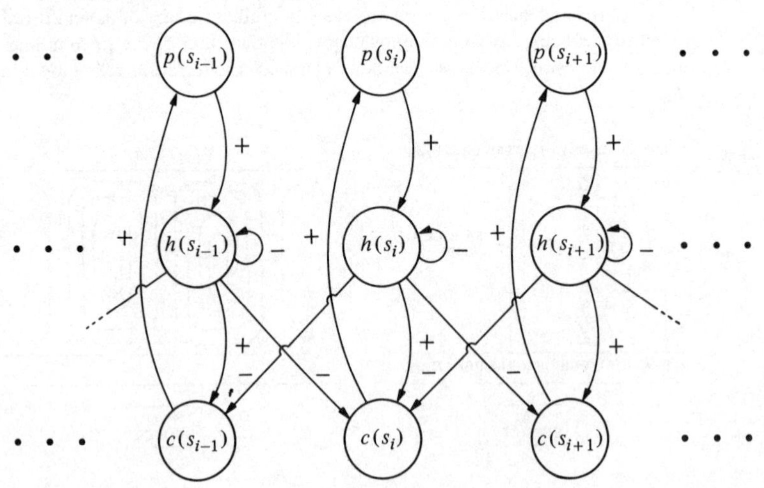

The second type of spatial behavior involving positive feedback is the hypothesized development of sharp spatial boundaries between competing species of vegetation (see Transeau, 1935; Oosting, 1955; Siccama, 1974; Yamamura, 1976; Shugart *et al.*, 1980). Consider two species competing along an environmental gradient. A pair of partial differential equations representing this situation are the following:

$$\frac{\partial X}{\partial t} = \frac{\partial}{\partial s}\left\{D_x(s)\frac{\partial X}{\partial s}\right\} + r_x(s)\{1 - X/K_x(s) - C_{xy}(s)\,Y\}\,X \qquad (6a)$$

$$\frac{\partial Y}{\partial t} = \frac{\partial}{\partial s}\left\{D_y(s)\frac{\partial Y}{\partial s}\right\} + r_y(s)\{1 - Y/K_y(s) - C_{yx}(s)\,X\}\,Y \qquad (6b)$$

where X and Y are the population densities of the competing species,

Fig. 6.10. Four qualitatively distinct configurations of isoclines of competing species. Configuration (*a*) is assumed to hold at one end of the environmental gradient, and (*b*) at the other end. The transition from (*a*) to (*b*) can proceed as in (*c*) or as in (*d*).

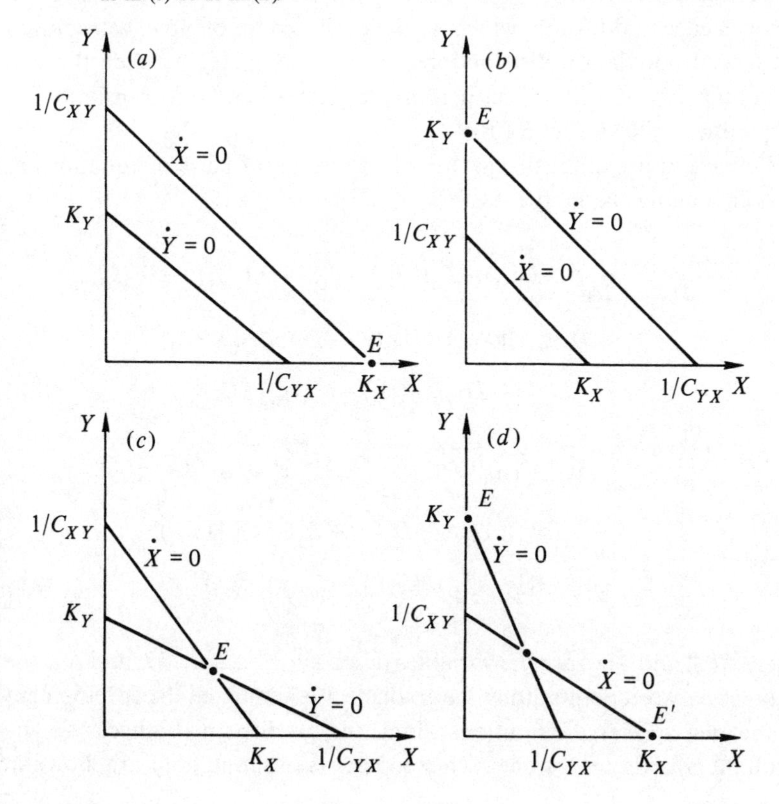

$C_{xy}(s)$ and $C_{yx}(s)$ are the competition coefficients, $D_x(s)$ and $D_y(s)$ are the diffusion coefficients due to seed dispersal, and s is the distance along the gradient.

Let us ignore the diffusion terms for the moment. Assume that the environmental conditions change gradually along this gradient, with species X being favored at small values of s, for which

$$1/C_{xy}(s) > K_y(s) \tag{7a}$$

$$K_x(s) > 1/C_{yx}(s), \tag{7b}$$

and Y being favored at large values of s, where the inequalities are reversed (Fig. 6.10*a, b*). A matter of importance is precisely how the transition from inequalities (7*a, b*) to the reverse inequalities proceeds as s increases. If this occurs as in Fig. 6.10(*c*), then there is a smooth transition from equilibrium point $(K_x, 0)$ to $(0, K_y)$. But if the transition occurs as in Fig. 6.10(*d*), there will be a region midway along the gradient in which two stable equilibrium points exist. Which point will be the solution at a given point in this region depends on the detailed conditions. The existence of diffusion, which we have ignored up till now, will smooth out jagged population distributions, but the boundary between the two populations will still be sharper in the case where the translation of inequalities follows Fig. 6.10(*d*).

We have approximated eqs. (6*a, b*) by a series of pairs of equations at 12 points along the gradient.

$$\begin{aligned}\frac{dX(s_i)}{dt} = & \frac{1}{2\Delta}\{[D_x(s_{i+1})+D_x(s_i)]\,X(s_{i+1})-[D_x(s_{i+1})+2D_x(s_i) \\ & +D_x(s_{i-1})]\,X(s_i)-[D_x(s_i)+D_x(s_{i-1})]\,X(s_{i-1})\} \\ & -r_x(s_i)[1-X(s_i)/K_x(s_i)-C_{xy}(s_i)]\,X(s_i)\end{aligned} \tag{8a}$$

$$\begin{aligned}\frac{dY(s_i)}{dt} = & \frac{1}{2\Delta}\{[D_y(s_{i+1})+D_y(s_i)]\,Y(s_{i+1})-[D_y(s_{i+1})+2D_y(s_i) \\ & +D_y(s_{i-1})]\,Y(s_i)-[D_y(s_i)+D_y(s_{i-1})]\,Y(s_{i-1})\} \\ & -r_y(s_i)[1-Y(s_i)/K_y(s_i)-C_{yx}(s_i)]\,Y(s_i)\end{aligned} \tag{8b}$$

where $X(s_i)$ and $Y(s_i)$ are the variables taken at $i = 1, 2, \ldots, 12$, and Δ is the distance between points along the gradient. We simulated the development of the two species distributions along the gradient in both cases. The resultant boundaries for one such comparison of simulations are shown in

Fig. 6.11 (*a*), (*b*). In the former case, each species population is distributed along the gradient largely as it would be if the other species did not exist. In the latter case, each species is virtually excluded from the region where the other species is dominant. This is a creation of spatial patterning due to positive feedback.

6.3 Summary

Recognition of both positive and negative feedback in biological systems goes back at least to Malthus, but a comprehensive view of the role of positive feedback in such systems only began to emerge since Maruyama (1963). In the present paper, positive feedback is viewed as not only a ubiquitous phenomenon of biological and ecological systems, but also a factor that plays a major role in ecosystem organization. As such, it influences the food web and community structure, temporal behavior, and spatial pattern observed in ecosystems. In particular, the following arguments are made:

1 Ecological food webs and communities contain a large number of positive feedback cycles. These may be direct linkages between competitive or mutualistic species, or longer loops involving

Fig. 6.11. Dynamics of populations of competing species started off as uniform distributions along the environmental gradient: (*a*) when the transition has the form shown in Fig. 6.10(*c*); (*b*) when the transition has the form shown in Fig. 6.10(*d*).

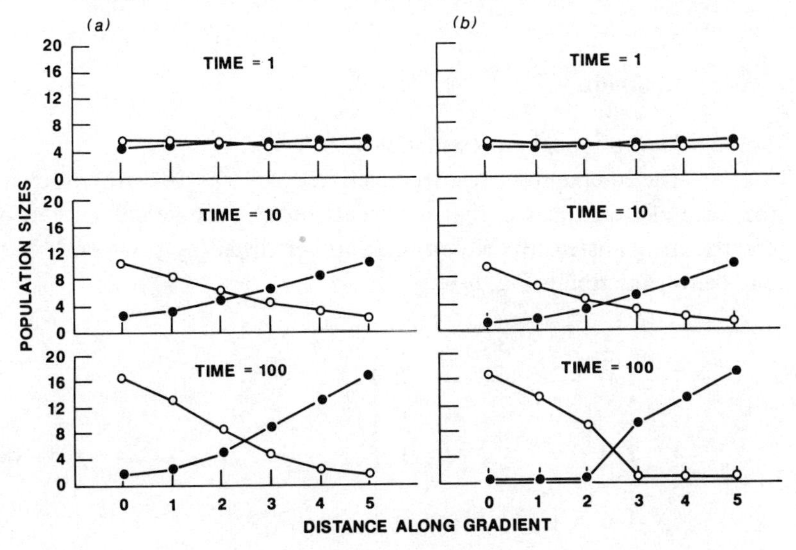

many species. Perturbations such as the removal of a species, can have large repercussions for the system as a whole.

2 Positive feedbacks generate temporal changes in populations. Such processes as Malthusian growth, depensatory mortality, and facultative colonization involve positive feedback loops. Temporal behaviors stimulated by positive feedbacks can include exponential growth, logistic growth, linear decline, and oscillations.

3 Spatial patterns can be generated or accentuated by positive feedbacks. In a simple mathematical model of termite colonies, for example, positive feedback reinforces initial colony development as a particular spatial location, and suppresses nearby colonies, leading to regular spacing of colonies. In plant communities, positive feedback between competitors sharpens spatial separation between plant types along environmental gradients.

Acknowledgement

Research supported by the National Science Foundation's Ecosystem Study Program, under Interagency Agreement No. BSR-81031810-A02 with the US Department of Energy under Contract No. DE-AC05-840R21400 with Martin Marietta Energy Systems, Inc. Environmental Sciences Division Publications No. 3426.

Appendix

A6.1 Mathematical representations

The interactions in a feedback network can be written down in the form of a sign matrix that represents the qualitative influence of one compartment (listed by column) on another (listed by row), such as the one below describing Fig. 6.1;

$$\begin{array}{c} \\ 1 \\ 2 \\ 3 \\ 4 \\ 5 \end{array} \begin{array}{c} \begin{array}{ccccc} 1 & 2 & 3 & 4 & 5 \end{array} \\ \begin{bmatrix} 0 & - & 0 & - & 0 \\ - & 0 & + & 0 & 0 \\ 0 & + & 0 & 0 & 0 \\ + & 0 & 0 & - & + \\ 0 & + & + & 0 & 0 \end{bmatrix} \end{array} . \tag{A1}$$

Many examples of systems exist in which a sign can be given to the interaction, but where the magnitude of the effects are difficult or impossible to determine. Qualitative analysis of the type introduced by Levins (1975) and May (1973*b*) can often provide information on the stability of such systems, even though only the sign of the interactions are known.

When the magnitudes of the interactions are known at equilibrium, the sign matrix can be replaced by a matrix incorporating this knowledge:

$$A = \begin{bmatrix} 0 & -a_{12} & 0 & -a_{14} & 0 \\ -a_{21} & 0 & -a_{23} & 0 & 0 \\ 0 & a_{32} & 0 & 0 & 0 \\ a_{41} & 0 & 0 & -a_{44} & a_{45} \\ 0 & a_{52} & a_{53} & 0 & 0 \end{bmatrix} \tag{A2}$$

where the a_{ij}'s are absolute magnitudes.

A problem of great interest is to determine whether, for a given matrix, A, the deviation-amplifying loops or the deviation-countering loops are most important in determining system behavior; that is, in determining whether the system is stable and will return to equilibrium following a perturbation to the X_i's or not. In general, this determination can be made by finding the eigenvalues, λ_1, λ_2, λ_3, λ_4, and λ_5 of A. If any eigenvalue has a real positive part, the system is unstable.

It is not always necessary to find the eigenvalues to determine this. When the system is of a special form where all feedback loops longer than unit length are positive, some mathematical theorems show that analytic criteria for the stability can be derived. Any matrix A of a system in which all feedback loops larger than unity are positive can be transformed into the form:

$$A' = \begin{bmatrix} a'_{11} & a'_{12} \cdots & a'_{1n} \\ a'_{21} & a'_{22} \cdots & a'_{2n} \\ \vdots & \vdots & \vdots \\ a'_{n1} & a'_{n2} \cdots & a'_{nn} \end{bmatrix} \tag{A3}$$

where $a'_{ij} \geqslant 0$ for all $j \neq i$. The transformation can be carried out by similarity transforms:

$$A' = SAS^{-1}. \tag{A4}$$

As an example, consider

$$A = \begin{bmatrix} 0 & a_{12} & -a_{13} \\ a_{21} & 0 & -a_{23} \\ -a_{31} & 0 & 0 \end{bmatrix}. \tag{A5}$$

The transform

$$S = \begin{bmatrix} 1 & 0 & 0 \\ 0 & 1 & 0 \\ 0 & 0 & -1 \end{bmatrix} \quad S^{-1} = \begin{bmatrix} 1 & 0 & 0 \\ 0 & 1 & 0 \\ 0 & 0 & -1 \end{bmatrix} \tag{A6}$$

converts A to

$$A' = \begin{bmatrix} 0 & a_{12} & a_{13} \\ a_{21} & 0 & a_{23} \\ a_{31} & 0 & 0 \end{bmatrix}. \tag{A7}$$

For matrices in the form A', very simple sufficient conditions for stability can be written. For A' to be stable, it is sufficient that the following conditions hold:

$$(-1)^k A'_k > 0 \quad (k = 1, 2, \ldots, n) \tag{A8}$$

where A'_k is the kth principal minor; e.g.

$$A'_1 = [a_{11}]$$

$$A'_2 = \det\begin{bmatrix} a_{11} & a_{12} \\ a_{21} & a_{22} \end{bmatrix}$$

$$A'_3 = \det\begin{bmatrix} a_{11} & a_{12} & a_{13} \\ a_{21} & a_{22} & a_{23} \\ a_{31} & a_{32} & a_{33} \end{bmatrix}$$

(see, for example, DeAngelis *et al.*, 1986). These criteria are much easier to calculate than is the computation of eigenvalues (which usually must be done numerically) and yield relationships among parameters to be satisfied, which can give insights.

References

Allee, W. C. (1938). *Cooperation Among Animals With Human Implications.* New York: W. W. Norton and Co.

Andrewartha, H. G. & Birch, L. C. (1984). *The Ecological Web. More on the Distribution and Abundance of Animals.* Chicago: The University of Chicago Press.

Berryman, A. A. (1981). *Population Systems: A General Introduction.* New York: Plenum Press.

Berryman, A. A. (1986). *Forest Insects. Principles and Practice of Population Management.* New York: Plenum Press.

Connell, J. H. & Orias, E. (1964). The ecological regulation of species diversity. *Am. Nat.*, **98**, 399–414.

Dawkins, R. (1986). *The Blind Watchmaker.* New York: W. W. Norton and Co.

DeAngelis, D. L., Post, W. M. & Travis, C. C. (1986). *Positive Feedback in Natural Systems.* Berlin: Springer-Verlag.

Deneubourg, J. L. (1977). Applications of order by fluctuations to description of some stages in building of termite nests. (French.) *Insectes Sociaux*, **23**, 329.

Diakonov, I. (ed.) (1969). *Ancient Mesopotamia.* Moscow: Nauka Publishing House.

Elton, C. S. (1966). *The Pattern of Animal Communities.* London: Chapman and Hall.

Gilbert, L. E. (1980). Food web organization and the conservation of neotropical diversity. In *Conservation Biology*, ed. M. E. Soule & B. A. Wilcox, pp. 11–33. Sunderland: Sinauer Associates, Inc.

Grasse, P.-P. (1959). La reconstruction du nid et les coodinations inter-individuelles chez Bellicositermes natalensis et Cubitermes spp. La theories de la stimergie: essai d'interpellation du comportement des termites constructeurs. *Insectes Sociaux*, **6**, 41–83.

Gutierrez, L. T. & Fey, W. R. (1980). *Ecosystem Succession: General Hypothesis and a Test Model of a Grassland.* Cambridge, MA: The MIT Press.

Horn, H. S. (1978). Optimal tactics of reproduction and life history. In *Behavioral Ecology: An Evolutionary Approach*, ed. J. R. Krebs & N. B. Davies, pp. 411–29. Sunderland: Sinauer Associates, Inc.

John, B. (ed.) (1979). *The Winters of the World.* New York: Halsted Press.

Levins, R. (1975). Evolution in communities near equilibrium. In *Ecology and Evolution of Communities*, ed. M. L. Cody & J. M. Diamond, pp. 16–55. Cambridge, MA: Belknap Press.

Lotka, A. J. (1925). *Elements of Physical Biology.* Baltimore: Williams and Wilkins. (Reprinted as *Elements of Mathematical Biology.* New York: Dover. 1956.)

Maruyama, M. (1963). The second cybernetics: Deviation amplifying mutual causal processes. *Am. Sci.*, **51**, 164–79.

May, R. M. (1973*a*). *Stability and Complexity in Model Ecosystems.* Princeton: Princeton University Press.

May, R. M. (1973*b*). Qualitative stability in model ecosystems. *Ecology*, **54**, 638–41.

Mayr, O. (1986). *Authority, Liberty, and Automatic Machinery in Early Modern Europe.* Baltimore: The Johns Hopkins University Press.

Odum, E. P. (1971). *Fundamentals of Ecology* (3rd edition). Philadelphia: W. B. Saunders Company.

Oosting, H. J. (1955). *The Study of Plant Communities: An Introduction to Plant Ecology* (2nd edition). San Francisco: W. H. Freeman and Company.

Patten, B. C. (1965). *Community organization and energy relationships in plankton.* ORNL/TM-3634. Oak Ridge: Oak Ridge National Laboratory.

Pimm, S. L. (1982). *Food Webs.* London: Chapman and Hall.

Puccia, C. J. & Levins, R. (1985). *Qualitative Modeling of Complex Systems.* Cambridge, MA: Harvard University Press.

Shugart, H. H., Emanuel, W. R. & DeAngelis, D. L. (1980). Environmental gradients in a simulation model of a beech-yellow-poplar stand. *Math. Biosci.*, **50**, 163–70.

Siccama, T. G. (1974). Vegetation, soil, and climate in the Green Mountains of Vermont. *Ecol. Monogr.*, **44**, 325–49.

Tanner, J. T. (1966). Effects of population density on the growth rates of animal populations. *Ecology*, **47**, 733–45.

Transeau, E. N. (1935). The prairie peninsula. *Ecology*, **16**, 423–37.

Travis, C. C. & Post, W. M. (1979). Dynamics and comparative statics of mutualistic communities. *J. Theor. Biol.*, **78**, 553–71.

Ulanowicz, R. E. (1986). *Growth and Development. Ecosystems Phenomenology.* New York: Springer-Verlag.

Underwood, A. J., Denley, E. J. & Moran, M. J. (1983). Experimental analysis of the structure and dynamics of mid-shore rocky intertidal communities in New South Wales. *Oecologia*, **56**, 202–19.

Van Dyne, G. M. (1966). Ecosystems, systems ecology and systems ecologists. TM-3957. Oak Ridge: Oak Ridge National Laboratory.

Whittaker, R. H. (1969). *Evolution of diversity in plant communities.* Symposium on Diversity and Stability in Ecological Systems. New York: Brookhaven National Laboratory.

Yamamura, N. (1976). A mathematical approach to spatial distribution and succession in plant communities. *Bull. Math. Biol.*, **38**, 517–26.

7

Structure, stability and succession of model competition systems

K. KAWASAKI, H. NAKAJIMA, N. SHIGESADA
AND E. TERAMOTO

7.1 Introduction

The relationship between the structure and stability of ecosystems is one of the most important problems in ecology. In this chapter, we introduce several models describing the dynamics of multi-species competition systems and examine the relationship between their structure and stability. We also analyze the effect of invasion of a new species on community structure, from which extremal principles concerning the successional process are derived. Although we restrict the scope of this chapter to competition systems, we first give a brief review of the previous studies that dealt with multi-species systems in general and were concerned with the following problems: (1) relationships between complexity and dynamical stability of ecosystems, (2) mathematical treatments of stability of multi-species systems and (3) ecosystem succession, especially the relevant extremal principles.

7.1.1 *Complexity and dynamical stability*

From both empirical and theoretical viewpoints, Elton (1958) argued that more complex communities are more stable. MacArthur (1955) proposed that stability is roughly proportional to the number of links between species in a food web. With a variety of trophic links, the flow of energy and nutrients would not change greatly when the population density of one species is raised or lowered abnormally. May's (1972) mathematical analysis of linear ecosystem models showed, however, that in random assemblages more complex systems tend to be less stable. He pointed out that ecological systems may be composed of blocks of species having more interactions within blocks than between blocks. The observations of McNaughton (1978), Rejmanek & Stary (1979), and

Moore & Hunt (1988) support this hypothesis; they indicated that real communities are organized as blocks of species, similar to *guilds* (Root, 1967), i.e. each species in a guild has weak or no interactions with species belonging to other guilds, and connectance, the number of links, decreases as the number of species increases. Gilpin (1975) and Goh & Jennings (1977) obtained the same result as May for the Lotka–Volterra system even when the condition of feasible equilibrium was imposed. DeAngelis (1975) analyzed the class of systems having a trophic structure in which the food supply to a consumer population is affected by the resource population, but not by the consumer population (donor-controlled system). According to his analysis, stability is either unaffected by or increases with complexity but it never decreases with complexity. Other investigations of stability for semi-randomly assembled systems based on real food webs were done by Gilpin (1975), Pimm & Lawton (1977), Yodzis (1981) and Pimm (1984). Pimm & Lawton (1977), in particular, found that omnivory makes the system less stable, and the system becomes less stable as the number of trophic levels increase.

In the above mathematical analyses, the concept of local stability is used mainly. Local stability means that the state of a system can return to an equilibrium state after it is moved *slightly* from this equilibrium state by an external perturbation. In contrast, global stability refers to the property that a system returns to its original state when subjected to *any* perturbation. Pimm (1979) defined another sort of stability: a system is called species-deletion stable, if all of the remaining species are retained at a locally stable equilibrium when one is removed. According to his computer simulations, more complex systems are more stable for the deletion of primary producers.

7.1.2 *Stability analysis of multi-species dynamical systems*

Local stability is mathematically tractable through the community matrix, whose elements a_{ij} represent the effect of species j upon species i near an equilibrium (see Section 7.2.2 for the precise mathematical definition). A system is locally stable if the real parts of eigenvalues of its community matrix are all negative. This stability condition can be also examined directly from the matrix elements using the Routh–Hurwitz criterion (Gantmacher, 1959). Several classes of local stability have been presented: D-stability, total stability (Svirezhev & Logofet, 1978), sign-stability (Quirk & Ruppert, 1965; May, 1973; Jeffries, 1974, 1975; Tansky, 1976) and connective stability (Šiljak, 1975). The mathematical relationships among these stability classes are discussed by Svirezhev & Logofet (1978).

Global stability properties have been studied intensively for a special class of Lotka–Volterra models. Consider the following Lotka–Volterra system:

$$\frac{\mathrm{d}x_i}{\mathrm{d}t} = \left(b_i + \sum_{i=1}^{n} a_{ij}\, x_j\right) x_i, \quad i = 1, 2, \ldots, n, \tag{1}$$

where x_i is the population density of the *i*th species, b_i is the intrinsic growth rate, and $[a_{ij}]$ is the interaction matrix. We call matrix $A = [a_{ij}]$ a *dissipative* matrix, if there exists a positive diagonal matrix C such that $CA + {}^{t}AC$ is negative definite (i.e. all eigenvalues of this matrix are negative). It has been shown that if A is dissipative and a positive equilibrium point of eq. (1) exists, the point is always globally stable, independent of the magnitudes of the growth rates b_i (Volterra, 1931; Harrison, 1979).

Sufficient conditions for the dissipativity have been studied by many authors (Goh, 1977, 1978, 1980; Takeuchi *et al.*, 1978; Harrison, 1979; Case & Casten, 1979; Takeuchi & Adachi, 1980; Ikeda & Šiljak, 1980; Solimano & Beretta, 1982). Ikeda & Šiljak (1980) showed that eq. (1) is dissipative if A is a diagonal dominant matrix, i.e. each diagonal element of the matrix is greater than the total magnitude of off-diagonal elements in the same row. A sign-stable matrix (Harrison, 1979; Solimano & Beretta, 1982) and an anti-symmetric matrix (Case & Casten, 1979) are also shown to be dissipative.

The Lotka–Volterra model may be analyzed more easily if we impose certain restrictions on the parameters. MacArthur & Levins (1967) presented a model for an exploitation–competition community in which the elements of the interaction matrix are given by the degree of resource-utilization overlap between the competing species, thus the matrix is symmetric. They showed that this system has a globally stable equilibrium point, at which the difference between the available resources and the resources actually used is minimized. Many authors have extended the MaçArthur & Levins model to explain niche partitioning and limiting similarity (MacArthur, 1968, 1970; May, 1973; Roughgarden, 1979). On the other hand, Shigesada *et al.* (1984, 1989) considered a Lotka–Volterra model for interference competition, in which each element of the interaction matrix is assumed to be the product of two factors: the strengths of interference and susceptibility. They analyzed the stability of all non-negative equilibrium points and the species composition of the community at stable equilibrium. Their analysis will be reviewed in Section 7.2.

7.1.3 *Succession of communities: extremal principles*

The second law of thermodynamics for physical and chemical irreversible processes asserts that entropy always increases in adiabatic processes. Succession and evolution of ecological systems are irreversible processes. Hypotheses for succession or evolution of ecological systems have been proposed in terms of increasing biological state functions. We call these hypotheses *extremal principles*. Lotka (1922*a*, *b*) asserted that the evolution of a system proceeds so as to maximize, compatible with the constraints, the total energy flow through the system. A similar proposal was made by H. T. Odum (1971), who emphasized the maximization of power (product of a force and the flow induced by this force). A fascinating hypothesis was given by Margalef (1968); he argued that the efficiency of an ecosystem, defined as the amount of biomass maintained per unit energy income to primary producers of the system, increases as succession or evolution proceeds. Recently, Ulanowicz (1986) proposed 'ascendency' (product of the total flow and the mutual information in a flow network) as an increasing quantity in ecological succession.

Extremal principles hold in a certain class of dynamical models in ecology. The total utilization function in MacArthur's (1970) exploitation–competition model always increases in ecological succession. Nakajima (1985) proposed a new method for deriving an extremal principle by means of a 'quasi-stationary method'. This method is described and applied to several models in Section 7.3.

7.2 Structure of interference competition communities

7.2.1 *Lotka–Volterra model of competition communities*

Consider an N-species competition community. Let x_i denote the population density of the ith species. We assume that the population dynamics is given by the following Lotka–Volterra equation:

$$\frac{\mathrm{d}x_i}{\mathrm{d}t} = \left(\varepsilon_i - \sum_{j=1}^{N} \mu_{ij} x_j\right) x_i, \quad i = 1, 2, \ldots, N \tag{2}$$

where $\varepsilon_i(>0)$ is the intrinsic growth rate of the ith species; $\mu_{ii}(>0)$ and $\mu_{ij}(>0, i \neq j)$ are the coefficients of intra- and interspecific competition, respectively.

Competition mechanisms are classified into two categories: exploitation and interference (Miller, 1967). The former refers to joint exploitation of limiting resources, and the latter to any activity that limits or interrupts the competitor's access to a necessary resource, such as a niche, food, defended space or particular habitat. A typical example of interference competition has been observed in communities of sessile organisms. Connell (1961) and Dayton (1971) reported that competition of sessile animals involved direct physical aggression: one species of barnacle outcompeted another for attachment space by undercutting or crushing individuals of the other species.

Here we focus on interference competition. To incorporate interference competition in system (2), we assume the following: the competitive interaction is factored into the intrinsic strength of interference to other individuals and the intrinsic ability to defend against attacks by other individuals. Thus the competition coefficients μ_{ij} are assumed to be given in the form:

$$\mu_{ij} = \begin{cases} \sigma_i \alpha_i & (i = j), \\ \sigma_i \beta_j & (i \neq j), \end{cases} \tag{3}$$

where β_j (or α_i) is an intrinsic factor of *interspecific* (*intraspecific*) *interference* of an individual of the jth (ith) species, and its effect on the ith species is reduced by a factor of $\sigma_i (0 < \sigma_i < 1)$ owing to the defensive ability of the ith species; σ_i is termed the *susceptibility*. With use of these notations, eq. (2) is rewritten as

$$\frac{\mathrm{d}x_i}{\mathrm{d}t} = \left(\varepsilon_i - \sigma_i \alpha_i x_i - \sum_{j=1 (j \neq i)}^{N} \sigma_i \beta_j x_j\right) x_i \equiv f_i(\boldsymbol{x}) x_i, \quad i = 1, 2, \ldots, N \tag{4}$$

where $\boldsymbol{x} = (x_1, x_2, \ldots, x_N)$.

Without loss of generality, we assign index numbers to the N species in decreasing order of ε_i/σ_i; thus we have

$$\frac{\varepsilon_1}{\sigma_1} > \frac{\varepsilon_2}{\sigma_2} > \ldots > \frac{\varepsilon_N}{\sigma_N}, \tag{5}$$

and we refer to the index number of each species in this ordering as its rank. We will see later that the magnitude of ε_i/σ_i and the sign of $\alpha_i - \beta_i$ play an important role in determining whether species i can survive or not. We call ε_i/σ_i *essential growth rate*, the species with $\alpha_i > \beta_i$ *auto-competitors* and the species with $\alpha_i < \beta_i$ *hetero-competitors*.

An equilibrium point of eq. (4) is obtained by setting $x_i = 0$ or $f_i(\boldsymbol{x}) = 0$ for all i, and this procedure provides 2^N equilibrium points in total. Each equilibrium point corresponds to a partitioning of the species set (index set) $I = \{1, 2, \ldots, N\}$ into two subsets, P and $I-P$, for which we set $f_i(\boldsymbol{x}) = 0$ for $i \in P$ and $x_i = 0$ for $i \in I-P$; the equilibrium for a particular partitioning, P and $I-P$, of the species set I is given by the following equations:

$$x_i^* = \begin{cases} \dfrac{(\varepsilon_i/\sigma_i) - C(P)}{\alpha_i - \beta_i} & \text{for all } i \in P \\ 0 & \text{for all } i \in I-P \end{cases} \tag{6}$$

where

$$C(P) = \frac{\sum\limits_{i \in P} \varepsilon_i \xi_i/\sigma_i}{1 + \sum\limits_{i \in P} \xi_i}, \quad \xi_i = \frac{\beta_i}{\alpha_i - \beta_i}. \tag{7}$$

In the following, we assume for mathematical simplicity (1) ε_i/σ_i ($i = 1, 2, \ldots, N$) are mutually distinct, (2) $\alpha_i \neq \beta_i$ for all i, and (3) $1 + \sum_{i \in P} \xi_i \neq 0$.

7.2.2 *Stability analysis of equilibrium states*

Among the equilibria given by eq. (6), biologically meaningful ones should be non-negative, that is, $x_i^* \geqslant 0$ for all i. The local stability of these non-negative equilibrium points can be analyzed by means of the linearization of system (4). From the basic theory of linearization analysis (Goh, 1980), for an equilibrium to be stable it is necessary that the real parts of all the eigenvalues of the coefficient matrix in linearization, namely, the community matrix, be negative or zero. Conversely, if all real parts of eigenvalues are negative, the equilibrium is stable. When some real parts of eigenvalues are zero, we cannot determine stability from the linearized equation. Hereafter, we call an equilibrium point *linearly stable* if the real parts of all the eigenvalues are negative. Whether each equilibrium of system (4) is linearly stable or not can be determined by the following theorem:

Theorem 1

An equilibrium point of system (4) is non-negative and linearly stable, if and only if it is either of the following types:

$$\text{(I)} \qquad \boldsymbol{x}^* = (x_1^*, x_2^*, \ldots, x_s^*, 0, 0, \ldots, 0), \tag{8a}$$

which is given by eq. (6) for $P = P_s \equiv \{1, 2, \ldots, s\}$, where s ($1 \leqslant s \leqslant N$) satisfies

$$\alpha_i > \beta_i \text{ for } i \in P_s \tag{8b}$$

$$\frac{\varepsilon_s}{\sigma_s} > C(P_s) > \frac{\varepsilon_{s+1}}{\sigma_{s+1}}. \tag{8c}$$

(II) $$x^* = (x_1^*, x_2^*, \ldots, x_s^*, 0, \ldots, 0, x_w^*, 0, \ldots, 0) \tag{9a}$$

which is given by eq. (6) for $P = P_{s,w} \equiv \{1, 2, \ldots, s\} + \{w\}$, where s and w $(0 \leqslant s < w \leqslant N)$ satisfy

$$\alpha_i > \beta_i \text{ for } i \in P_s, \; \alpha_w < \beta_w \tag{9b}$$

$$\frac{\varepsilon_s}{\sigma_s} > C(P_{s,w}) > \frac{\varepsilon_{s+1}}{\sigma_{s+1}} \tag{9c}$$

$$1 + \sum_{i \in P_{s,w}} \xi_i < 0. \tag{9d}$$

For $s = 0$ in case (II), $P_{s,w} \equiv \{w\}$, and all the elements of x^* except x_w^* are zero.

Note that if there exists no integer s that satisfies (8*b*) and (8*c*), the system does not possess any linear stable equilibrium of type (I). Similarly, if an integer s satisfying (9*b*)–(9*d*) for a given w does not exist, the system does not have any equilibrium of type (II) with a surviving hetero-competitor of rank w. We show an outline of the proof in Appendix A7.1. For more detail, see Shigesada *et al.* (1984). Theorem 1 indicates that the system in general has multiple domains of attraction; the stable equilibrium that is attained at infinite time depends upon the initial state of the system.

Before discussing the biological interpretation of Theorem 1, we supplement it by the following theorem:

Theorem 2

(a) The system (4) has neither chaotic solutions nor limit cycles.

(b) If a linearly stable equilibrium point of system (4) is positive (i.e. $x_i^* > 0$ for all $i \in I$), it is globally stable.

(c) The number of stable equilibrium points of type (I) in Theorem 1 is at most one, and the number of stable equilibrium points of type (II) for a given $w(\in I)$ is at most one.

We prove Theorem 2(a) by using the following Lyapunov function:

$$Q(x(t)) = -2\sum_{i=1}^{N} \varepsilon_i \beta_i x_i / \sigma_i + \sum_{i=1}^{N} \alpha_i \beta_i x_i^2 + \sum_{i=1}^{N} \sum_{j=1(j \neq i)}^{N} \beta_i \beta_j x_i x_j. \tag{10}$$

The time derivative of $Q(x(t))$ along any trajectory of system (4) is given by

$$\frac{\mathrm{d}}{\mathrm{d}t} Q(x(t)) = -2\sum_{i=1}^{N} \beta_i x_i \{f_i(x)\}^2 / \sigma_i \leqslant 0, \tag{11}$$

where $\mathrm{d}Q/\mathrm{d}t = 0$ holds at the equilibrium points. Since $\mathrm{d}Q/\mathrm{d}t \leqslant 0$, every solution $\boldsymbol{x}(t)$ tends to one of the equilibrium points as $t \to +\infty$ (LaSalle & Lefshetz, 1961). Therefore, system (4) has neither chaotic solutions nor limit cycles; the community structure at infinite time is determined as one of the equilibrium points given by Theorem 1. The proofs of Theorem 2(b) and (c) are given in Appendix A7.2.

7.2.3 *Structure of interference competition systems*

To discuss the biological meaning of these results, we schematically illustrate Theorem 1. Assume that system (4) contains N species among which κ species are hetero-competitors of ranks specified by $w_1, w_2, \ldots, w_i, \ldots, w_\kappa$ $(1 \leqslant w_1 < w_2 < \ldots < w_i < \ldots < w_\kappa \leqslant N)$, and that the rest of the species are auto-competitors. If we denote by ○ a surviving auto-competitor, by ● a surviving hetero-competitor, and by × an extinct species, then stable non-negative equilibrium states given by (I) and (II) in Theorem 1 are depicted as (*a*) and (*b*), respectively, in Fig. 7.1. Characteristic properties of the system revealed from these diagrams may be summarized as follows:

(i) Among the auto-competitors $(\alpha_i > \beta_i)$, the surviving species are determined ordinally, i.e. the species with a higher essential growth rate has a higher priority in survival.

(ii) The number of surviving hetero-competitors $(\alpha_i < \beta_i)$ is at most one.

(iii) The ranks of surviving auto-competitors $(\alpha_i > \beta_i)$ are higher than rank w_1, which is the highest rank among the hetero-competitors. Therefore, the number of surviving species is at most w_1. If $w_1 = 1$, only one hetero-competitor of rank 1 can survive.

Furthermore, Theorem 2(c) implies that

(iv) The number of stable equilibria of system (4) is at most $\kappa + 1$.

We now derive conditions for all N species to coexist in the system. Putting $s = N$ in (8*c*), and $s = N-1$ and $w = N$ in (9*c*) of Theorem 1 leads to, respectively,

$$\sum_{i=1}^{N-1} \xi_i \left(\frac{\varepsilon_i/\sigma_i}{\varepsilon_N/\sigma_N} - 1 \right) < 1,$$

and

$$\sum_{i=1}^{N-1} \xi_i \left(\frac{\varepsilon_i/\sigma_i}{\varepsilon_N/\sigma_N} - 1 \right) < 1 < \sum_{i=1}^{N} \xi_i \left(\frac{\varepsilon_i/\sigma_i}{\varepsilon_{N-1}/\sigma_{N-1}} - 1 \right).$$

Since $\xi_i > 0$ and $\varepsilon_i/\sigma_i > \varepsilon_N/\sigma_N$ for all i except N, we have the following conditions on the coexistence of all species:

(a) If the constituent species of a system are all auto-competitors, then $\varepsilon_i/\sigma_i \approx \varepsilon_N/\sigma_N$ and/or $\alpha_i \gg \beta_i$ for all i except N are necessary conditions for coexistence.

(b) If the constituent species of a system are both auto- and hetero-competitors, then $\alpha_N \approx \beta_N$, in addition to the above conditions (a), are necessary for their coexistence.

We can see that, in general, the conditions for all species to coexist are highly restrictive.

Gilpin & Case (1976) studied system (2) by means of computer simulations in which they assigned randomly chosen values to parameters μ_{ij}. They found that the system generally has multiple stable equilibrium states (multiple domains of attraction), and that only 2 or 3 species on average ultimately survive, even when a large number of species exist at the initial stage.

Let us consider randomly assembled communities for system (4). If we assign randomly chosen values to α_i and β_i in our model (4), each species will be either an auto- or a hetero-competitor with equal probability. Thus, the probability is $(1/2)^w$ that the species of ranks 1 through $w-1$ are all auto-competitors and the species of rank w is a hetero-competitor. Furthermore, the total number of surviving species at any equilibrium is

Fig. 7.1. Stable equilibrium states of eq. (4). N species are ranked in decreasing order of ε_i/σ_i. Species are further classified into auto-competitors ($\alpha_i > \beta_i$) and hetero-competitors ($\alpha_i < \beta_i$). Ranks of hetero-competitors are indicated by w_1, $w_2, \ldots, w_k$. Other symbols are: ○, surviving auto-competitor; ●, surviving hetero-competitor; ×, extinct species. Starting from arbitrary initial population densities, the community always approaches either type (*a*) or type (*b*). The number of surviving auto-competitor species, s, is determined for each community type from the conditions in Theorem 1. If there is no such integer s, the corresponding equilibrium does not exist.

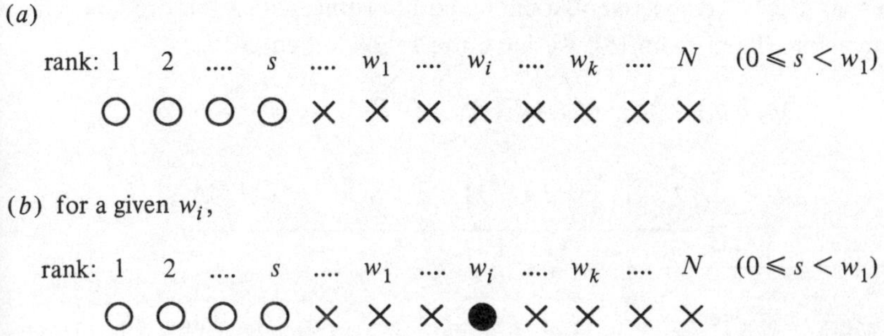

(*a*)

rank: 1 2 .... s .... w_1 .... w_i .... w_k .... N $(0 \leqslant s < w_1)$

○ ○ ○ ○ × × × × × × × ×

(*b*) for a given w_i,

rank: 1 2 .... s .... w_1 .... w_i .... w_k .... N $(0 \leqslant s < w_1)$

○ ○ ○ ○ × × × ● × × × ×

almost w (as remarked in (iv)). Taking these into account, we can estimate that the average number of surviving species is less than 2. Thus, the expected number of surviving species at each stable state roughly accords to that obtained from the randomly assembled Lotka–Volterra model of Gilpin & Case (1976).

7.2.4 *Invasion and succession in interference competition communities*

We now proceed to the problem of invasion into communities. We consider the case where a small propagule of a new species migrates to a given area where pre-occupant species have already reached a stable state. We say that a new species *succeeds in invasion* if, after competitive processes, the community reaches a new stable state in which the new species is a constituent member. Under the assumption that only one new species can invade in a single invasion process, a criterion for successful invasion is given by the following:

Theorem 3 (Invasion criterion)
Consider a community that has already reached a stable positive equilibrium point $\boldsymbol{x}^* = (x_1^*, x_2^*, \ldots, x_N^*)$, where $x_i^* > 0$ for all i. If a new species μ immigrates at low density into the community, the following condition is necessary (or sufficient) for the species to succeed in invasion:

$$\frac{\varepsilon_\mu}{\sigma_\mu} \geqslant C(I) \quad \left(\text{or } \frac{\varepsilon_\mu}{\sigma_\mu} > C(I)\right), \tag{12}$$

where $I = \{1, 2, \ldots, N\}$.

The proof of Theorem 3 is given in Appendix A7.3. When a new species satisfying condition (12) invades, the structure of the community will attain a new stable state. By comparing the pre-existing community with the community resulting from a single successful invasion process, we obtain changes in the population densities of the pre-occupant species. Let $\boldsymbol{x}^*$ and $\boldsymbol{x}^{**}$, respectively, denote equilibrium points before and after invasion. Then from (6), we have the following equation:

$$(\alpha_i - \beta_i)(x_i^{**} - x_i^*) = C(P^*) - C(P^{**})$$

$$= \frac{\sum_{k \in P^*} \xi_k \left(\frac{\varepsilon_k}{\sigma_k} - C(P^*)\right) - \sum_{k \in P^{**}} \xi_k \left(\frac{\varepsilon_k}{\sigma_k} - C(P^*)\right)}{1 + \sum_{k \in P^{**}} \xi_k},$$

where $P^* = \{i \mid x_i^* > 0\}$ and $P^{**} = \{i \mid x_i^{**} > 0\}$. By using eq. (6) and taking account of the invasion condition (12) and stability conditions (8) and (9), we can evaluate whether each species increases or decreases after invasion. We illustrate the results in Fig. 7.2, with signs + and − to represent an increase and decrease, respectively, in the population densities. In all cases except (i) and (ii), the population densities of hetero-competitors increase ($x_w^* < x_w^{**}$), while the population densities of auto-competitors decrease ($x_i^* > x_i^{**}$), and there is a possibility that some auto-competitors with the lowest ranks become extinct, as marked by crosses. In case (i), they change oppositely to the above case. In case (ii), the direction of change depends on the sign of $C(P^*)\{(-\xi_N)+\xi_\mu\}-\{(\varepsilon_N/\sigma_N)(-\xi_N)+(\varepsilon_\mu/\sigma_\mu)\xi_\mu\}$. If the sign is plus, the population densities of auto-competitors decrease and the population density of the hetero-competitors increases, and vice versa if the sign is minus.

Fig. 7.2. Change of community structure by species invasion. $\overset{a}{\downarrow}$ represents invasion of auto-competitor, $\overset{h}{\downarrow}$ invasion of hetero-competitor, ○ a surviving auto-competitor, ● a surviving hetero-competitor, and × an extinct species. The signs +/− represent the direction of change in population densities. In all cases except (i) and (ii), population densities of auto-competitors decrease, some may go extinct, and population density of the hetero-competitor increases. In (i), all pre-occupant species survive; in (ii), all pre-occupants, except the hetero-competitor, survive.

(*a*) Case without pre-existent hetero-competitor

before invasion

after invasion

(*b*) Case with one pre-existent hetero-competitor

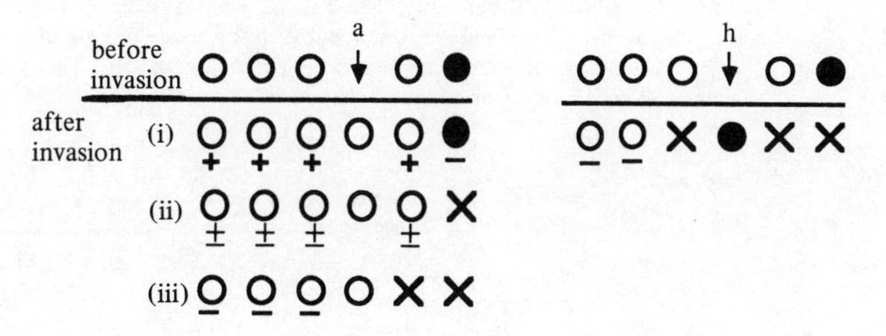

Theorem 3 also suggests that species with higher values of essential growth rate, ε_i/σ_i, can more easily invade the community. Furthermore, when a hetero-competitor invades, the larger the value of ε_i/σ_i is for the invading species, the more the species richness in the resultant community decreases.

7.3 Succession of model competition systems

In the previous section, the invasion rules for a Lotka–Volterra competition system were analyzed and the direction of succession obtained. That analysis was based on the stability of dynamical systems.

The relationship between the stability of population dynamics and community structure or succession is also of interest. Many analytical results have been obtained from a similar viewpoint. According to May (1972), complexity leads to instability of randomly connected food webs. This implies that natural communities must have structure. Structure is a consequence of the historical development of the system during a time period that is long in comparison with the ecological period during which its populations vary. During this long time-scale system development, changes in the character of species or species replacement occurs. The former is evolution, the latter is succession.

We now focus our attention on succession, where an invasion by a species into a community changes the species composition; the structure of the system is made up through a series of invasions. We will consider a simple succession model that represents intra- and interspecific interactions by differential equations. We derive results on the direction of succession and the structure of the climax state by using a quasi-stationary method.

Our succession model is shown in Fig. 7.3. The successional process is divided into two stages. One is an 'invasion' period, during which the

Fig. 7.3. Model of succession. Random invasions by new species into steady-state system are indicated by arrows. Species succeed or fail to invade the system, depending on the stability of that steady state. System state changes from the invaded steady state to a new stable steady state during ecological change period according to the dynamical population equations. The system attains a climax state when no species can invade successfully.

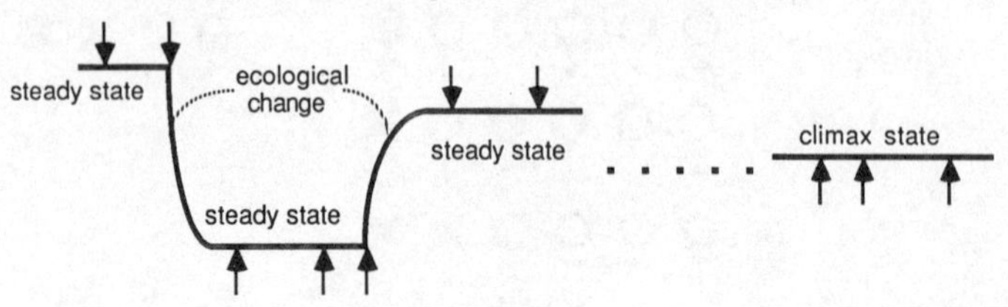

system stays in a steady state while new species repeatedly and often unsuccessfully attempt to invade the system. New species come to the system at random, i.e. the invasion process is stochastic. Whether the population density of an invading species will increase or decrease in the system depends on the stability of the pre-existent steady state of the entire system, including the new species. An incoming species whose growth rate is negative in this state will soon become extinct. In this case, the system remains unchanged. On the other hand, a species whose growth rate is positive, can join the community. The invasion period continues until one species succeeds in invasion.

The second stage is a period of 'ecological change', during which the state of the system changes, according to the dynamical equations, from the previous steady state to a new stable steady state. This period continues until the system reaches the new steady state. Our succession model is an alternating sequence of these two stages. The system may finally attain a state that cannot be invaded by any species. We call this state the climax state.

7.3.1 *Lyapunov function and characteristic quantities representing the direction of succession*

First, we shall analyze the following Lotka–Volterra competition system with symmetric coefficients:

$$\frac{\mathrm{d}x_i}{\mathrm{d}t} = r_i\left(1-\sum_{j=1}^{N} a_{ij}x_j/K_i\right)x_i, \quad i=1,2,\ldots,N \tag{13a}$$

$$a_{ii}=1, \quad i=1,2,\ldots,N, \quad a_{ij}=a_{ji}, \quad i,j=1,2,\ldots,N \tag{13b}$$

where x_i, r_i and K_i are population density, intrinsic growth rate and carrying capacity of species i, respectively, and a_{ij} is the competition coefficient between species i and j. MacArthur (1970) derived eq. (13) for resource-consumer systems using Lotka–Volterra equations with an appropriate approximation. Equation (13) is called the exploitative competition model. Furthermore, he found a Lyapunov function for this system, which is given by

$$G=\sum_{i=1}^{N} K_i x_i - \frac{1}{2}\sum_{i,j=1}^{N} a_{ij}x_i x_j, \tag{14}$$

the derivative of which, along a trajectory of system (13), satisfies

$$\frac{\mathrm{d}G}{\mathrm{d}t}=\sum_{i=1}^{N} r_i K_i\left(1-\sum_{j=1}^{N} a_{ij}x_j/K_i\right)^2 x_i \geqslant 0. \tag{15}$$

Function G may be interpreted as the total resource utilization in MacArthur's (1970) resource-consumer system. He obtained the result that the total resource utilization of an exploitative competition system always increases, and the final state is characterized by the maximum of that value.

We can qualitatively analyze the succession of system (13) by considering eqs. (14) and (15). Let G^* denote the value of G at the steady state $\{x_i^*\}$. Then, from the steady-state relation,

$$\left(K_i - \sum_{j=1}^{N} a_{ij} x_j^*\right) x_i^* = 0, \quad i = 1, 2, \ldots, N, \tag{16}$$

we get

$$G^* = \frac{1}{2} \sum_{i=1}^{N} K_i x_i^*. \tag{17}$$

Consider the following situation: a new species, species $N+1$, comes to the system at steady state $\{x_i^*\}$, the population density of species $N+1$ grows, and then the system reaches finally a new steady state $\{x_i^{**}\}$. If the population density of species $N+1$ is very small when it invades, then the value of G is equal to G^* at that moment of invasion. According to eq. (15), function G increases until the system goes to the new steady state $\{x_i^{**}\}$. Therefore, we obtain the following inequality,

$$G^* < G^{**}, \tag{18}$$

where G^{**} denotes the value of function G at $\{x_i^{**}\}$. Even if the population density of the invading species is not sufficiently small at the initial stage, we can derive this inequality for the case where the competition coefficient matrix is positive definite. Thus, the new steady state is globally stable.

The change of G is the sum of the alterations in density x_i weighted by K_i. Inequality (18) does not mean that the total population density increases necessarily in succession. The irreversibility of this succession is shown by eq. (18). The climax state of system (13) is characterized by the maximum of G^*.

We can also derive a Lyapunov function for an extended Lotka–Volterra system with semi-symmetric interactions. This is shown in Appendix A7.4. For that extended system, we have similarly a state function that increases during succession.

Like the special case of extended Lotka–Volterra system with semi-symmetric interaction, there is a quantity satisfying inequality (18) for the interference competition system (4). Let us define function G as follows,

$$G(\boldsymbol{x}) \equiv \sum (\varepsilon_i/\sigma_i)\beta_i x_i. \tag{19}$$

This quantity represents the sum total of the products of the essential growth rates and the intensities of interference competition of all species. Consider the Lyapunov function $Q(\boldsymbol{x})$ defined by eq. (10). Because $Q(\boldsymbol{x})$ decreases along any trajectory $x(t)$ with time t, we have

$$Q(\boldsymbol{x}^*) > Q(\boldsymbol{x}^{**}). \tag{20}$$

The quantity $Q(\boldsymbol{x}^*)$ can be rewritten as follows:

$$\begin{aligned} Q(\boldsymbol{x}^*) &= -2 \sum_{i\in P^*} \varepsilon_i \beta_i x_i^*/\sigma_i + \sum_{i\in P^*} \sum_{j\in P^*-\{i\}} \beta_i \beta_j x_i^* x_j^* \\ &\quad + \sum_{i\in P^*} \alpha_i \beta_i x_i^{*2} \\ &= -\sum_{i\in P^*} \varepsilon_i \beta_i x_i^*/\sigma_i + \sum_{i\in P^*} (\beta_i/\sigma_i)\{-\varepsilon_i + \sigma_i \alpha_i x_i^* \\ &\quad + \sum_{j\in P^*-\{i\}} \sigma_i \beta_j x_j^*\} x_i^* \\ &= -\sum_{i\in P^*} \varepsilon_i \beta_i x_i^*/\sigma_i = -G(\boldsymbol{x}^*) \end{aligned} \tag{21}$$

where we used the relation $f_i(\boldsymbol{x}^*)\, x_i^* = 0$. Similarly, we have

$$Q(\boldsymbol{x}^{**}) = -\sum (\varepsilon_i/\sigma_i)\beta_i x_i^{**} = -G(\boldsymbol{x}^{**}). \tag{22}$$

Thus, it follows from relation (20) that $G(\boldsymbol{x})$ increases monotonically with each invasion process.

7.3.2 *Quasi-stationary process*

A simple Lyapunov function exists for Lotka–Volterra competition systems with constrained coefficients, such as symmetric coefficients. We cannot find, however, a Lyapunov function for general Lotka–Volterra systems; in the extended case shown in Appendix A7.4, the implicit symmetry plays an important role in obtaining the Lyapunov function. Note that if we can find relationships between the values of a state function at two steady states such as eq. (18), it would provide us with a quantitative means for analyzing the irreversible process of succession and characterizing the climax state of the system by the maximum of such a function; it is not necessary to find a Lyapunov function, which monotonically increases along non-stationary change, like function (14) for system (13).

Succession from one steady state to another is actually non-stationary. But by very slow alteration of system parameters, we can change a competition system from one steady state to another in such a way that the system follows a series of steady states. We shall call such a process a 'quasi-stationary process'. It is analogous to quasi-static processes in thermodynamics, where external parameters are changed slowly so as to keep a system close to equilibrium. In the following, we present a new method, established upon this quasi-stationary process, to derive

Fig. 7.4. Comparison of non-stationary and quasi-stationary processes. Two-species Lotka–Volterra competition systems are considered. Solid lines and broken lines indicate isoclines, $dx_1/dt = 0$ and $dx_2/dt = 0$, respectively, and dots indicate state of the system: (*a*) a non-stationary process where state of system changes from unstable steady state $(K_1, 0)$ to stable one $(0, K_2)$ by invasion of second species. (*b*) a quasi-stationary process where invasion of second species causes no non-stationary change, because the steady state is stable for invasion of the second species. The quasi-stationary process occurs by slow changes of parameters, e.g. K_2.

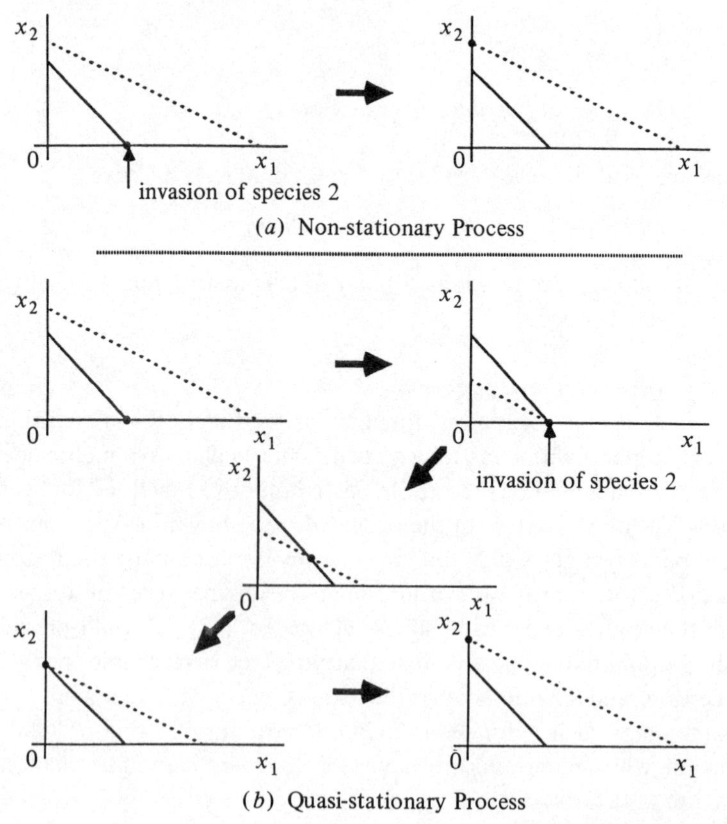

characteristic quantities indicating the direction of succession; this offers a means alternative to, and easier than, one using a Lyapunov function. First, we illustrate the idea of a quasi-stationary process with a simple example.

Fig. 7.4 compares a non-stationary change and the corresponding quasi-stationary change in a two-species Lotka–Volterra competition system. Assume that

$$K_1 < \frac{K_2}{a_{21}} \text{ and } \frac{K_1}{a_{12}} < K_2. \tag{23}$$

Then, the steady state $(K_1, 0)$ (i.e. the population of species 1) is unstable for an invasion of species 2, while the steady state $(0, K_2)$ (population 2) is stable for an invasion of species 1. Thus, an invasion of species 2 at steady state $(K_1, 0)$ would cause a non-stationary change into a new steady state $(0, K_2)$, as depicted in Fig. 7.4(*a*). On the other hand, we can make the same state change from $(K_1, 0)$ to $(0, K_2)$ along a quasi-stationary process as follows: at first, we decrease K_2 to $K_1 a_{21}$. After this parameter change, steady state $(K_1, 0)$ becomes stable; therefore, the state $(K_1, 0)$ would remain unchanged by an invasion of species 2. When we increase K_2 to K_1/a_{12} the system traces the continuous line of steady states in state space from $(K_1, 0)$ to $(0, K_1/a_{12})$. During this process, the two species coexist, until the first species dies out at the final stage. Finally, we change the value of K_2 from K_1/a_{12} to its original value, which brings the system to $(0, K_2)$. As a result, the state is altered from $(K_1, 0)$ to $(0, K_2)$, and the system follows a series of steady states throughout the entire process.

7.3.3 *General competition systems*

Consider a competition system given in the following general form,

$$\frac{dx_i}{dt} = f_i(x_1, \dots, x_N; K_i)\, x_i, \quad i = 1, 2, \dots, N. \tag{24}$$

We assume

$$\frac{\partial f_i}{\partial x_i} \leqslant 0, \quad i, j = 1, \dots, N, \tag{25a}$$

and

$$f_i(0, \dots, K_i, \dots, 0; K_i) = 0, \quad i = 1, 2, \dots, N, \tag{25b}$$

i.e. K_i is equal to the population density of the ith species in the situation where all other species are absent. We call K_i carrying capacity, which depends on the organic and inorganic environmental factors for the ith species. We also assume that,

$$\frac{\partial f_i}{\partial K_i} > 0. \tag{26}$$

Inequality (26) has a reasonable biological meaning; if the environment becomes more favorable for the ith species, then its growth rate becomes greater, and vice versa.

The quasi-stationary process of system (24) is given in Appendix A7.5. We get the following theorem using the idea of quasi-stationary process for a general competition system (24).

Theorem 4

If there exists a state function G such that

$$\frac{\partial G}{\partial K_i} > 0, \quad i = 1, 2, \ldots, N, \tag{27}$$

where independent variables of G are population densities x_i, $i = 1, \ldots, N$, then the value of function G at steady state always increases during succession, i.e. the inequality,

$$G(x_1^*, \ldots, x_N^*) < G(x_1^{**}, \ldots, x_N^{**}), \tag{28}$$

holds where succession proceeds from a steady state $(x_1^*, \ldots, x_N^*)$ to another state $(x_1^{**}, \ldots, x_N^{**})$.

The proof of this theorem, for which the quasi-stationary process idea is central, is shown in Appendix A7.7. The following is a corollary of this theorem; its proof is in Appendix A7.8.

Corollary

If there exists a state function G, such that

$$\frac{\partial G}{\partial K_i} > 0, \quad \text{for } x_i > 0, \quad i = 1, 2, \ldots, N, \tag{29a}$$

$$\frac{\partial G}{\partial K_i} = 0, \quad \text{for } x_i = 0, \quad i = 1, 2, \ldots, N, \tag{29b}$$

where independent variables of G are population densities x_i and carrying capacities K_i, $i = 1, \ldots, N$, then the value of function G increases monotonically during succession.

As an example of this result, we now show that a state function of system (13) defined by eq. (17), with independent variables x_i^*'s and K_i's, satisfies above conditions (29). Differentiating eq. (17) with respect to K_l, we have the following equations from the steady state relation (16):

$$\frac{\partial}{\partial K_l}\left(\tfrac{1}{2}\sum_{i=1}^{N} K_i x_i^*\right) = \tfrac{1}{2}x_l^* + \tfrac{1}{2}\sum_{i=1}^{N} K_i \frac{\partial x_i^*}{\partial K_l}$$

$$= \tfrac{1}{2}x_l^* + \tfrac{1}{2}\sum_{i=1}^{N}\sum_{j=1}^{N} a_{ij} x_j^* \frac{\partial x_i^*}{\partial K_l}$$

$$= \tfrac{1}{2}x_l^* + \tfrac{1}{2}\sum_{j=1}^{N} x_j^* \frac{\partial}{\partial K_l}\left(\sum_{i=1}^{N} a_{ji} x_i^*\right)$$

$$= \tfrac{1}{2}x_l^* + \tfrac{1}{2}\sum_{j=1}^{N} x_j^* \delta_{jl} = x_l^*, \quad l = 1, 2, \ldots, N. \tag{30}$$

These equations show that function G in eq. (17) satisfies conditions (29). Therefore, from the above corollary, we can conclude inequality (18) without using a Lyapunov function.

7.3.4 *Suppressing competition systems*

Competition systems satisfy inequalities (25*a*): the growth rate of each species is reduced by the increase in the population densities of other species, i.e. each member species of the competition system has a negative effect on the growth rate of other species.

We now consider the effect of each member species of a competition system on the others from a different viewpoint: the effect of K_i on the population densities at a steady state. Hereafter, we denote by x_i the stationary population density of species i. It is assumed that x_i increases as K_i increases, i.e. the environment for species i is improved:

$$\frac{\partial x_i}{\partial K_i} > 0, \quad i = 1, 2, \ldots, N. \tag{31}$$

Note that the inequality

$$\frac{\partial x_i}{\partial K_j} \leqslant 0, \quad \text{for } i \neq j, \quad i, j = 1, 2, \ldots, N \tag{32}$$

does not always hold in competition systems. For example, consider a three species competition system, in which the competition between species 1 and species 2 and between species 2 and species 3 are much stronger than the competition between species 1 and species 3. When K_1

increases, x_1 increases according to eq. (31). Then x_2 decreases, because of a negative interaction between species 1 and species 2. The effect of a decrease of x_2 is to increase x_3. If the magnitude of this indirect positive effect of K_1 on x_3 is great enough to dominate the direct negative effect of K_1 on x_3, then x_3 increases under these effects. Therefore, $\partial x_3/\partial K_1 > 0$ in this case.

When inequality (32) holds for a competition system, we call that system a 'suppressing' competition system. Inequality (32) means that all member species of this system suppress each other at their stationary population densities, and no indirect positive effect overcomes the corresponding direct negative effect.

We obtain the following theorem for suppressing competition systems (the proof of this theorem is given in Appendix A7.9).

Theorem 5

In a suppressing competition system that satisfies inequalities (26) and (32), the population density of each species always decreases with an invasion of other species.

In suppressing competition systems, the population density of a species increases only when that species invades into the system. Once a species becomes established in the community, its population density always decreases with subsequent invasions. Succession of these systems is simpler than the exploitative competition system represented by eq. (13).

In the previous section, several properties are analyzed for the class of Lotka–Volterra competition systems that have competition coefficients a_{ij} given by the product of two constants, i.e. $a_{ij} = \sigma_i \beta_j$ $(i \neq j)$. It is easily shown that inequality (32) holds for this class of systems, if the conditions $a_{ii} > \sigma_i \beta_i$, $i = 1, \ldots, N$, are satisfied, i.e. all member species are auto-competitors. Such systems are in the class of suppressing competition systems.

7.4 Discussion

Pomerantz & Gilpin (1979) analyzed by computer simulation the effect of community matrix structure on the number of alternative stable subsets of species (domains of attraction) and the number of species per domain of attraction. They normalized eq. (2) by setting ε_i and μ_{ii} equal to one, and constructed community matrices of three covariance types: positive, negative and zero $\operatorname{cov}(\mu_{ij}, \mu_{ji})$. A matrix with positive $\operatorname{cov}(\mu_{ij}, \mu_{ji})$ corresponds to a community dominated by exploitative competition. A matrix with negative $\operatorname{cov}(\mu_{ij}, \mu_{ji})$ corresponds to a community dominated by interference competition (transitive competitive relationships), and one with zero $\text{cov}(\mu_{ij}, \mu_{ji})$ corresponds to a community dominated by gratuitous interference (Case & Gilpin, 1974). Pomerantz & Gilpin (1979) showed that, regardless of the covariance, more domains of attraction and fewer species per domain of attraction occur with increasing average competition among community members.

Although our model (4) is based on different assumptions, it is related to Pomerantz & Gilpin's model in the following way; if ε_i, α_i and σ_i in eq. (4) are set to one, then the species with larger β_i is stronger in interspecific competition. Furthermore, if β_i is chosen randomly, then the expected value of $\text{cov}(\beta_i, \beta_j)$ is zero, and a community with gratuitous interference competition results. Because a species with β_i greater than 1 is a hetero-competitor, a larger value of the average β_i implies a community with more hetero-competitors. From remarks (iii) and (iv), and the discussion in Section 7.2, it is clear that a community with more hetero-competitors has a higher number of stable equilibrium states (domains of attraction), and each stable equilibrium state consists of fewer auto-competitors, plus a single hetero-competitor. This corresponds to Pomerantz & Gilpin's (1979) conclusion.

We formulated a simple model for succession consisting of a sequence of two stages: an invasion period and an ecological change period. New species come from outside of the system during an invasion period. If one species invades successfully, then the state of the system changes, according to the dynamical equations of the populations, from a previous steady state to a new one. We assumed in our model that the organic and inorganic environment of the system remains unchanged throughout succession. If the environments of communities change during succession, we need to incorporate into our model environmental factors as state variables.

Dynamical models have been analyzed to explain system properties, for example, stability, catastrophic change of state, and periodic change of population size. In these analyses, attention focused on changes of state with time, stability of steady state, and change of stable steady state through parameter change. We now have a new application of dynamical models to the succession of communities or ecosystems. Dynamical equations of populations represent interactions among member species of a system. Therefore, the invasion rules obtained from the stability of steady states of a dynamical system reflect the interactive structure among species in the system under consideration. The invasion rules and thus, the interactive structure, determine partially the direction of succession and the climax state.

In our analysis of succession, we need not analyze dynamical behavior beyond steady state relations and their local stabilities. We need no dynamical Lyapunov function to obtain a state function that increases with the successional process. A state function with this property indicates the irreversibility of succession, and the climax state is characterized by the maximum of this state function. If this state function has only one maximum, then there might be an unique climax state toward which succession proceeds independent of the order of invasion. On the other hand, when that function has several maximum points, there might be more than one climax state. Which state will be attained depends on the order of invasion.

Inherent in the definition of suppressing competition systems is the notion that changing the carrying capacity of one species has both direct and indirect effects on other species population densities. Most dynamical models represent interactions among species by the dependence of the growth rate function of one species on that of another species. However, the effects represented in growth functions are direct and pairwise. We propose different expressions for interactions: the influence of one species on another species at a steady state. This expression contains indirect effects as well as direct effects. Patten (1983) has argued for the significance of indirect effects in ecosystems, and Higashi & Patten (1989) quantitatively analyzed indirect influence in flow networks. Indirect effects are determined by not only direct interaction of two species, but also the interactive structure of the system.

7.5 Summary

1. The structure of communities and ecosystems are determined by succession and evolution. We focus on the structure of competition systems formed by succession. Interactions among species are described in terms of growth-rate functions of dynamical equations. First, invasion rules are obtained from the stability condition of dynamical systems. The direction of succession is then obtained from the invasion rules. Finally, the structures of the climax states are determined by the direction of succession.

2. Interference competition systems are analyzed to gain insight into the species composition of competition systems as a result of succession. This analysis is based on a Lotka–Volterra model in which the competition coefficients are constrained to be the products of two factors: the intrinsic force of interference to other individuals and the susceptibility to interference from other individuals. In this model, survival or extinction of each species at a non-negative steady state, crucially depends both on

the rank of the magnitude of essential growth rate [(intrinsic growth rate)/(susceptibility to interference)], and on whether the species is an auto-competitor (i.e. a species with more severe intra-specific interference than inter-specific) or a hetero-competitor (the opposite to auto-competitor). The priority for survival among auto-competitors is also determined by their rank.

3. If the system contains a number of hetero-competitors, it may have more than one stable steady state, and the multiplicity of steady states increases with the number of hetero-competitor species. Only one hetero-competitor, if any, can survive at each stable equilibrium, occupying a rank lower than that of any surviving auto-competitor. Therefore, although a system consisting of a large number of species has generally many stable steady states, each stable steady state has only a small number of species with positive population densities. This analytical result agrees with the results of Pomerantz & Gilpin (1979) from computer simulations.

4. Theorem 3 gives the criterion for successful invasion of a species; it must have a greater essential growth rate. If a hetero-competitor with a high essential growth rate invades, all auto-competitors with lower essential growth rates should become extinct, thereby drastically reducing species richness.

5. A proposed model of succession has two stages: invasion and ecological change. In the former, the system remains in steady state, while new species attempt to invade the system. An invading species can enter the system only when its growth rate is positive at the invading steady state. In the stage of ecological change, the system changes from this steady state to a new stable steady state according to the dynamical population equations.

6. In exploitative competition systems, resource utilization of a system increases during succession. Resource utilization is a dynamical Lyapunov function in these systems. In general, the direction of succession of competition systems can be analyzed as a quasi-stationary process. The quasi-stationary process is analogous to a quasi-static process in equilibrium thermodynamics. Even when no dynamical Lyapunov function is known, this new method can show the irreversibility of succession under a certain condition.

7. In suppressing competition systems, a special class of competition systems for which an increase in carrying capacity of one species suppresses that of other populations at a steady state, it is proved that each population density decreases when any species invades. This result is also derived by the quasi-stationary process method. A special case of

interference competition systems in which all species are auto-competitors falls into the class of suppressing competition systems.

Appendix

A7.1 Outline of proof of Theorem 1

Let $\boldsymbol{x}^* = (x_1^*, \ldots, x_N^*)$ be a non-negative equilibrium point of system (4), and let $P = \{i \mid x_i^* > 0\}$ and $I - P = \{i \mid x_i^* = 0\}$. From eq. (6) and $x_i^* > 0$ for $i \in P$, we have the following relations: for $i \in P$,

$$\frac{\varepsilon_i}{\sigma_i} > C(P), \quad \text{if } \alpha_i > \beta_i, \tag{A1}$$

$$\frac{\varepsilon_i}{\sigma_i} < C(P), \quad \text{if } \alpha_i < \beta_i. \tag{A2}$$

Next, we will obtain the conditions for a non-negative equilibrium $\boldsymbol{x}^*$ to be linearly stable. As we defined in Section 2, the condition for linear stability is that all eigenvalues λ of a community matrix A, which are given by the solution of the characteristic equation $|A - \lambda I| = 0$, have negative real parts. The characteristic equation of the community matrix of system (4) is given by the following polynomial equations:

$$|A - \lambda I| = \{\prod_{i \in P} (\lambda - b_i) + \sum_{j \in P} a_j \prod_{i \in P(i \neq j)} (\lambda - b_i)\} \prod_{i \in I-P} (\lambda - c_i) = 0, \tag{A3}$$

where $a_i = \sigma_i \beta_i x_i^*$, $b_i = \sigma_i(\beta_i - \alpha_i) x_i^*$ and $c_i = f_i(\boldsymbol{x}^*)$. From (A3), we can show that all the eigenvalues are negative (i.e. $\boldsymbol{x}^*$ is linearly stable) if the following conditions are satisfied (see Shigesada *et al.* (1984) for details):

For $i \in P$,

(a) $\alpha_i > \beta_i$ for all i, or (A4)

(b) only one species satisfies $\alpha_i < \beta_i$ and all other species satisfy $\alpha_i > \beta_i$, and

$$1 + \sum_{i \in P} \xi_i < 0; \tag{A5}$$

for $i \in I - P$,

$$f_i(\boldsymbol{x}^*) < 0, \quad \text{that is,} \quad \frac{\varepsilon_i}{\sigma_i} < C(P). \tag{A6}$$

Let s be the largest integer in P that satisfies $\alpha_i > \beta_i$. Then, (A1), (A2) and (A6) together with (5) from the text imply that all the species ranked from

1 through s belong to P and are auto-competitors. Thus, from (A4) and (A5), respectively, we get conditions (I) and (II) of Theorem 1.

A7.2 Proof of Theorem 2(b)

Let $\boldsymbol{x}^* = (x_1^*, \ldots, x_N^*)$ be a positive equilibrium point of system (4), (i.e. $x_i^* > 0$ for all i). Consider the following positive definite function:

$$V(\boldsymbol{x}) = \sum_{i=1}^{N} \frac{\varepsilon_i}{\sigma_i}\left(x_i - x_i^* - x_i^* \ln \frac{x_i}{x_i^*}\right). \tag{A7}$$

By differentiating $V(\boldsymbol{x}(t))$ with respect to t, we have

$$\frac{\mathrm{d}}{\mathrm{d}t} V(\boldsymbol{x}(t)) = -\sum_{i=1}^{N} \sum_{j=1(j \neq i)}^{N} \beta_i \beta_j (x_i - x_i^*)(x_j - x_j^*) - \sum_{i=1}^{N} \alpha_i \beta_i (x_i - x_i^*)^2. \tag{A8}$$

If all eigenvalues of the symmetric matrix,

$$Q_{ij} = \begin{cases} -\alpha_i \beta_i & (i = j) \\ -\beta_i \beta_j & (i \neq j) \end{cases} \tag{A9}$$

are negative, then eq. (A8) becomes negative definite, so that the equilibrium point $\boldsymbol{x}^*$ is globally stable. The characteristic equation of (A9) becomes identical to (A3), if $\sigma_i x_i^*$ is substituted for β_i in (A9). Since (A4) and (A5), which give the conditions for all real parts of solutions of (A3) to be negative, do not depend on $\sigma_i x_i^*$, the eigenvalues of the matrix (A9) have the same signs as those of the community matrix of system (4). Thus, if the equilibrium point $\boldsymbol{x}^*$ are linearly stable, all the eigenvalues of the matrix (A9) are negative, and $\boldsymbol{x}^*$ is globally stable.

A7.3 Proof of Theorem 2(c)

First we consider the steady states of type (I) in Theorem 1. Assume that there exist two $s_1, s_2 \in I$ $(s_1 < s_2)$ that satisfy (8b) and (8c). Then, the following relations are fulfilled:

$$\varepsilon_{s_1}/\sigma_{s_1} > C(P_{s_1}) > \varepsilon_{s_1+1}/\sigma_{s_1+1}, \tag{A10a}$$

$$\varepsilon_{s_2}/\sigma_{s_2} > C(P_{s_2}) > \varepsilon_{s_2+1}/\sigma_{s_2+1}, \tag{A10b}$$

$$\alpha_i > \beta_i \text{ for } i \in P_{s_2}, \tag{A10c}$$

where $P_{s_1} = \{1, 2, \ldots, s_1\}$ and $P_{s_2} = \{1, 2, \ldots, s_2\}$. When $1 + \sum_{i \in P_s} \xi_i$ and $1 + \sum_{i \in P_s} \xi_i + \xi_{s+1}$ have the same sign, we can show that if

$$C(P_s) > \frac{\varepsilon_{s+1}}{\sigma_{s+1}},$$

then

$$C(P_{s+1}) > \frac{\varepsilon_{s+1}}{\sigma_{s+1}} > \frac{\varepsilon_{s+2}}{\sigma_{s+2}}. \tag{A11}$$

Because $\xi_i = \beta_i/(\alpha_i - \beta_i) > 0$ for all $i \in P_{s_2}$, substituting $s = s_1, s_1+1, \ldots, s_2-1$ in (A11) and using (A10a), we obtain that

$$C(P_{s_2}) > \varepsilon_{s_2}/\sigma_{s_2}.$$

This inequality contradicts (A10b). Thus s must be unique. For the case (II) of Theorem 1, we get the uniqueness of s by replacing P_s with $P_{s,w}$, where $P_{s,w} = P_s + \{w\}$.

A7.4 Proof of Theorem 3

Because we assume that the pre-occupant N species have already reached a stable equilibrium state $\boldsymbol{x}^*$, and that a small number of species μ invade the community, the growth rate $f_\mu(\boldsymbol{x}, x_\mu)$ of species μ is approximately given by $f_\mu(\boldsymbol{x}^*, 0)$. Therefore, it is necessary for successful invasion that $f_\mu(\boldsymbol{x}^*, 0) \geqslant 0$, that is, $\varepsilon_\mu/\sigma_\mu \geqslant C(P)$.

We next show that $f_\mu(\boldsymbol{x}^*, 0) > 0$ is sufficient for successful invasion of species μ. Since $f_\mu(\boldsymbol{x}^*, 0) > 0$, the population of the invading species initially increases and the system finally reaches a new equilibrium state $(\boldsymbol{x}^{**}, x_\mu^{**})$. However, it is not obvious whether the species finally becomes a constituent member of the community or not, that is, whether $x_\mu^{**} > 0$ or $x_\mu^{**} = 0$. So, we must show that $x_\mu^{**} > 0$. Since Theorem 2(b) assures that $\boldsymbol{x}^*$ is globally stable in the original community (i.e. the $\boldsymbol{x}$-space), we obtain, from the property of the Lyapunov function (10), that

$$Q(\boldsymbol{x}^*, 0) < Q(\boldsymbol{x}, 0) \text{ except for } \boldsymbol{x} = \boldsymbol{x}^*. \tag{A12}$$

Furthermore, it follows from eq. (11) that $Q(\boldsymbol{x}(t), x_\mu(t))$ always decreases as t increases. Thus we obtain

$$Q(\boldsymbol{x}^{**}, x_\mu^{**}) < Q(\boldsymbol{x}^*, 0). \tag{A13}$$

Substituting $\boldsymbol{x}^{**}$ for $\boldsymbol{x}$ in the relation (A12), and combining the resultant relation with (A12) leads to the following:

$$Q(\boldsymbol{x}^{**}, x_\mu^{**}) < Q(\boldsymbol{x}^{**}, 0). \tag{A14}$$

If the invading species becomes extinct at the final stage, that is $x_\mu^{**} = 0$, it contradicts (A14). Thus species μ must be a constituent member in the resultant community.

A7.5 Lyapunov function of an extended Lotka–Volterra competition system

We consider the following competition system:

$$\frac{dx_i}{dt} = r_i\left(1 - \frac{v_i(x_i) + \sum_{j=1,(j\neq i)}^{N} \varphi_{ij} u_j(x_j)}{v_i(K_i)}\right)x_i, \quad i = 1, 2, \ldots, N, \tag{A15}$$

where

$$\varphi_{ij} = \varphi_{ji}, \quad i, j = 1, 2, \ldots, N, \tag{A16}$$

$$u_i(0) = 0, \frac{du_i(x)}{dx} > 0 \text{ and } v_i(K_i) > 0, \quad i = 1, 2, \ldots, N. \tag{A17}$$

We define function G as

$$G = \sum_{i=1}^{N} \int_0^{x_i} \left\{v_i(K_i) - v_i(x)\right\} \frac{du_i(x)}{dx} dx - \frac{1}{2}\sum_{i=1}^{N} \sum_{j=1,(j\neq i)}^{N} \varphi_{ij} u_i(x_i) u_j(x_j). \tag{A18}$$

Differentiating eq. (A18) with respect to time t, we have

$$\frac{dG}{dt} = \sum_{i=1}^{N} r_i \frac{du_i(x)}{dx} v_i(K_i)\left(1 - \frac{v_i(x_i) + \sum_{j=1,(j\neq i)}^{N} \varphi_{ij} u_j(x_j)}{v_i(K_i)}\right)^2 x_i. \tag{A19}$$

In the above derivation, we used symmetry condition (A16). It follows from (A17) that $dG/dt > 0$. Thus, function G is a Lyapunov function of system (A15) with conditions (A16) and (A17). From eq. (A15), we have the steady state relation,

$$\left\{v_i(K_i) - v_i(x_i^*) - \sum_{j=1,(j\neq i)}^{N} \varphi_{ij} u_j(x_j^*)\right\} x_i^* = 0, \quad i = 1, 2, \ldots, N. \tag{A20}$$

Let G^* denote the value of G at the steady state $\{x_i^*\}$. From eqs. (A18) and (A20), we obtain

$$G^* = \sum_{i=1}^{N} \left\{ \tfrac{1}{2}(v_i(K_i) - v_i(x_i^*))\, u_i(x_i^*) + \int_0^{x_i} u_i(x) \frac{dv_i(x)}{dx} dx \right\}. \tag{A21}$$

A7.6 Quasi-stationary process in general competition systems

Let state $(x_1^*, \ldots, x_{N-1}^*, 0)$ be a steady state of system (24). Assume that species N invades into the system at this steady state, and that the population density of this new species can increase. Then, according to eq. (24) and the instability condition for an invasion, we have

$$f_i(x_1^*, \ldots, x_{N-1}^*, 0; K_i) = 0, \quad i = 1, 2, \ldots, N-1 \tag{A22a}$$

$$f_N(x_1^*, \ldots, x_{N-1}^*, 0; K_N) > 0. \tag{A22b}$$

By the invasion of species N, the state of the system changes to a new steady state $(x_1^{**}, \ldots, x_N^{**})$. In this new steady state, all N species co-exist or some species die out. Let s be the number of species becoming extinct, then $0 \leqslant s \leqslant N-1$. Without loss of generality, species 1 through s die out. Thus we have

$$x_i^{**} = 0 \text{ for } 1 \leqslant i \leqslant s. \tag{A23}$$

According to eq. (A23) and the stability condition of the new steady state, we obtain

$$f_i(0, \ldots, 0, x_{s+1}^{**}, \ldots, x_N^{**}; K_i) \leqslant 0, \quad \text{for } 1 \leqslant i \leqslant s, \tag{A24a}$$

$$f_i(0, \ldots, 0, x_{s+1}^{**}, \ldots, x_N^{**}; K_i) = 0, \quad \text{for } s < i \leqslant N. \tag{A24b}$$

To bring about the same state change in the system by a quasi-stationary process, we perform the following three steps:

(i) Decrease the value of K_N to $K_{N'}$, where

$$f_N(x_1^*, \ldots, x_{N-1}^*, 0; K_{N'}) = 0, \tag{A25}$$

From eqs. (26), (A22b), and (A25) we obtain

$$K_N > K_{N'}. \tag{A26}$$

(ii) Let species N invade into the system at this state, then increase the value of K_N to its original value with K_i $(i \leqslant s)$ increased to $K_{i'}$, where

$$f_i(0, \ldots, 0, x_{s+1}^{**}, \ldots, x_N^{**}; K_{i'}) = 0, \quad \text{for } 1 \leqslant i \leqslant s. \tag{A27}$$

From eqs. (26), (A24a), and (A27) it is clear that

$$K_i \leqslant K_{i'}, \quad \text{for } 1 \leqslant i \leqslant s. \tag{A28}$$

Species i $(i \leqslant s)$ become extinct at the final state of this step.

(iii) Finally, we decrease K_i $(i \leqslant s)$ to their original values.

In step (i), the population densities of all species in this system remain unchanged. At the final state of step (i), the steady state becomes stable for the invasion of species N. In step (ii), all N species co-exist in the steady state, and the population densities x_i are functions of K_i, $i = 1, 2, \ldots, N$. The values of carrying capacities K_i $(i \leqslant s)$ and K_N increase in this step. At the final state of this step, species 1 through s become extinct. In the step (iii), the population densities of all species remain unchanged. As a result, the K_i return to their original values, and the state moves from $(x_1^*, \ldots, x_{N-1}^*, 0)$ to $(0, \ldots, 0, x_{s+1}^{**}, \ldots, x_N^{**})$ along the quasi-stationary process.

A7.7 Proof of Theorem 4

Let G be a state function with independent variables x_i, $i = 1, 2, \ldots, N$. We denote by ΔG the difference in the value that G takes between two steady states $\{x_i^*\}$ and $\{x_i^{**}\}$. This difference is evaluated by the following integral:

$$\Delta G = \int_{\text{step (i)+step (ii)+step (iii)}} \mathrm{d}G \tag{A29}$$

Function G does not change in step (i) and step (iii) because the population densities x_i, $i = 1, 2, \ldots, N$, remain unchanged in these steps. We thus get

$$\Delta G = \int_{\text{step (ii)}} \mathrm{d}G = \int_{\text{step (ii)}} \sum_{i=1}^{N} \frac{\partial G}{\partial K_i} \mathrm{d}K_i. \tag{A30}$$

In step (ii), K_i does not decrease for any i, and increases for $i = 1, 2, \ldots, s$ and N, and $\partial G/\partial K_i$ are positive by assumption. Therefore, ΔG is positive, and we obtain the result of Theorem 4.

A7.8 Proof of Corollary of Theorem 4

Let G be a state function with independent variables x_i and K_i, $i = 1, 2, \ldots, N$. In step (i), $\mathrm{d}K_i = 0\ (1 \leqslant i \leqslant N-1)$, and $\mathrm{d}K_N < 0$ but $\partial G/\partial K_N = 0$ because $x_N = 0$; function G thus remains unchanged in this step. In step (iii), $\mathrm{d}K_i = 0$ for $i = s+1, \ldots, N$ and $\partial G/\partial K_i = 0$ for $i = 1, \ldots, s$ because of assumption (29*b*). Therefore, the value of G is also constant in step (iii). The change in the value of function G comes only from step (ii). It can be easily shown that the change of G in step (ii) is positive, using the same reasoning as in the proof of Theorem 4. Hence the total change in the value of function G is positive, which concludes the proof.

A7.9 Proof of Theorem 5

We apply the quasi-stationary process method to this system. No population density changes during steps (i) and (iii). We denote the change of x_i from the steady state $\{x_i^*\}$ to $\{x_i^{**}\}$ by Δx_i. We have

$$\Delta x_i = \int_{\text{step (ii)}} \mathrm{d}x_i = \int_{\text{step (ii)}} \sum_{j=1}^{N} \frac{\partial x_i}{\partial K_j} \mathrm{d}K_j. \tag{A31}$$

From the suppressing condition (32) and $dK_j \geqslant 0$ for $j = 1, \ldots, N$, we can prove that Δx_i is negative.

References

Case, T. J. & Casten, R. G. (1979). Global stability and multiple domains of attraction in ecological systems. *Am. Nat.*, **113**, 705–14.

Case, T. J. & Gilpin, M. E. (1974). Interference competition and niche theory. *Proc. Natl. Acad. Sci.* (*USA*), **71**, 3073–7.

Connell, J. H. (1961). The influence of interspecific competition and other factors on the distribution of the barnacle *Chthamalus stellatus*. *Ecology*, **42**, 710–23.

Dayton, P. K. (1971). Competition, disturbance and community organization: the provision and subsequent utilization of space in a rocky intertidal community. *Ecol. Monogr.*, **41**, 351–89.

DeAngelis, D. L. (1975). Stability and connectance in food web models. *Ecology*, **56**, 238–43.

Elton, C. (1958). *The Ecology of Invasion by Animals and Plants*. London: Methuen.

Gantmacher, F. R. (1959). *Applications of the Theory of Matrices*. New York, London, Sydney: John Wiley & Sons.

Gilpin, M. E. (1975). Stability of feasible predator-prey systems. *Nature*, **254**, 137–9.

Gilpin, M. E. & Case, T. J. (1976). Multiple domains of attraction in competition communities. *Nature*, **261**, 40–2.

Goh, B. S. (1977). Global stability in many-species systems. *Am. Nat.*, **111**, 135–43.

Goh, B. S. (1978). Sector stability of a complex ecosystem model. *Math. Biosci.*, **40**, 157–66.

Goh, B. S. (1980). *Management and Analysis of Biological Populations*. New York: Elsevier.

Goh, B. S. & Jennings, L. S. (1977). Feasibility and stability in randomly assembled Lotka–Volterra models. *Ecol. Mod.*, **3**, 63–71.

Harrison, G. W. (1979). Global stability of food chains. *Am. Nat.*, **114**, 455–7.

Higashi, M. & Patten, B. C. (1989). Dominance of indirect causality in ecosystems. *Am. Nat.*, **133**, 288–302.

Ikeda, M. & Šiljak, D. D. (1980). Lotka–Volterra equations: decomposition, stability, and structure. *J. Math. Biol.*, **9**, 65–83.

Jeffries, C. (1974). Qualitative stability and digraphs in model ecosystems. *Ecology*, **55**, 1415–19.

Jeffries, C. (1975). Stability of ecosystems with complex food webs. *Theor. Popul. Biol.*, **7**, 149–55.

LaSalle, J. & Lefshetz, S. (1961). *Stability by Liapunov's Direct Method with Applications*. New York, London: Academic Press.

Lotka, A. J. (1922*a*). Contribution to the energetics of evolution. *Proc. Natl. Acad. Sci.* (*USA*), **8**, 147–51.

Lotka, A. J. (1922*b*). Natural selection as a physical principle. *Proc. Natl. Acad. Sci. (USA)*, **8**, 151–4.

MacArthur, R. H. (1955). Fluctuations of animal populations, and a measure of community stability. *Ecology*, **36**, 522–6.

MacArthur, R. H. (1968). The theory of the niche. In *Population Biology and Evolution*, ed. R. C. Lewontin, pp. 159–76. New York: Syracuse University Press.

MacArthur, R. H. (1970). Species packing and competitive equilibrium for many species. *Theor. Popul. Biol.*, **1**, 1–11.

MacArthur, R. H. & Levins, R. (1967). The limiting similarity, convergence, and divergence of coexisting species. *Am. Nat.*, **101**, 377–85.

Margalef, R. (1968). *Perspectives in Ecological Theory*. Chicago: University of Chicago Press.

May, R. (1972). Will a large complex system be stable? *Nature*, **238**, 413–14.

May, R. (1973). *Stability and Complexity in Model Ecosystems*. Monographs In Population Biology, 6. Princeton: Princeton University Press.

McNaughton, S. J. (1978). Stability and diversity of ecological communities. *Nature*, **274**, 251–2.

Miller, R. S. (1967). Pattern and process in competition. *Adv. Ecol. Res.*, **4**, 1–74.

Moore, J. C. & Hunt, H. W. (1988). Resource compartmentation and the stability of real ecosystems. *Nature*, **333**, 261–3.

Nakajima, H. (1985). Succession of model competition system. *Mem. of the Res. Inst. of Sci. and Eng. Ritsumeikan Univ.*, **44**, 1–12.

Odum, H. T. (1971). *Environment, Power and Society*. New York: Wiley.

Patten, B. C. (1983). On the quantitative dominance of indirect effects in ecosystems. In *Analysis of Ecological Systems: State-of-the-Art in Ecological Modeling*, ed. W. K. Lauenroth & G. V. Skogerboe, pp. 27–37. Amsterdam: Elsevier.

Pimm, S. L. (1979). The structure of food webs. *Theor. Popul. Biol.*, **16**, 144–58.

Pimm, S. L. (1984). *Food Webs*. New York: Chapman and Hall.

Pimm, S. L. & Lawton, J. H. (1977). Number of trophic levels in ecological communities. *Nature*, **268**, 329–31.

Pomerantz, M. J. & Gilpin, M. E. (1979). Community covariance and coexistence. *J. Theor. Biol.*, **79**, 67–81.

Quirk, J. P. & Ruppert, R. (1965). Qualitative economics and the stability of equilibrium. *Rev. Econ. Stud.*, **32**, 311–26.

Rejmanek, M. & Stary, P. (1979). Connectance in real biotic communities and critical values for stability of model ecosystems. *Nature*, **280**, 311–13.

Root, R. (1967). The niche exploitation patterns of the blue-grey gnatcatcher. *Ecol. Monogr.*, **37**, 317–50.

Roughgarden, J. (1979). *Theory of Population Genetics and Evolutionary Ecology: An Introduction*. New York: MacMillan.

Shigesada, N., Kawasaki, K. & Teramoto, E. (1984). The effect of interference competition on stability, structure and invasion of a multi-species system. *J. Math. Biol.*, **21**, 97–113.

Shigesada, N., Kawasaki, K. & Teramoto, E. (1989). Direct and indirect effects of invasions of predators on a multiple-species community. *Theor. Popul. Biol.* **36**, 311–38.

Šiljak, D. D. (1975). When is a complex ecosystem stable? *Math. Biosci.*, **25**, 25–50.

Solimano, F. & Beretta, E. (1982). Graph theoretical criteria for stability and boundedness of predator-prey systems. *Bull. Math. Biol.*, **44**, 579–85.

Svirezhev, Yu. M. & Logofet, D. O. (1978). *Stability of Biological Communities.* Moscow: Mir Publishers.

Takeuchi, Y. & Adachi, N. (1980). The existence of globally stable equilibria of ecosystems of the generalized Volterra type. *Math. Biol.*, **10**, 401–15.

Takeuchi, Y., Adachi, N. & Tokumaru, H. (1978). Global stability of ecosystems of the generalized Volterra type. *Math. Biosci.*, **42**, 119–36.

Tansky, M. (1976). Structure, stability, and efficiency of ecosystem. *Prog. Theor. Biol.*, **4**, 205–62.

Ulanowicz, R. E. (1986). *Growth and Development: Ecosystems Phenomenology.* New York: Springer Verlag.

Volterra, V. (1931). *Leçons sur la théorie mathématique de la lutte pour la vie.* Paris: Gauthier-Villars.

Yodzis, P. (1981). The stability of real ecosystem. *Nature*, **289**, 674–6.

8

Hierarchical evolution in ecological networks: environs and selection

T. P. BURNS, B. C. PATTEN AND M. HIGASHI

8.1 Introduction

Most ecologists conceive ecosystems as complex and describe them so. Ecosystems are more than energy and nutrient flows, trophic webs, and competition communities. They are the full interrelations among coexistent living organisms and their nonliving milieu (e.g. Tansley, 1935).

Implicit in this causal model of the world are the ideas of network and hierarchy. Where there is only matter heterogeneously distributed in four dimensions, there is no network and no hierarchy. Choosing to recognize discontinuities in space-time as entities – ourselves for instance – means we must deal with their temporal dynamics and their hierarchical and interactive structure. Evolutionary theory has for years debated the necessity of a hierarchy of evolving entities, and still largely ignores the network of ecological interactions to which living entities belong. Ecology, long concerned with evolutionary processes in the hierarchy from organism to the biosphere, has not developed a general theory of order-preserving temporal change. It remains, for both disciplines, a worthwhile goal to determine how and why ecological entities interacting at the recognized hierarchical levels change, or not, through time.

Riedl (1978) takes a systems perspective to the study of morphological evolution. He presents four forms of morphological order: (1) standard part, (2) hierarchy, (3) interaction, and (4) traditive inheritance. The order of standard parts is the constraint that complex forms are built up from multiple copies of a single form – an ensemble of similarly derived individuals. Hierarchical order is the constraint that the whole plays on the parts and vice versa. Interactive order is the constraint on the possible variations of any entity to the subset that is compatible with others with which it interacts as part of a larger entity. Traditive inheritance is the constraining force of the information contained in an entity; the 'burden' of familial history which is most evident in cells and organisms. These four impose constraints upon evolutionary possibilities. As a result we observe the structural order of living organisms and the temporal order of phylogeny. Riedl's classification will provide a framework for considering how supra-organismal levels of biological order change, i.e. do they evolve? It requires a hierarchical conception of the natural world, an understanding of the interactive networks that living entities form, and a general model of order-preserving change.

Ensembles of entities change in composition through time as a result of differential reproduction, survival, or both, of the individual entities that compose them. A simple and general model is presented in Fig. 8.1. This general model of change in composition through time includes what is usually considered Evolution[1], changes in gene frequency in an inter-

Fig. 8.1. A general model of replication (with error) and selection over a discrete time step (T_0, T_1).

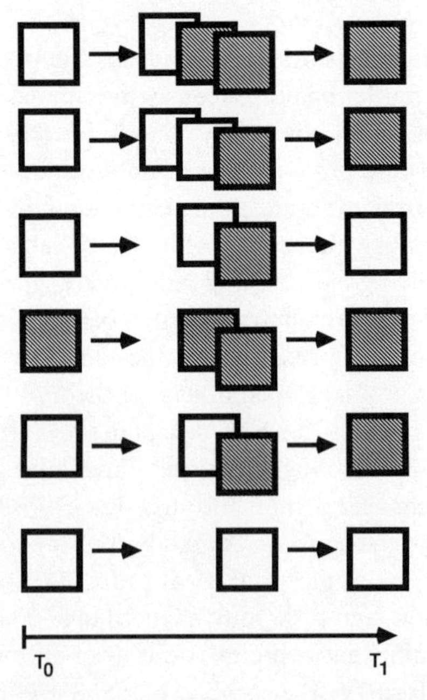

1 We use 'evolution' and 'selection', respectively, for change in composition and choice among variants. Evolution (capitalized) and natural selection refer, respectively, to population gene frequency change, and differential survival and reproduction of organisms.

breeding population of organisms. The future composition of any ensemble of living entities is constrained, because they are parts of larger interactive systems (ecological networks). This chapter describes how networks of entities and their within-system interactions, are the context in which evolution occurs.

8.1.1 ***Evolution and ecology: A review***

8.1.1i *Darwin and beyond*

Three principles of Darwin's thesis pertain to this discussion. They are: natural selection among different organisms results in the origin and evolution of species; natural selection is 'by' the sum of total environmental influences; and variations originate from within organisms in some unknown way and these accumulate gradually in the population.

Populations evolve, in part, because some subset of the organisms in it are selected, i.e. survive long enough to produce more offspring than others. Selection is by the environment in the sense that it determines what traits are more adaptive and lead to greater reproductive success. Biologists have since undertaken the task of uncovering those environmental influences that were most important to the evolution of any trait (phenotype) of interest. Too often this has neglected the constraints and counterbalancing forces on adaptations. Mayr (1966) justified an ecological systems approach to Evolution when he wrote:

> Evolution is the change of systems, the modification of one extremely complex system under the impact of extremely complex sets of selective and random forces. . . .

Darwinism is a hypothesis about the evolution of populations of organisms. The *Origin of the Species* (Darwin, 1859) proposed that many small steps were necessary for the modification of character distributions to the extent that once interbreeding populations could become reproductively isolated and result in two related but distinct species. Provine (1982) summarized Darwin's view:

> ...varieties were potentially incipient species. Natural selection drove organisms into ecological niches..., creating new varieties. In time, some, but not all of these varieties gradually diverged in morphology and became more sterile in crossing so that taxonomists would then rank them as species.

There was in Darwinism no essential concern with evolution at other levels of the biological hierarchy.

Neo-Darwinism can be characterized as the synthesis of Mendelian genetics and classic Darwinian selection. Genetics provided Darwinism with the inheritance mechanism and source of variation it lacked. The quantitative population-genetics models of Fisher, Haldane, and Wright verified that 'a very small selection rate could alter the hereditary constitution of a population' (Provine, 1982). Whether this process can be extrapolated to phyletic diversification (speciation) has been the subject of recent debate. The modern synthesis 'focuses on a single level of evolutionary dynamics and claims that the neo-Darwinian paradigm is both necessary and sufficient to explain all evolutionary phenomena' (Mayr, 1980). Eldredge & Salthe (1984) among others have 'questioned both the effectiveness and propriety of viewing the population-level processes of natural selection and genetic drift as the sole important determinants in evolution.'

A small group of evolutionary scientists has recently moved beyond considering only the evolution of species, i.e. populations of organisms. It has identified entities at many different spatial and temporal scales (Hull, 1980; Vrba & Eldredge, 1984). Punctuated equilibrium theory questions the sufficiency of the neo-Darwinian paradigm to explain the pattern of macro-evolution observed in the fossil record (Eldredge & Gould, 1972; Gould & Eldredge, 1977; Gould, 1982). This does not mean that changes at lower levels (e.g. gene frequencies) are not necessary conditions; they are with respect to the usual grouping of organisms based on genetic make-up. Nevertheless, they hypothesize that evolutionary processes at different levels result from different mechanisms (e.g. Raup *et al.*, 1973; Stanley 1975, 1979, 1985; Vrba, 1980; Cracraft, 1982). Their concern is usually with evolution above the species-population level, but they recognize a need for more general theories of evolution (Gould, 1980, 1985; Vrba, 1982; Vrba & Eldredge, 1984).

Existing theories of evolution as a general process of ordered change have come not from biology, but from physics and general systems theory (Allen, 1981; Haken, 1981; Csanyi, 1982; Jantsch, 1981). In addition, a great deal of corroborating evidence is accumulating in the study of chemical reaction systems (Eigen, 1971; Haken, 1977; Prigogine, 1980), life's origin (Fox & Dose, 1972), epigenetic systems (Ho & Saunders, 1979, 1984), cell evolution (Schwemmler, 1984) and the biosphere (Margulis & Lovelock, 1974) that there is a common and fundamental description of self-organizing change in far-from-equilibrium systems. What these theories share is a recognition that entities are systems evolving within still larger interactive systems, entities with environments both modified by and constraining their evolution.

8.1.1ii *Ecology and evolution theory*

Ecology, the study of the entity-environment relationship, has been concerned with adaptations of organisms (physiological ecology), dynamics of single populations (population ecology), populations as parts of competition communities and food webs (community ecology) and networks of entities transferring energy-matter, nutrients and information at many scales (ecosystem ecology). It has yet to develop a general systems foundation to unite these perceptually distinct concerns with one another and a general model of evolution.

A formal hierarchical approach is a necessary part of a general systems theory of ecology and evolution. It helps avoid the confounding of issues at different levels (Allen & Starr, 1982; O'Neill *et al.*, 1986). For example, coevolution theory (Roughgarden, 1979) has been criticized for explaining organismal adaptations as the result of competition among populations (Connell, 1980). Strong (1984) suggests that these traits can often be best explained as adaptations to autecological factors, not synecological ones. But where interspecific competition was historically a strong force, attributes that increased the survival of the population, such as dispersal ability (which, of course, depends on the mobility of individuals or their propagules) might have evolved, as well as those increasing the survival or fecundity of individual organisms (e.g. feeding adaptations). Ecology and evolutionary biology are beginning to recognize a hierarchy of entities and spatio-temporal scales at which variation, selection, and evolution, or evolution-like processes occur (Stanley, 1975; Wade, 1977; Wilson, 1980; Gould, 1982).

Hierarchies are not new in the ecological literature, but evolution has not been generally the primary concern. Miller's (1978) 'living systems' theory is one attempt to bring a general systems framework to the ecological hierarchy. At each level, this viewpoint focuses on the entity and its subsystems (e.g. reproducer, ingestor), which are evolving in response to energy-matter and information exchange with the environment. MacMahon *et al.* (1978) presented a multi-hierarchical scheme centered on the organism, the only entity common to the anatomical, phylogenetic, coevolutionary, and energy-matter exchange hierarchies that they recognized.

A more strictly evolution-centered approach distinguishes between two equally important and mutually dependent hierarchies – a genealogical one and an ecological one – which interact to provide, respectively, the players and the stage upon which evolution takes place (Eldredge & Salthe, 1984; Salthe, 1985). Eldredge & Salthe (1984) provide an excellent

Table 8.1. *Two conceptions of the hierarchy of living entities. On the left Eldredge & Salthe's* (1984) *distinguishes between evolutionary and ecological entities. On the right a unitary hierarchy of evolving entities as systems, as developed in this chapter, is presented. This table is not meant to suggest a necessary equivalence between aligned entries or that these lists are comprehensive.*

Genealogical	Ecological	Evolutionary
Codons	Enzymes	Epigenetic system
Genes	Cells	Physiologic system
Organisms	Organisms	Organism
Demes	Populations	Species-population
Species	Local ecosystems	Ecosystem
Monophyletic taxa	Biotic region	
(Special case: all life)	Biosphere	Biosphere

discussion of the nature of hierarchies based in the state-space and general system theories. They review and develop the justification for distinct levels above and below that of organisms.

> Each particular instance of these classes of individuals – base pairs, genes, organisms, demes, species, and monophyletic taxa – is a spatiotemporally restricted entity... It is our contention that each level of the genealogical hierarchy constitutes a quasi-discrete level of the evolutionary process. Specifically, change at any level reflects the shuffling of properties among the next lower level individuals within any individual. (Eldredge & Salthe, 1984).

Their synthesis of evolutionary theory and ecology is an inspiring development, but it unnecessarily fosters the separation between entities and their environments. Eldredge & Salthe (1984) state, '...the genealogical hierarchy is certainly the seat of both replication and evolution, while the ecological hierarchy is certainly the seat of interaction.' In contrast, we see in nature a hierarchy of systems (Table 8.1). The entities of their genealogical hierarchy have environments, which makes them ecological entities. There is no separation between the object and everything else. Every entity and its environment are mutually defining and unitary (Patten, 1978; Odling-Smee, 1988).

It is important to realize that these entities form networks of selective interactions. They also receive input from (can be influenced by) and export to (can influence) the environment beyond the system's boundaries, as defined by the observer. An obvious example for populations in ecosystems is the immigration from and emigration to the larger species pool. Possibilities at each level of entification are limited by the variants produced in lower-level processes (law-limited) and constrained by the selective 'environment' of higher levels (rule-limited) (Allen *et al.*, 1984). The hierarchical constraint of Allen & Starr (1982) and Salthe's (1985) 'triadic' model is thus an integral part of evolutionary networks (Fig. 8.2). What is gained from this perspective is the explicit realization that evolving entities at any hierarchical level, especially populations of interbreeding organisms (i.e. species), are embedded in interactive ecological networks.

8.2 Evolution in ecological networks

8.2.1 *Evolutionary networks and environs*

Darwin, in *The Origin of Species*, expressed his sense of system as follows:

> Almost every part of every organic being is so beautifully related to its complex conditions of life...plants and animals remote in the scale of nature, are bound together by a web of complex relations...I can see no limit...to the beauty and complexity of

Fig. 8.2. A hierarchical network conception of ecological systems.

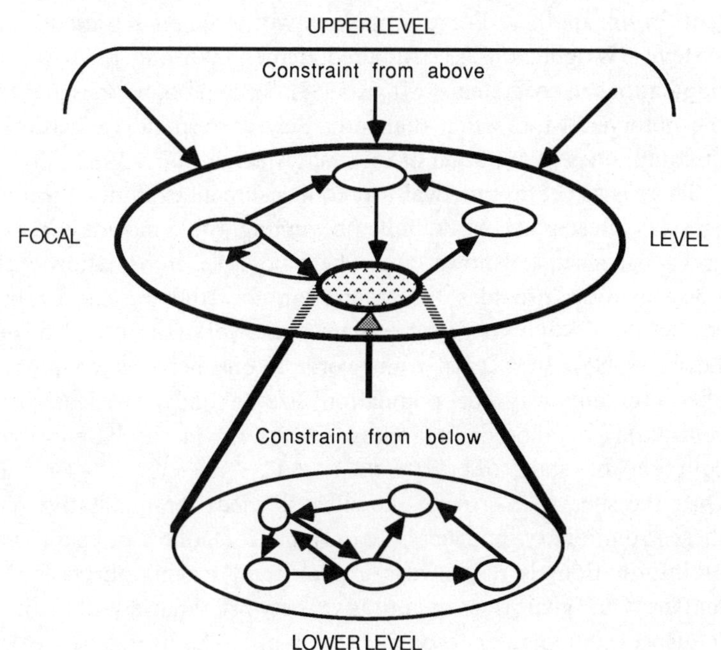

> the co-adaptations between all organic beings, one with another and with their physical conditions of life.... (Darwin, 1859)

Darwin's organisms were co-adapted with their living environment through a set of closely evolved relationships. Evolutionary biology does not lack this sense, but has failed to formalize theory that captures it. Wilson (1980) agreed with this assessment:

> Traditional theory does not sufficiently emphasize the extent to which environment is created by the activities of organisms, and how completely the fitness of an individual can depend on other members of the community. A bark beetle does not simply eat trees. [It]...modifies the structure of the environment, which affects the fitness of many other species.

What evolutionary biology must deal with eventually is the network of evolutionary influences of ecological entities on each other in a variety of spatiotemporally scaled ecological systems.

Patten *et al.* (1976) presented a general systems formalism to represent hierarchical networks. A digraph representation of an abstract ecological system is shown in Fig. 8.3(*a*). For this discussion of evolution the network's nodes, represented by open circles, are entities at any defined spatial and temporal scale. The arcs, directed lines connecting nodes, indicate ecological interactions and evolutionary influence of one node (entity) on another. For example, networks might represent epigenetic systems (Wright, 1982), structured demes (Wilson, 1980), or species-populations in ecosystems (Mayr, 1963). System boundaries are drawn by the observer. More often than not, they correspond to spatiotemporal discontinuities at the scale of interest (Allen & Starr, 1982).

There is as yet no practical way to measure all evolutionary influences, especially in ecosystems. A limited beginning would include energy-matter and biogeochemical flows, and where possible, information exchanges. Loop analysis provides the inspiration for studying the evolution of entities in the context of their ecological network (Levins, 1975; Puccia & Levins, 1985). In Levins' framework, if one network component (X_j) affects in any way the population size (abundance, density or concentration) of another, it appears as a term in the latter's dynamical equation of state ($\mathrm{d}X_i/\mathrm{d}t = f(X_i, \ldots, X_j, \ldots) = aX_i + \ldots + (-c)X_j + \ldots$). Only the sign, $(+, -, 0)$, of the effect is used; the qualitative nature of these evolutionary influences (i.e. whether it is a flow of energy-material, or information) is not always explicit, and for his purposes does not matter. Our goal is a quantitative network analysis of evolutionary influences in complex ecological systems. The formalism exists; the challenge is now to quantify objectively with a common currency as many of the diverse evolutionary influences as is possible (Odling-Smee, 1988).

Ecological systems of evolving entities are open to exogenous inputs, that is inputs from beyond the system boundaries (Patten *et al.*, 1976). Few would argue that endogenous, within-system causes are sufficient for evolution. Ultimately, ecological systems are far from thermodynamic equilibrium and must be open to the influx of energy and other ordering forces ($\mathrm{d}_e S/\mathrm{d}t < 0$) to counteract the always positive internal production of entropy ($\mathrm{d}_i S/\mathrm{d}t > 0$). Evolution, as a process opposing the spontaneous tendency toward disorder ($\mathrm{d}S/\mathrm{d}t < 0$), is otherwise impossible. This is an expression of the Second Law of Thermodynamics applied to far-from-equilibrium systems: $\mathrm{d}S/\mathrm{d}t = \mathrm{d}_e S/\mathrm{d}t + \mathrm{d}_i S/\mathrm{d}t$ (Prigogine, 1967).

Fig. 8.3. Networks and environs: (*a*) directed-graph representation of a hypothetical network model of an open ecosystem; (*b*) the input and output environs of node H_i of the system depicted in 8.3(*a*) showing its direct and proximal indirect interaction sequences.

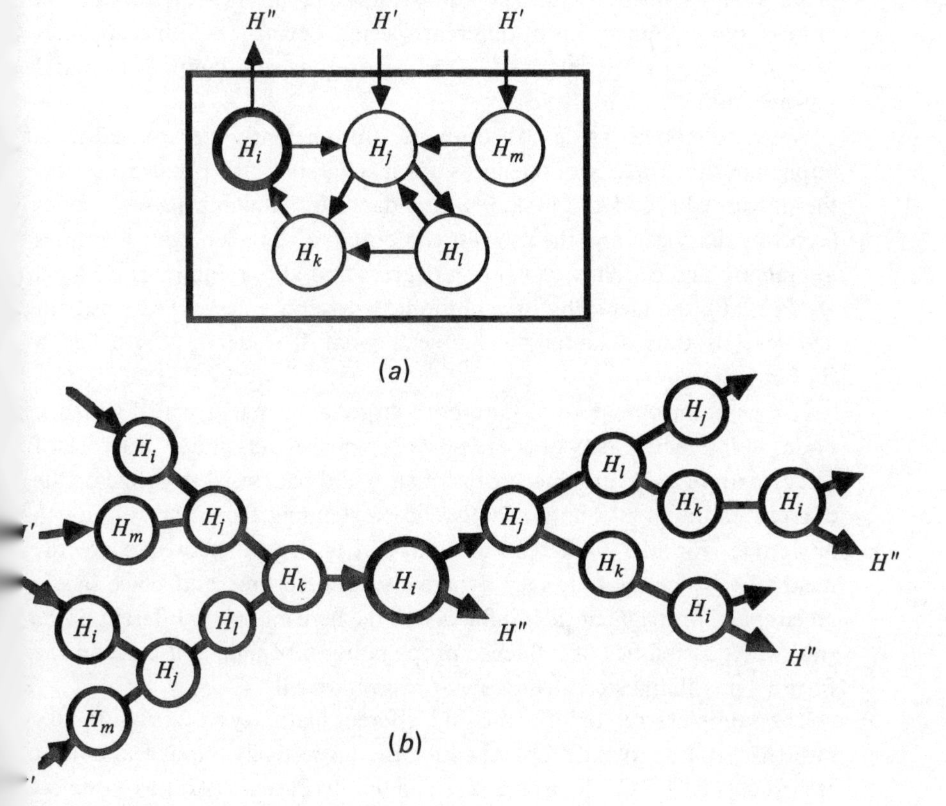

When ecosystem networks include only biotic entities, all abiotic influences, both those general to all members of the system and those specific to one or more, must be represented as inputs to the nodes. The quality and quantity of incoming energy was and is a major evolutionary influence in ecosystems. Other exogenous evolutionary influences at different ecological levels might include mutagens, anthropogenic impacts, shifting weather systems (e.g. El Nino events), and asteroids striking the earth. Whereas these are usually considered 'disturbances' leading to disorder, atmospheric and hydrologic circulation patterns, seasonal cycles, the day/night cycle and other informative inputs from higher-level processes act as constraining or ordering forces on ecological systems.

Although Fig. 8.3(*a*) does not include loops (self-influences), it could. Loops represent evolutionary processes at lower levels (Vrba & Eldredge, 1984; Salthe, 1985). There is no question that ecological entities influence their own evolution. For example, population ecology studies the interactions among individuals or phenotypes of the same species and this evolutionary influence would be represented by loops in a digraph model of interacting populations of different species. Levins (1975) incorporates exogenous inputs into his 'self-loops', as well as self-damping (−) and self-promoting (+) influences.

Network models, as described above, are inherently hierarchical. It is implicit in the property of openness, for it suggests a larger scaled, rest-of-the-universe beyond the system's boundary. By recognizing that nodes (evolving entities) are themselves composed of smaller scaled entities interacting and evolving to varying degrees and at varying rates (O'Neill *et al.*, 1986), the hierarchy of evolutionary systems emerges. An entity at any level is thus a hierarchical object, what Koestler (1967) called a 'holon'.

Just as holons are doorways to both larger and smaller scaled systems, nodes in interactive networks are portals in space and time. A network of directed arcs can be unfolded both forward and backward to delineate the efferent and afferent temporal sequences extending from and leading to any node (Fig. 8.3). Afferent pathways in ecological networks are the means by which evolutionary dynamics, at some time and place in the recent past, propagated their influence to the here and now. Efferent paths trace the possibilities of influence propagating to a time and place in the future – possibilities created anew at each moment.

The complete set of afferent and efferent pathways of evolutionary influence leading to and from an entity are, respectively, input and output environs (Fig. 8.3*b*). Environs were defined by Patten (1978) as 'a holon together with its associated within-system environment.' They are defined by the ecologist's or evolutionary biologist's network model of interactions and influences among the entities of concern. These temporal sequences are truncated at the boundary of the entity's ecological system (Patten, 1982) as they diverge backward to their origin and forward to extinction. For a complete description of environs the reader is referred to Patten (1978, 1982).

Environment, evolution theory's all powerful, but vague selective agent, becomes input environs, the set of direct and indirect (historical) influences of evolving entities on other evolving entities that coexist as parts of ecological systems and the abiotic inputs to each member. Instead of reducing Darwin's 'web of complex relations' to the direct, and most obvious interactions, environs provide an entity-centered, holistic systems perspective and description of each entity–environment unit limited only by knowledge and ability to model. System structure – the network of entities and their within-system interactions – is the context in which evolution generally, and Evolution specifically, occurs.

The structural information in environs is unambiguous only while the network's configuration remains unchanged. We might expect network structure to be constant on ecological time scales (e.g. the lifespan of the fully formed higher-level system represented by the network), but not on longer evolutionary time scales. For example, many if not most structural changes at the epigenetic level (mutations) are deleterious to the organism; allowable genetic changes are constrained directly and indirectly by other genomic elements. Instead, '...changes in organismal integration systems appear to provide most of the impetus to morphological evolution' (Raff & Kaufman, 1983). A link in an ecosystem level network means that the two populations will interact with probability $p > 0$ in some interval of time, Δt, sufficiently long, but less than the lifetime of the population or the ecosystem itself. We argue below, as a specific example of a more general hierarchical property, that changes in ecosystem organization (network structure) are significant in the evolution of their species populations, and vice versa. Environs provide a formal, although (for explicit implementation) model dependent, framework in which to view evolutionary and coevolutionary processes in their ecological setting.

8.2.2 *Environ-mediated selection*

At least two different modes of selection act on the entities in an interactive network: selection of the node (the whole set of individual entities) and selection of variants (non-empty subsets) at the node. The former selection is familiar to ecologists from the study of biogeography (MacArthur & Wilson, 1967), succession (Margalef, 1968), and community assembly (Diamond, 1975; Yodzis, 1981; Sugihara, 1982). The problem is one of survival and positive population growth versus extinction for an 'introduced' entity, or that of the resident in the face of invasion or an unstable environment (see Kawasaki *et al.*, this volume). The determinant of extinction or survival is the whole network of direct interactions (e.g. competition, predation, disease, parasitism) and their indirect consequences (Levine, 1976; Vandermeer, 1980; Vandermeer *et al.*, 1985; Kerfoot & Sih, 1987), inputs to the system when the abiotic environment is unsuitable, or the stability of the larger system (Levins, 1975). With the extinction of a local population, individuals and all lower-level entities die, but it is the structure of the larger entity (e.g. species, ecosystems) that is affected by the loss. Possible outcomes of this mode of selection are addition or replacement of nodes without changes in interactive structure, or new structure and rates of interactions. At the focal level of the node, it is not a question of relative fitness of the individuals which compose it or the ensemble's evolution. For the whole node the question is, can it coexist with the other entities; that is 'Does it fit in the system?'

Ecological systems also play a role in the 'survival of the fittest' or relative fitness of individuals. A non-homogeneous or 'structured' node in an evolutionary network (i.e. one composed of varying organisms, or trait groups) is subject to selection due to its interactions with other members of the network. This can occur in two qualitatively different ways. First, we define the single environs mode to be that where all members of the node set have identical interactions with the other system components (Fig. 8.4*a*). Alternatively, individuals could vary as to which members of the ecosystem network they actually interact with, defining a structured environs mode (Fig. 8.4*b*). It is hypothesized that either could occur at any level in the hierarchy of interactive systems where a node is composed of varying individuals.

The single input-environs model, where all individuals are exposed to the same set of potential causal agents (prey, predators and parasites, competitors, etc.), but vary in a trait affecting their relative success in the interaction, captures the usual understanding of Darwinian selection of varying organisms. The logic has been applied also to selection of genes (Dawkins, 1976, 1982), groups (Wright, 1947; Wynne-Edwards, 1987) and species (Stanley, 1979). Here it is explicit that the ecological network is the context for this selection, i.e. it is not only the competitors, or predators and prey that determine what traits are adaptive and to what extent, but everything in an entity's input environ.

The structured input-environs model is diagrammed in Fig. 8.4(*b*). As an example of varying environs at the species-population level, imagine a benthic marine invertebrate with two larval phenotypes: one negative, the

Fig. 8.4*a*. For caption see p. 224.

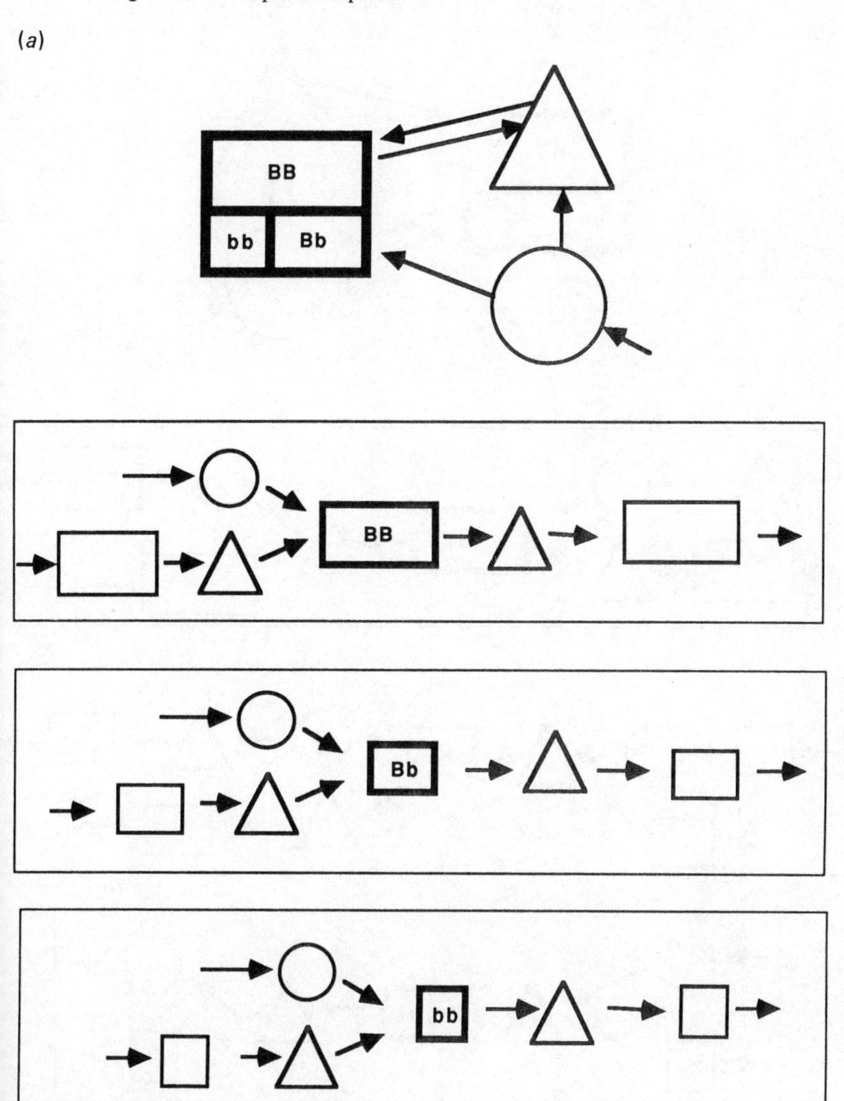

other positive phototaxic. The two larval types might interact with and be subject to selective pressures from different species or guilds (e.g. benthic fishes vs. arborescent cnidarians). The two larval trait groups have different input environs. The differential success of one could eventually

Fig. 8.4. Diagrammatic examples of (*a*) single environs and (*b*) structured environs models of selection. In 8.4(*b*), without recognition of the substructure of H_i the network appears maximally connected. The lower diagrams indicate that subsets of H_i can actually have different input and output environs.

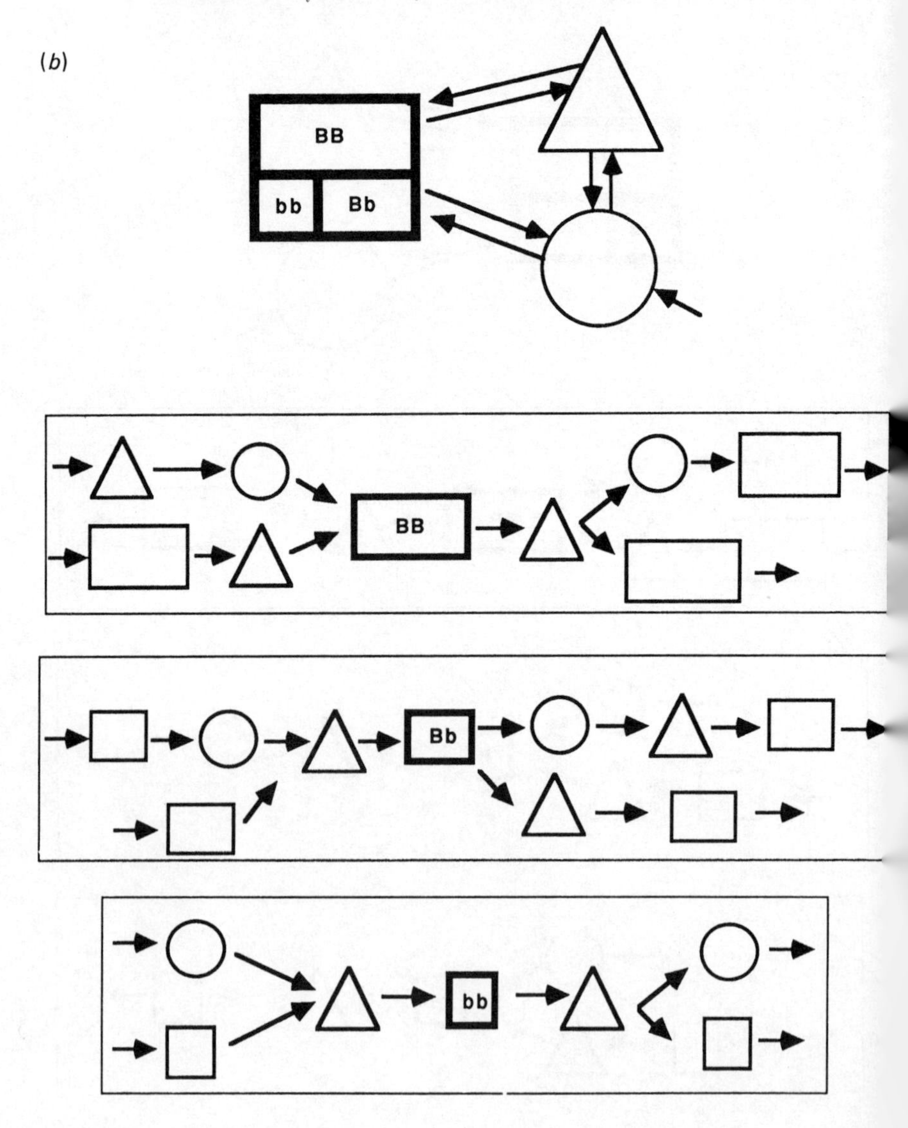

result in a single type and a single input environ for the whole population, but there appears to be no reason that this is a necessary outcome. Alternatively, we might hypothesize that different input environs could lead to divergence of variants at a node. Sympatric speciation is no longer in vogue, but an analogous process might be important at levels above and below the species-population where greater spatial scales and stronger interactive strengths, respectively, might promote differentiation of coexistent variants. More generally, this mode of selection can be observed to result in a change in network structure, unlike the single environs mode which can only result in a change in rate or strength of interactions.

8.2.3 *The selection of output environs*

We argue above that selective influence acts through the input environs of an ecological entity. Different system structure or interaction strengths result from, respectively, selection among environ or trait variants at the nodes. A similar situation appears when looking at this process from the output perspective. Selective influences, from outside the system for instance, act directly upon a node and then have indirect consequences on entities in that node's output environ. Variants at each step along these efferent pathways can be selected in response to the new influence. Selection of structured output environs is also a possibility if variants at a node have distinct subnetworks of interrelationships. Thus, selection of entities becomes selection of environs. Here is a simple hypothetical example; a new variant of phytoplankton arises that is able to use light at lower intensities and expand its population earlier in the spring. As a result, a new trophic network could arise from that subset of the available species pool able to take advantage of this new resource. Selection acting upon the varying entities at a single node, as it is normally perceived to do in Evolution by natural selection, selects the output environ. This sequence of selective events propagating through the network determines the structure and function of ecological systems.

The structured input- and output-environs models highlight the interesting possibility that ecological networks, at hierarchical levels where variants at nodes can be identified, select their own interactive structure through selection among these variants. Given the definition of environment as the ecological subnetwork of interactions and the concept of entity–environment unity, it is almost trivial to conclude that system structure can be determined (selected) by selection acting upon the entity. What is not trivial is understanding that whole systems are selected through selective forces acting upon the parts, and that the system itself is often the selective force. Natural selection is significant in the determination of the structure of ecosystems (i.e. networks of species-populations) and vice versa, the structure of ecosystems is a significant selective force on species-populations. If only the abiotic environment selects from among internally produced variants, order in the natural world is highly improbable. It is only through interaction that persistent assemblages of living, evolving entities become self-organizing, or autogenic (Jantsch, 1982; Csanyi, 1985).

8.2.4 *Circular causal networks and coevolution*

Empirical interest in coevolution usually results from the natural history study of a single species, with the subsequent realization that another figures significantly in its survival and adaptations. There can be both direct (pairwise and diffuse) and indirect coevolution in networks of evolving entities. Direct coevolution is the type most commonly acknowledged by evolutionary biologists. Symbioses, competition, parasitism and predator–prey interactions are easily recognized, and their possibly coevolved nature has been modeled in detail (for a review see Thompson, 1982). Such pairs are, however, embedded in much larger community networks. More recently, ecologists with ecosystem perspectives have observed significant indirect influences (e.g. Montague, 1980; Inouye, 1981; Davidson *et al.*, 1984; Sterner, 1986). The theory of coevolution in ecosystem networks is still developing, but indirect relationships have been modeled (Levine, 1976; Vandermeer, 1980; Vandermeer *et al.*, 1985; Puccia & Levins, 1985), and the potential for indirect influences to exceed direct ones in cyclical networks is clear (Patten, 1982; Higashi & Patten, 1989).

This broader view of coevolution has its beginnings in the perception of ecosystems as circular causal systems (Hutchinson, 1948). This gave ecology a holistic perspective on the organism-environment relationship. For the same reasons that Hutchinson said, 'the biological cycles of the elements must be regarded as a unity,' so must circular causal networks of ecological interactions. Causation includes evolutionary processes and circular refers to closure of the pathways of evolutionary influence. Patten (1982) argued that path closure is both a necessary and sufficient condition for coevolution in causal networks. The details of this argument are presented in Appendix A8.1.

Closure in a network of evolutionary influence means the entities forming the cycle exist in one another's input and output environs (Fig.

8.5). Selective pressures acting on any one cycle member will propagate through the pathways of interactions to others and back again. Selection operates on the set of individuals that comprise a node, but does so via the other nodes in its input environs. Eigen & Schuster (1978) discovered similar coevolutionary principles at work in molecular-genetic hypercycles. Odling-Smee (1988) describing the coevolution of the organism-environment unit (Patten, 1982), distinguishes between natural selection, the environment's evolutionary influence on the organism, and 'niche construction', the organism's influence on the environment. Odling-Smee emphasizes the abiotic environment, but by focusing on the other living entities in the environ, entities with genetic memories (traditive inheritance), one can more easily see the truly coevolutionary nature of the entity–environment unit.

If selective influences propagate through interactive networks as described, it is obvious that ecological entities can influence their own evolution indirectly. Environ-mediated selection of self (Fig. 8.6) is distinct from lower-level evolutionary mechanisms, which were described previously (e.g. production of variants, sexual selection). Indirect self-selection would appear, from an atemporal perspective, to be selective influence from the entity's environment (e.g. direct interactions of competition or predation), but in fact, the initial selective event or change could have originated earlier at the node itself. It is no simple matter to isolate the agent ultimately responsible for change in complex ecological networks – a cycle has no beginning, no end. It is possible, within the

Fig. 8.5. The environs of node H_i of the Fig. 8.3(a) system with part of the indirect or environ-mediated coevolutionary relationship $\{H_i, H_j\}$ highlighted.

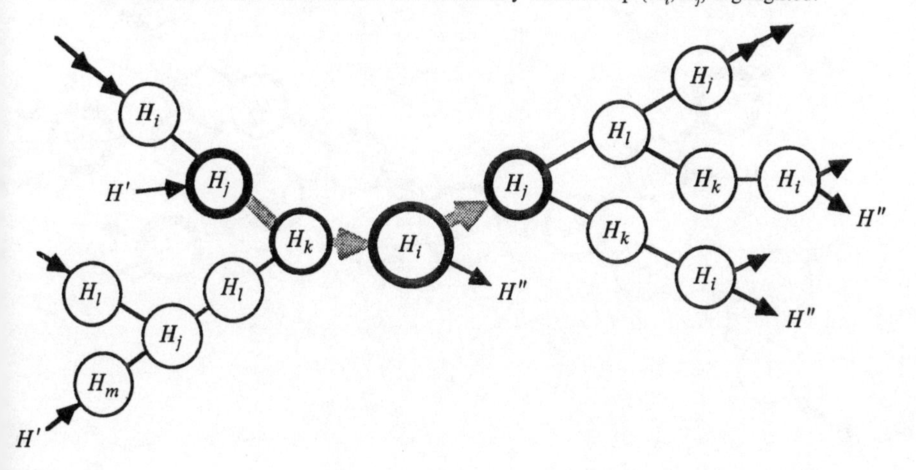

limits of knowledge about the system, to articulate the pathways and the intermediaries of influence or causation, because this information is contained in the network structure.

- With the recognition of cyclic structure comes the realization that the entity and its within-system environment are an evolutionary unit. Genealogists know the unity of a lineage – shared genetic history. Shared interactive history, influence propagating from past to present to future generations through nondirectly related interacting entities, brings unity to ecological networks on evolutionary time scales. Humans will not survive as a species if we do not learn soon that interactions among nations and with the natural world have indirect consequences for our children and grandchildren.

We have come full circle. The ecological network in which evolving entities are embedded is a significant source of selective influence on each entity (Sect. 8.2.3), and in turn, the network's structure is determined by selection acting upon the network's entities. Because ecological networks are cyclic (circular causal networks), they are self-determining. But, do they evolve?

8.3 The evolution of ecological systems

Up to this point, our thesis should be uncontroversial. It deals with the evolution of entities like species (populations of reproducing individuals and elements of ecological networks), and how they and their

Fig. 8.6. The environs of node H_i of the Fig. 8.3(a) system with part of the indirect or environ-mediated self-influencing relationship $\{H_i, H_i\}$ highlighted.

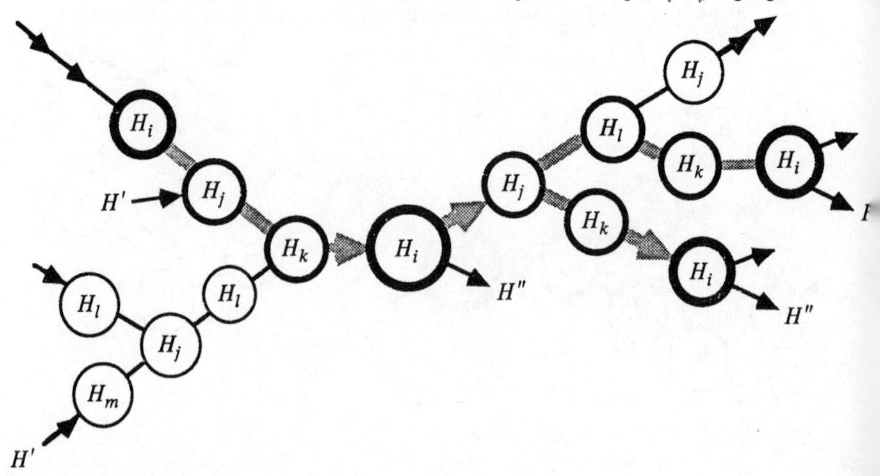

environs are mutually determining. It suggested that this was a specific example of a general model of change at different levels in a hierarchy of ecological networks. The following discussion, however, is more in the nature of speculation about the differences and similarities between organisms (and species-populations) and higher-level ecological networks, such as ecosystems. Ecosystems are not organisms, but like organisms, they are ordered dynamic entities.

First, we ask whether ecological systems must possess Riedl's (1978) four forms of order to be evolving entities? Ecosystems, like organisms, are made up of parts, but to what extent these parts are standardized is questionable. Oak trees in a forest ecosystem might be no more variable than cells in the human liver. We might conclude that ecosystems possess the order of standard parts to some degree. We have said throughout this chapter that all ecological systems are also inherently hierarchical and interactive structures. These three might be necessary, but not sufficient conditions for evolution; machines possess standard part, hierarchical, and interactive order, but they do not reorganize or change anti-entropically independently of man. And so far, machines do not possess traditive inheritance. Traditive inheritance, on the other hand, may be both necessary and sufficient to deduce the existence of a living evolving entity.

What does it mean to suggest that supra-organismal entities, for instance ecosystems, possess traditive inheritance? To possess traditive inheritance means that an entity is able to pass on the information contained in it to the future, information that will allow for the construction and functioning of a like entity. The information content of ecosystems – the successive orderings of structure in atomic arrangements, chemical configurations, the arrangement of base pairs, gene frequencies, genotype frequencies, structured demes, species relative abundances, confamilial species complexes/functional groups and network structure – extends through time, but not unchanged. Over a time interval long enough to include the demise and replacement of every living lower-level entity, the ecological system becomes a completely new entity which has inherited the information in all existing lower levels.

This does not mean that ecosystems make more of themselves. Multiplication is spatial replication (Csanyi, 1985). Ecological systems, including organisms, replicate temporally in a process of self-renewal. Any ecosystem belongs, therefore, to the set of all ecosystems possible, given the available components and environmental context. Through time the set of all possible ecosystems changes composition with the appearance of new components. In addition, ecosystem interactive structure (i.e. the interactions between species-populations) can vary, even though the species composition remains unchanged, which is possible with the structured environs mode of selection. Is this long term temporal change equivalent to what is usually meant by the 'evolution' of organisms: the generation to generation changes in morphology, instinctual behavior, biochemistry, etc., observed in phylogenies? It is probably more accurate to speak of the 'anagenesis' of ecosystems, progressive (sequential) change in organization (Jantsch, 1982), rather than their evolution.

Riedl's system of choice was the organism. As ecological networks of interacting cellular subsystems, organisms possess all four forms of order: traditive inheritance, interaction, hierarchy, and standard part. Nevertheless, most evolution theorists claim organisms do not not evolve: they come into existence, live for some time, and then die. Only species-populations, sets of genetically related and potentially interbreeding organisms, evolve. Therefore, traditive inheritance and the other forms of order must be properties of the members of a set of related individuals (e.g. organisms in a species-population), but only the set can evolve. Evolution is unmistakably a hierarchical process; a set transformation emerging from the properties of the lower level entities. Although ecosystems and other supra-organismal levels in the hierarchy of ecological networks possess Riedl's forms of order, this is not sufficient to qualify as evolving entities. Ecosystems, as far as we know, are probably not members of a set of entities that fulfills the above criteria; ecosystems do not evolve in a way homologous to species.

At the other end of the biological hierarchy, theories of cell evolution via endosymbiosis (Margulis, 1981; Schwemmler, 1984) strongly suggest that what were once ecosystem-like interactive networks of discrete living entities (e.g. heterotrophic prokaryotes, mitochondria, chloroplasts and spirochete-like organisms), became individuals at a new level of organization. It is not incredible that these entities were capable of replicating temporally and spatially. Csanyi's (1982, 1985) general theory of evolution posits that, given enough time, ecological systems become reproducing entities, thereby fulfilling the requirements to be truly evolutionary entities. Whether there will be enough time or space on this planet for what we usually recognize as ecosystems to reach this point is doubtful. It is more possible that within the lifespan of the biosphere highly organized ecological networks at much smaller spatial and interactive scales (e.g. myxotricha, coral-zooxanthellae, lichen, sponge-inhabitant and sargassum-epiphyte microecosystems) have or will become discrete evolutionary entities.

8.4 Summary

The principal concepts of this chapter are reiterated below. This chapter is our first exploration and exposition of evolution from the perspective of an explicitly hierarchical network approach to ecosystem ecology. We fully expect criticism and corrections, and offer the following conclusions in this spirit.

1 The underlying logic of the Darwinian process describing Evolution of species-populations by the natural selection among varying organisms is applicable to evolution at other levels where a set of entities and a selective set transformation can be identified.

2 Ecology and evolution science must be concerned with developing a general systems foundation that recognizes biological entities as systems evolving as parts of larger interactive systems. The genealogical hierarchy need not be abstracted from the hierarchy of evolving ecological systems.

3 Ecological network models can capture at any scale and moment our knowledge of the structure of evolutionary influences biotic and abiotic entities exert on one another. Network systems are open to exogenous evolutionary forces and they are inherently hierarchical.

4 The evolving entity and its within-system subnetwork of interactions are a unit (environs), and together they determine which units survive and which are more fit. The environ is the ecological unit of evolution.

5 Two types of selection operating via environs are identified. First, ensembles of individual entities (e.g. species-population of organisms) do or do not survive in the larger network. Secondly, some variants survive and are more fit than others because of some trait. If the selective trait involves interactions with other network components, a subnetwork of interactors leading to an entity (environs) can be selected.

6 Selection acts directly upon an evolving entity and then indirectly upon all others in the former's output environ. Subnetworks in ecological systems can be selected when an ecological variant is selected.

7 Coevolution can only occur in circular causal networks, e.g. ecosystems (Appendix A8.1). Closure in evolutionary networks might be sufficient, as well as necessary for coevolution.
8 Biological entities influence the evolution of their descendants through interactions with unrelated entities in ecological networks (indirect self-selection).
9 All ecological systems replicate temporally (they are finitely self-renewing), and possess Riedl's four forms of order. Like organisms, ecosystems may undergo anagenesis, but they do not evolve as do species-populations of organisms. What we recognize today as evolutionary entities might have evolved from ecological systems that at some point in their anagenesis became spatially replicating.

Acknowledgements

Our thanks to T. F. H. Allen, R. V. O'Neill, H. R. Pulliam and C. Benkman for their criticism and helpful comments.

Appendix

A8.1 Formal conditions for coevolution

The following formal conditions for coevolution between two evolvable entities were given by Patten (1982). Let H be an arbitrary evolutionary entity, that is, a set whose members can vary and be more or less successful in surviving or producing new members of like character such that the subset of like members is non-empty in subsequent time periods. Let H^* be the environment of H. For convenience, H can be referred to as 'species'; environment will be taken to mean the entire ecosystem of collected abiotic factors and other 'species' interacting directly and indirectly with H.

With Z and Z^* the respective sets of admissible inputs, X and X^* the respective state spaces, and Y and Y^* the respective outputs of H and H^*, a determinate (unique) input–output relationship is established by:

$$H: \begin{array}{l} \rho: Z \times X \rightarrow Y \\ \tau: Z \times X \rightarrow X, \end{array} \qquad H^*: \begin{array}{l} \rho^*: Z^* \times X^* \rightarrow Y^* \\ \tau^*: Z^* \times X^* \rightarrow X^*, \end{array} \qquad \text{(A1)}$$

where ρ and ρ^* are response functions, and τ and τ^* are state transition functions on some appropriate time interval T. Interactions between H and H^* are expressed by coupling constraints:

$$y^* \equiv z, y \equiv z^*, \tag{A2}$$

where $z \in Z, z^* \in Z^*, y \in Y, y^* \in Y^*$.

Expressions (A1) and (A2) together define a closed species–environment relationship $\{H, H^*\}$, where:

$$H: \begin{array}{l} \rho: Y^* \times X \to Y \\ \tau: Y^* \times X \to X, \end{array} \qquad H^*: \begin{array}{l} \rho^*: Y \times X^* \to Y^* \\ \tau^*: Y \times X^* \to X^*. \end{array} \tag{A3}$$

In other words, output from H is input to H^* and vice versa.

An open relationship can be produced by separating environment into two non-overlapping components, input environment H' and output environment H'' (Patten *et al.*, 1976; Patten, 1978). Then, establishing as before the sets Z', Z, and Z'', X', X, and X'', and Y', Y, and Y'', and the interaction constraints:

$$y' \equiv z, y \equiv z'', \tag{A4}$$

determinate models for these environments are:

$$H': \begin{array}{l} \rho': Z' \times X' \to Y' \\ \tau': Z' \times X' \to X', \end{array} \qquad H'': \begin{array}{l} \rho'': Z'' \times X'' \to Y'' \\ \tau'': Z'' \times X'' \to X''. \end{array} \tag{A5}$$

An open species–environment system $\{H', H, H''\}$ is realized by:

$$H': \begin{array}{l} \rho': Z' \times X' \to Y' \\ \tau': Z' \times X' \to X', \end{array} \qquad H: \begin{array}{l} \rho: Y' \times X \to Y \\ \tau: Y' \times X \to X, \end{array}$$

$$H'': \begin{array}{l} \rho'': Y \times X'' \to Y'' \\ \tau'': Y \times X'' \to X'', \end{array} \tag{A6}$$

where Z' is understood to come from and Y'' to go to the system-level environment.

In the closed $\{H, H^*\}$ system, let the environment, H^*, change. This means a change in one or more of ρ^*, τ^*, Z^*, X^* or Y^*, causing a different set of outputs Y^{*1} to be generated. Four possibilities exist for the relation of Y^{*1} to the original Y^*: $Y^{*1} \cap Y^* \neq \varnothing$ (the empty set), $Y^{*1} \cap Y^* = \varnothing$, $Y^* \supseteq Y^{*1}$, and $Y^{*1} \supseteq Y^*$. The first case is most likely; it implies a slight to substantial, but not complete, change in conditions. The second case signifies a catastrophic environmental change to totally new conditions. The third denotes a narrowing and the fourth case an expansion of the original environment. Except in the case of contraction where $Y^* \supseteq Y^{*1}$, some or all of the new outputs, $Y^{*1} - Z'$, will be inadmissible inputs to H. Others, $Z^1 = Y^{*1} - Z$, will be admissible, but the set of admissible inputs will have shrunk compared to the original, $Z^1 \subset Z$, leading to a reduced set of species responses also, $Y^1 \subset Y$. This obviously is also true for the case where $Y^* \supseteq Y^{*1}$, leading to $Z \supseteq Z^1$ and $Y \supseteq Y^1$, and $Y^{*1} \subset Y^*$ produces $Y^1 \subset Y$. Thus, H becomes restricted in its behavior:

$$H^1: \begin{array}{l} \rho: Z^1 \times X \to Y^1 \\ \tau: Z^1 \times X \to X. \end{array} \tag{A7}$$

With $Y^1 \subset Y$, Z^* becomes restricted to $Z^{*1} \subset Z^*$, hence H^* becomes further narrowed:

$$H^*: \begin{array}{l} \rho^*: Z^{*1} \times X^* \to Y^{*2} \\ \tau^*: Z^{*1} \times X \to X^*, \end{array} \tag{A8}$$

where $Y^{*2} \subset Y^{*1}$. The original environmental change has initiated a contracting process which, with each cycle around the $\{H, H^*\}$ loop, progressively attenuates the behavioral range, i.e. the input and output sets, of both H and H^*. To interrupt this process, H must alter its input set to admit the new behaviors Y^{*1} generated by the initial change in H^*. This is where evolution occurs. A 'mutant' H^1, with an expanded input set $Y^{*1} \supseteq Z^1$ (rather than $Z^1 = Y^{*1} - Z$), is required. Thus, H and H^* must coevolve, or their input and output sets will contract to the vanishing point. The conclusion follows:

> *Hypothesis 1.* Closure of an entity-environment system is sufficient to induce system coevolution; coevolution is a necessary concomitant of closure, and absence of coevolution indicates absence of closure.

In the case of the open or acyclic interaction system $\{H', H, H''\}$, if H'' undergoes a change, then its output set will change to Y''^1 and output environment will be different. But since there is no feedback from H'' to H', there will be no further effect within the system. As above, the new response set will be reduced compared to the original, $Y''^1 \subset Y''$, but now persistent. If H' changes, a reduced output set $Y'^1 \subset Y'$ results, reducing the input set of H to $Z^1 \subset Z$, and this reduces the species' response set, $Y^1 \subset Y$. The input set of H'' is thereby narrowed, $Z''^1 \subset Z''$, leading finally to a restricted set of system outputs, $Y''^1 \subset Y''$, which are not fed back to Z'. Thus, the behavior of H', H and H'' all become narrowed in response to the original change in H'. This stenotopic condition is stable, and may lead to an abundance of specialized forms, and perhaps endemism, in ecosystems. A contracting process is not initiated, and there is no impetus for an expanded input set mandating a coevolutionary response in the $\{H', H, H''\}$ open system. The following conclusion is warranted:

> *Hypothesis 2.* Lack of closure in a species–environment system is sufficient to preclude system coevolution; closure is a necessary condition for coevolution, and thus coevolution is sufficient to indicate closure.

Hypotheses 1 and 2 establish, respectively, sufficient and necessary relationships between closure and coevolution of entire species–environment systems encompassed by such closure.

Expanded rather than contracted output sets are also possible following perturbation of H^* or H' in the cases where $Y^* \subset Y^{*1}$ and $Y' \subset Y'^1$. This does not necessarily alter the input set of H. The new environmental conditions $Y^{*1} - Y^*$ and $Y'^1 - Y'$ can simply remain inadmissible, or H can, by a drift process, evolve a wider input set, $Z \subset Z^1$, and achieve a more eurytopic relationship, $Y \subset Y^1$, with the new environment.

References

Allen, P. M. (1981). The evolutionary paradigm of dissipative structures. In *The Evolutionary Vision*. AAAS Selected Symp. No. 61, ed. E. Jantsch, pp. 25–72. Boulder: Westview Press.

Allen, T. F. H. & Starr, T. B. (1982). *Hierarchy: Perspectives for Ecological Complexity*. Chicago: The University of Chicago Press.

Allen, T. F. H., O'Neill, R. V. & Hoekstra, T. W. (1984). *Interlevel Relations in Ecological Research and Management: Some Working Principles from Hierarchy Theory*. Fort Collins: U.S. Dept. of Agriculture.

Connell, J. (1980). Diversity and the coevolution of competitors, or the ghost of competition past. *Oilkos*, **35**, 131–8.

Cracraft, J. (1982). A non-equilibrium theory for the rate-control of speciation and extinction and the origin of macroevolutionary patterns. *Syst. Zool.*, **31**, 348–65.

Csanyi, V. (1982). *General Theory of Evolution. Studia Biol. Acad. Sci. Hungaricae*. Vol. 18. Budapest: Akademiai Kiado.

Csanyi, V. (1985). Autogenesis: the evolution of self-organizing systems. In *Dynamics of Macrosystems*, ed. J-P. Aubin, D. Saari, & K. Sigmund. Lect.

Notes in Economics and Mathematical Systems, pp. 253–67. Berlin: Springer-Verlag.

Darwin, C. (1859). *On the Origin of Species by Means of Natural Selection, or the Preservation of Favoured Races in the Struggle for Life*. London: John Murray.

Davidson, D. W., Inouye, R. S. & Brown, J. H. (1984). Granivory in a desert ecosystem: experimental evidence for indirect facilitation of ants by rodents. *Ecology*, **65**, 1780–6.

Dawkins, R. (1976). *The Selfish Gene*. Oxford: Oxford University. Press.

Dawkins, R. (1982). *The Extended Phenotype*. San Francisco: W. H. Freeman and Co.

Diamond, J. M. (1975). Assembly of species communities. In *Ecology and Evolution of Communities*, ed. M. L. Cody & J. M. Diamond, pp. 342–4. Cambridge, MA: Belknap Press.

Eigen, M. (1971). Self-organization of matter and the evolution of biological macromolecules. *Naturwissenschaften*, **58**, 465–523.

Eigen, M. & Schuster, P. (1978). The hypercycle: a principle of natural self-organization. Part C: The realistic hypercycle. *Naturwissenschaften*, **65**, 341–69.

Eldredge, N. & Gould, S. J. (1972). Punctuated equilibria: an alternative to phyletic gradualism. In *Models in Paleobiology*, ed. T. T. Schopf, pp. 82–115. San Francisco: Freeman, Cooper and Co.

Eldredge, N. & Salthe, S. N. (1984). Hierarchy and evolution. *Oxford Surveys in Evol. Biol.*, **1**, 184–208.

Fox, S. W. & Dose, K. (1972). *Molecular Evolution and the Origin of Life*. San Francisco: W. H. Freeman Co.

Gould, S. J. (1980). Is a new and general theory of evolution emerging? *Paleobiology*, **6**, 119–30.

Gould, S. J. (1982). The meaning of punctuated equilibrium and its role in validating a hierarchical approach to macroevolution. In *Perspectives on Evolution*, ed. R. Milkman, pp. 83–104. Sunderland: Sinauer.

Gould, S. J. (1985). The paradox of the first tier. *Paleobiology*, **11**, 2–12.

Gould, S. J. & Eldredge, N. (1977). Punctuated equilibria: the tempo and mode of evolution reconsidered. *Paleobiology*, **3**, 115–51.

Haken, H. (1977). *Synergetics: An Introduction. Nonequilibrium Phase Transitions and Self-Organization in Physics, Chemistry, and Biology*. 2nd. ed. 1978. Heidelberg: Springer-Verlag.

Haken, H. (1981). Synergetics: is self-organization governed by universal principles? In *The Evolutionary Vision*. AAAS Selected Symp. No. 61, ed. E. Jantsch, pp. 15–24. Boulder: Westview Press.

Higashi, M. & Patten, B. C. (1989). Dominance of indirect causality in ecosystems. *Am. Nat.*, **133**, 288–302.

Ho, M. W. & Saunders, P. T. (1979). Beyond neo-Darwinism: an epigenetic approach to evolution. *J. Theor. Biol.*, **78**, 573–91.

Ho, M. W. & Saunders, P. T. (1984). *Beyond Neo-Darwinism*. London: Academic Press.

Hull, D. L. (1980). Individuality and selection. *Ann. Rev. Ecol. Syst.*, **11**, 311–32.

Hutchinson, G. E. (1948). Circular causal systems in ecology. *Ann. N. Y. Acad. Sci.*, **50**, 221–4.

Inouye, R. S. (1981). Interactions among unrelated species: granivorous rodents, parasitic fungus, and a shared prey species. *Oecologia*, **49**, 425–7.

Jantsch, E. (1981). Unifying principles. In *The Evolutionary Vision*, AAAS Selected Symp. 61, ed. E. Jantsch, pp. 83–115. Boulder: Westview Press.

Jantsch, E. (1982). From self-reference to self-transcendence: The evolution of self-organization dynamics. In *Self-organization and Dissipative Structures: Applications in the Physical and Social Sciences,* ed. W. C. Schieve & P. M. Allen, pp. 344–53. Austin: University of Texas Press.

Kerfoot, W. C. & Sih, A. (1987). *Predation: Direct and Indirect Impacts on Aquatic Communities.* Hanover: University Press of New England.

Koestler, A. (1967). *The Ghost in the Machine.* New York: MacMillan.

Levine, S. H. (1976). Competitive interactions in ecosystems. *Am. Nat.*, **110**, 903–10.

Levins, R. (1975). Evolution in communities near equilibrium. In *Ecology and Evolution of Communities*, ed. M. L. Cody & J. Diamond, pp. 16–50, Cambridge, MA: Belknap Press.

MacArthur, R. H. & Wilson, E. O. (1967). *The Theory of Island Biogeography.* Princeton: Princeton University Press.

Margalef, R. (1968). *Perspectives on Ecology.* Chicago: The University of Chicago Press.

Margulis, L. (1981). *Symbiosis in Cell Evolution.* San Francisco: W. H. Freeman and Co.

Margulis, L. & Lovelock, J. E. (1974). Biological modulation of the Earth's atmosphere. *Icarus*, **21**, 471–89.

Mayr, E. (1963). *Animal Species and Evolution.* Cambridge, MA: Belknap Press.

Mayr, E. (1966). Evolutionary challenges to the mathematical interpretation of evolution. In *Mathematical Challenges to the Neo-Darwinian Theory of Evolution.* ed. P. S. Moorehead & M. M. Kaplan, Wistar Inst. Symp. 5, pp. 117–20. Philadelphia: The Wistar-Institute Press.

Mayr, E. (1980). Prologue. In *The Evolutionary Synthesis*, ed. E. Mayr & W. B. Provine, pp. 1–48. Cambridge. MA: Harvard University Press.

MacMahon, J. A., Phillips, D. L., Robinson, J. V. & Schimpe, D. J. (1978). Levels of biological organization: an organism-centered approach. *BioScience*, **28**, 700–704.

Miller, J. G. (1978). *Living Systems.* New York: McGraw-Hill.

Montague, C. L. (1980). *The Net Influence of the Mud Fiddler Crab,* Uca pugnax, *on Carbon Flow Through a Georgia Salt Marsh: The Importance of Work by Macroorganisms to the Metabolism of the Ecosystem.* Ph.D. Dissertation, University of Georgia, Athens, Georgia.

Odling-Smee, F. J. (1988). Niche constructing phenotypes. In *The Role of Behavior in Evolution*, ed. H. C. Plotkin, pp. 73–132. Cambridge, MA: MIT Press.

O'Neill, R. V., DeAngelis, D. L., Waide, J. B. & Allen, T. F. H. (1986). *A Hierarchical Concept of Ecosystems.* Monographs in Population Biology, 23. Princeton: Princeton University Press.

Patten, B. C. (1978). A systems approach to the concept of environment. *Ohio J. Sci.*, **78**, 206–22.

Patten, B. C. (1982). Environs: relativistic elementary particles for ecology. *Am. Nat.*, **119**, 179–219.

Patten, B. C., Bosserman, R. W., Finn, J. T. & Cale, W. G. (1976). Propagation of cause in ecosystems. In *Systems Analysis and Simulation in Ecology*, vol. 4, ed. B. C. Patten, pp. 457–579. New York: Academic Press.

Prigogine, I. (1967). *Introduction to Thermodynamics of Irreversible Processes.* New York: Wiley.

Prigogine, I. (1980). *From Being to Becoming: Time and Complexity in Physical Sciences.* San Francisco: W. H. Freeman.

Provine, W. B. (1982). Influence of Darwin's ideas on the study of evolution. *BioScience*, **32**, 501–6.

Puccia, C. J. & Levins, R. (1985). *Qualitative Modeling of Complex Systems: An Introduction to Loop Analysis and Time Averaging.* Cambridge, MA: Harvard University Press.

Raff, R. & Kaufman, T. C. (1983). *Embryos, Genes and Evolution.* New York: Macmillan Publ. Co.

Raup, D. M., Gould, S. J., Schopf, T. J. M. & Simberloff, D. (1973). Stochastic models of phylogeny and the evolution of diversity. *J. Geol.*, **81**, 525–42.

Riedl, R. (1978). *Order in Living Organisms.* New York: Wiley.

Roughgarden, J. (1979). *Theory of Population Genetics and Evolutionary Biology: An Introduction.* New York: Macmillan.

Salthe, S. N. (1985). *Evolving Hierarchical Systems: Their Structure and Representation.* New York: Columbia University Press.

Schwemmler, W. (1984). *Reconstruction of Cell Evolution: A Periodic System.* Boca Raton: CRC Press, Inc.

Stanley, S. M. (1975). A theory of evolution above the species level. *Proc. Natl. Acad. Sci. (USA)*, **72**, 646–50.

Stanley, S. M. (1979). *Macroevolution: Pattern and Process.* San Francisco: W. H. Freeman.

Stanley, S. M. (1985). Macroevolution and the fossil record. *Evolution*, **36**, 460–73.

Sternier, R. W. (1986). Herbivores' direct and indirect effects on algal populations. *Science*, **231**, 605–7.

Strong, D. R., Jr. (1984). Natural variability and the manifold mechanisms of ecological communities. In *Ecology and Evolutionary Biology*, ed. G. W. Salt, pp. 56–80. Chicago: University of Chicago Press.

Sugihara, G. (1982). *Niche Hierarchy: Structure, Organization and Assembly in Natural Communities.* Ph.D. Dissertation. Princeton University, Princeton.

Tansley, A. G. (1935). The use and abuse of vegetational concepts and terms. *Ecology*, **16**, 284–307.

Thompson, J. N. (1982). *Interaction and Coevolution.* New York: Wiley-Interscience.

Vandermeer, J. (1980). Indirect mutualism: Variations on a theme by Stephen Levine. *Am. Nat.*, **116**, 441–8.

Vandermeer, J., Hazlett, B. & Rathcke, B. (1985). Indirect facilitation and mutualism. In *The Biology of Mutualism*, ed. D. Boucher, pp. 326–43. London: Croom Helm.

Vrba, E. S. (1980). Evolution, species, and fossils: how does life evolve? *S. Afr. J. Sci.*, **76**, 61–84.

Vrba, E. S. (1982). Darwinism in 1982: the triumph and challenges. *S. Afr. J. of Sci.*, **78**, 275–8.

Vrba, E. S. & Eldredge, N. (1984). Individuals, hierarchies and processes: towards a more complete evolutionary theory. *Paleobiology*, **10**, 146–71.

Wade, M. J. (1977). An experimental study of group selection. *Evolution*, **31**, 134–53.

Wilson, D. S. (1980). *Natural Selection of Populations and Communities.* Menlo Park: Benjamin/Cummings.

Wright, S. (1949). Adaptation and selection. In *Genetics, Paleontology, and Evolution*, ed. G. L. Jepsen, E. Mayr, and G. B. Simpson, pp. 365–89. Princeton: Princeton University Press.

Wright, S. (1982). Character change, speciation, and the higher taxa. *Evolution*, **30**, 427–43.

Wynne-Edwards, V. C. (1987). *Evolution Through Group Selection.* Oxford: Blackwell Scientific Publications.

Yodzis, P. (1981). The structure of assembled communities. *J. Theor. Biol.*, **92**, 103–17.

9

Control theory in the study of ecosystems: a summary view

B. HANNON AND J. BENTSMAN

9.1 **Introduction**

In spite of the efforts of numerous researchers for over a century, ecosystem modeling is still in its infancy. Coupled with modern technological developments, it provides more challenges for biologists, ecologists and system theorists than ever before. The enormous complexity of the subject and its major role in the very existence of our civilization have caused several theories to be proposed and approaches to be tried. However, there has not yet been produced a satisfactory predictive dynamical or informational account of ecosystem behaviour; hence new theories should be created.

The classical approach to this problem in ecological theory has been to model the interaction of two organisms with detailed, complex, nonlinear differential equations. Using the same methodology for cases involving more than two interacting organisms results in formulations that are often intractable. Therefore, approaches should be sought that constitute a trade-off between the complexity of the individual behavior and the ability of the model to describe the interactions among large numbers of species on a given territory, as well as the interactions among various territories. A summary of the development of an approach with such features is the goal of this chapter.

In recent years, a modeling theory known as flow analysis (to be discussed below) has been developed and demonstrated to provide a reasonable accounting framework for the functioning of a broad variety of ecosystems (Hannon, 1985*a*). As with any new methodology, the simplest setup – deterministic linear time invariant ordinary differential equations – has been adopted as a formal basis for the construction of flow models. Despite demonstrated relevance of this modeling framework to real ecosystems it has become obvious that further refinements of the theory are necessary to better explain observed phenomena, predict effects of disturbances, and provide a more reliable basis for possible ecosystem management. Here several issues are of primary importance.

The first and foremost issue is ecosystem resiliency (stability) to environmental as well as internal stresses (disturbances). A straightforward application of the flow accounting technique yields consistent but inherently unstable models which describe the flow changes between the various ecosystem processes. The actual ecosystems, however, do possess an ability to avoid collapse when perturbed (stability); hence, the question naturally arises as to how this stability should best be achieved in the context of flow models. The application of the concepts of control theory, a good account of which is given by Wonham (1984), provides a logical way to find the answer to this question. Indeed, from a system–theoretic viewpoint, a stable ecosystem can be thought of as a living system whose individual components possess the necessary mechanisms or controls capable of maintaining the system stability and coping with disturbances. Thus, formulation of testable hypotheses regarding operation of such controls and actual testing on real systems should provide vital insights into ecosystem stability.

The second issue is the relation of the model complexity to the observed and expected phenomena. Clearly, complete reality can hardly be squeezed into any mathematical formalism. However, good models should be simple enough and still account for the main features of system behavior. In this regard an inclusion of time delays into ecosystem models has been often called for. Indeed, the ecosystem accomplishes its normal functioning, including information exchanges in its controls, via exchange of energy and matter among its own components and between itself and surroundings. That such exchange does not occur instantaneously requires that time delays be incorporated into the model.

The third issue is the adaptability of ecosystems. Under environmental stress, the ecosystem parameters may vary in a broad range, therefore the ecosystem functioning might be expected to severely deteriorate. In many situations (e.g. hurricanes and coastal ecosystems), this obviously does not happen. Such response to stress indicates that ecosystem controls possess a very important feature – adaptability. Flow accounting models and adaptive control principles should provide a basis for identification of some of the adaptation mechanisms.

The last issue to be discussed here is optimality. Many physical phenomena, such as motion of a particle in a potential field, are extremal in a certain sense. This means that an optimality criterion can be formed and it can be shown that the behavior of a particle is directed toward minimization of this functional. The question, still unanswered at present, is whether an ecosystem behaves in such a manner as to minimize (optimize) some meaningful criteria. At present, consensus seems to be that it does not, but proofs are lacking. However, such criteria, if discovered could serve as principles of ecosystem behavior.

9.2 Current state of the theory

In the following discussion, we summarize the ecosystem modeling theory known as flow analysis, show its extension from static to dynamic behavior where it becomes unstable and sketch how a variety of controls can be theoretically constructed to stabilize the system.

9.2.1 *The static model*

The first step in any system modeling program is the development of an accounting framework in which every conceivable type of ecosystem flow of a given type (e.g. energy or nitrogen) has an unambiguous location. Over the last decade, a group of theorists has been struggling with this problem (Hannon, 1973, 1979, 1985*a*; O'Neill, 1979; Herendeen, 1981; Ulanowicz, 1986; Levine, 1977, 1980; Finn, 1976; Patten *et al.*, 1976; Barber *et al.*, 1979; Costanza & Hannon, 1990). The consensus seems to be that the arrangement shown in Fig. 9.1 meets these requirements.

In Fig. 9.1, $n \times n$ matrix P is called the production–consumption matrix. This matrix represents n processes which consume and produce n commodities. By process, we mean an aggregation of similar consumers–producers (Hannon, 1985*a*) that is viewed as a single ecosystem component. By commodities, we mean the substances produced and consumed by the components of the ecosystem. The elements of the ith

Fig. 9.1. Steady state ecosystem flow accounting diagram.

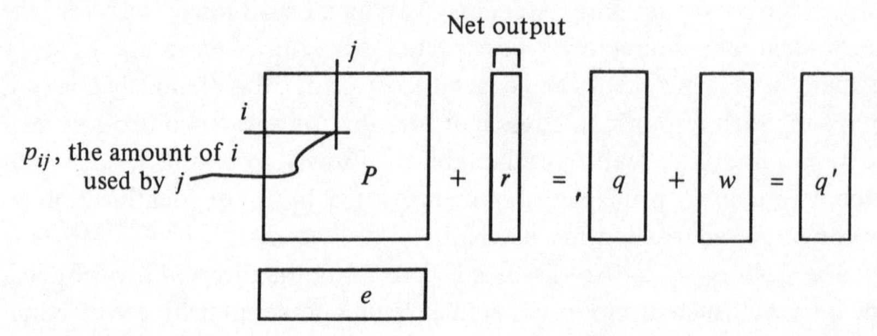

column represent the breakdown of the main part of the consumption of the *i*th process. The elements of the *i*th row describe the breakdown of the main part of the production by the same process. Therefore, each element of P is the amount of commodity (row number) that is used by process j (column number) in the given time period. For example, p_{ij} could be the daily amount of algal biomass (commodity i) consumed by a particular class of herbivores (process j). This is a multicommodity system as commodities listed along any of the rows are non-commensurable with commodities in any other row. Therefore, the row sums may be calculated as they are all the same commodity and, we assume, possess the same nutritional qualities for all consumers. (The exception to this rule is the non-basal heat of respiration, which by definition has zero value to any component in the ecosystem.) But, in general, the column sums cannot be formed because a common measure of the value of each element along the columns may not exist. Commodities of different qualities, even though measured in the same units (e.g. g-carbon) cannot be meaningfully added together. The inputs to omnivores and detritivores, for example, are of different qualities (relative value to the consumer), both chemically and in nutritional meaning, to the consumer.

The diagonal elements in P are the self-use terms, which are, for example, own-waste consumption by rabbits, the consumption of decomposers by decomposers and cannibalism.

The full output vector q' is the sum of the vector of the non-basal heat w given off by each of the components and the total output vector q.

The system in Fig. 9.1 is without joint products, that is, each process (column) is assumed to produce a community of only one type. The joint product case is discussed in Hannon (1985*a*) and Costanza & Hannon (1990). The joint input–output configuration is a more complete but also more complex one. However, the end result of the joint product theory is the linear transformation that reduces the joint configuration to the single output form described in Fig. 9.1. In this chapter, the theory is developed for the simpler case, with the understanding that it will apply easily to the more complicated description of ecosystem flows.

The relationship to the external environment of the measurable quantities in the ecosystem modeled in Fig. 9.1 is summarized in Table 9.1. The features of each quantity in this table are identified by the letters in the corresponding boxes. The table shows two vectors: r and e. The net output vector r is composed of three types of flows: exports (A & D), imports (D & E) and the heat of basal metabolism (C). By imports we mean those quantities that can be produced by the ecosystem but enter the

Table 9.1. *The description of the quantities forming the net output and non-produced input.*

			Net output r			Non-produced input e
			Exports	Imports	Basal heat	
Commodities that the ecosystem is capable of producing	Produced by the system	Leaves the system	A		C	
		Stays				
	Not produced by the system	Leaves	D	D		
		Stays		E		
Commodities that the ecosystem is incapable of producing						F

system from the external environment. Exports are those quantities that can be produced by the ecosystem for the external environment. The letter D in the import and export columns indicates those measured quantities passing through the ecosystem in the given time period; the quantity A – E is the net export. The system is perturbed by the externally induced change of the net export. The heat of basal metabolism (basal respiration) is that given off by the organism at rest. We take the heat of basal metabolism (C) as a surrogate for the commodity flows that are used in rebuilding the stocks metabolized during the given period. By stocks we mean the accumulated output quantities in each of the components in the system. The stocks are, for example, the amount of biomass of algae that has accumulated in the producer (sun capturing) component of an aquatic ecosystem. Stocks at this time in our model development are assumed to be fully productive at all times (on average), i.e. there is no concept of 'idle stock' in our present formulation of the ecosystem model. In reality, biological stocks may require some time of growth before they begin to be productive (e.g. the few weeks that trees spend each spring in leaf production). In our current model formulation, no such 'lag time' has been incorporated, but that feature significantly influences actual system behavior. The quantity F (vector e) stands for those input commodities that the ecosystem is incapable of producing (e.g. sunlight), but that are necessary for ecosystem functioning. It is also a possible external source of perturbation to the ecosystem; in the present state of model development, these disturbances are limited to changes in the net exports portion of vector r.

9.2.2 *The dynamic linear model*

Next we combine the flow definitions above with the possibility of a growth in the stock of process j during the given time period Δt. These flows are graphically shown in Fig. 9.2 for the individual process. The consumption flows p_{ij}, production flows p_{jk} and the storage flow $\Delta s_j/\Delta t$ are internal to the ecosystem boundary, whereas the net output flows r_j, the non-basal respiration flow w_j and the non-produced input flow e_j cross the ecosystem boundary. The non-basal respiration flow (e.g. the energy used in chasing prey, avoiding predators, food-searching and reproduction) cannot be utilized further by the ecosystem, and it is therefore considered a waste. The r_j consists of the net export of the process (export minus import) and the stock replacement (basal respiration). The net input vector e is assumed to cause no restriction to the level of q_j at the current stage of the model development.

The total outflow q'_j is defined for the steady state ecosystem as

$$q'_j \stackrel{\Delta}{=} \sum_{k=1}^{n} p_{jk} + r_j + w_j. \qquad (1)$$

Fig. 9.2. The definition of the input and output flows of a typical process (j).

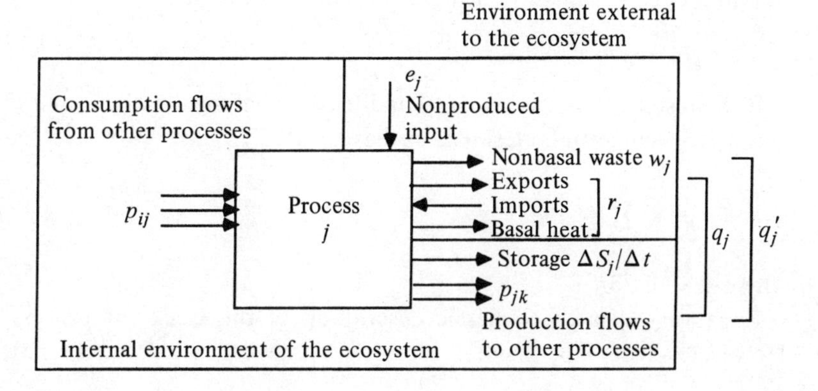

To take into account a growth in stock, Δs_j, over the time period Δt when the system is not in the steady state, definition (1) is augmented as

$$q'_j \stackrel{\Delta}{=} \sum_{k=1}^{n} p_{jk} + r_j + w_j + \frac{\Delta s_j}{\Delta t}. \tag{2}$$

Three important simplifying assumptions are now made for the ecosystem shown in Fig. 9.2 with q'_j defined in (2):

(i) commodity weighting or importance factors are assigned to each of the commodities produced in the system. The weight for each commodity is independent of where this commodity goes in the system. A weight of zero is given to the non-basal heat of respiration, w. The weights may vary as the q'_j vary. The vector w disappears from the formulation and the element q_j can be formed by the simple addition of all the elements along the jth row of matrix P, the rate of the jth stock growth and the jth element of vector r.

(ii) the inputs to process j, p_{ij}, form a constant ratio with the output of process j, q_j. Thus, $p_{ij}/q_j = g_{ij} =$ constant. The constants g_{ij} are determined from the data on the ecosystem at its steady state and they are assumed to remain constant for the dynamic form of our model presented below. These constants represent the internal behavior of the jth process. The g_{ij} incorporate the consumption flows into the model by locking them into a constant relationship with the output of the receiving process. Thus, the problem of summing the consumption flows (see Fig. 9.2) is avoided.

(iii) the stock (s_j) of any process (j) stays in constant proportion to the total output (q_j) of this process. That is: $b_{jj} = s_j/q_j =$ constant, forming a diagonal matrix $\mathrm{B} = \mathrm{diag}\{b_{11}, \ldots, b_{nn}\}$. This assumption allows us to obtain a balance equation using definition (2) since now

$$q_j = s_j/b_{jj}. \tag{3}$$

If the results of assumptions (i) and (ii) are combined with (2) and (3), and if Δt becomes infinitesimal, we have

$$\frac{s_j}{b_{jj}} = \sum_{k=1}^{n} g_{jk}\frac{s_k}{b_{kk}} + r_j + \dot{s}_j, \tag{4}$$

where $\dot{s}_j = \mathrm{d}s_j/\mathrm{d}t$.

Equation (4) is the dynamic description of the stock for process j. However, most experimental ecosystem data are presented as flows.

Therefore, we change (4) into a dynamic description of the flows for process j. Substituting (3) and its time derivative into (4) yields

$$b_{jj}\dot{q}_j = q_j - \sum_{k=1}^{n} g_{jk} q_k - r_j, \quad \forall j, 1 \leqslant j \leqslant n, \tag{5}$$

or in matrix form

$$\dot{q} = Nq - B^{-1}r, \quad N \triangleq B^{-1}(I-G). \tag{6}$$

This time invariant ordinary differential eq. (6) is in the 'standard' form for the flow analysis approach. The assumptions made for its derivation could cause subtle modeling difficulties. First, the assumption that the ratio of stock to the total output of a process remains constant requires that the current stock be fully involved in production. This is clearly not the case for a tree in winter. However, with careful averaging of the production data over a sufficiently long period (e.g. a year), this 'idle stock' problem could be minimized. Second, the manner in which the stock change was incorporated into the model does not allow for a 'lag time' between the initial stages of stock building and the onset of its productive activity. The study of the effect of lag times on output and ecosystem stability, however, is an important issue already mentioned earlier. Third, the existence of a constant vector of weights (see assumption (i) above) for the commodities produced in the system may not be appropriate. Even if the weights were varied appropriately (see Hannon, 1985*c*, for the equation of dynamic ecosystem 'prices' or weights), the lag time between the initiation of the stock building process and time of the full contribution of that stock to output means that two sets of weights may be needed, one set for the current output and another set for the output that was needed earlier for the production of the current stock. This complexity is the cost of weighting the commodities to reflect their (changing) value to various consumers in the ecosystem as the stocks and output rates of the various processes change.

The stability properties of the behavior of q when the system is subjected to a step change in r depend entirely on the matrix N in (6). If the real parts of all the eigenvalues of N were negative, the system would respond in a stable manner (Luenberger, 1979, p. 158). However, in (6) the sum of the eigenvalues of matrix N is always positive. Therefore, the system will always respond to 'sufficiently rich' changes in r in an unstable manner. If the system (6) is to accurately represent the functioning of an ecosystem, the equations must be judiciously modified to include stabilizing flows. Next, we introduce the techniques of ecosystem control theory and apply some of them to an actual ecosystem data set.

9.2.3 ***The system control strategies***

An early attempt to incorporate control theory into ecosystem management is found in Olson (1961), who used analog computers to model the flow of trace nuclides through ecosystems. Lowes & Blackwell (1975) and Boling & Van Sickle (1975) describe, in general, the application of engineering control theory to ecosystems. Mulholland & Sims (1976) used the stable compartmental model to demonstrate in a general way how control theory could be used to achieve desired flow levels with linear and nonlinear models. Kercher (1983) applied control theory to stable, dynamic, linear models to find their frequency response to harmonic input. Goh (1979) demonstrates why nonlinear control theory can be applied to systems with only one or two variables.

There are several problems with these historic approaches to ecosystem modeling. One problem is that the basic models used are always stable. Thus, they cannot reflect complete reality as ecosystems are known to be occasionally unstable. More important, these stable modeling approaches require summing of non-commensurable flows, and therefore do not seem to be completely consistent. Our approach is more appropriate as it allows for ecosystem instability and also more consistent, since non-commensurable quantities are not lumped together.

From an ecological viewpoint, a positive r represents an output of the ecosystem (for example, the amount of fish caught in the annual season). From the control theory viewpoint, however, this output represents an *input* to the system or a *control* action. For example, the amount of fish caught directly affects the rate of (re)production of fish *and* many other quantities produced in the ecosystem, which in turn, also affect later fishing success. At this point, we note that a population biologist might model fish reproduction in great detail, but he would miss its dependence on the other processes in the rest of the ecosystem. While we model the behavior of each component (e.g. fish) in less detail, we can account for the effect of changes in fishing success on the rest of the ecosystem and, most important, the return effect of these changes on the abundance of fish.

Feedback, feedforward and periodic controls are assumed to exist in some of the components in a stable ecosystem. Since the controls need not be expressed in averaged measurements taken on actual ecosystems, they might be thought non-existent. Dynamic testing of a specific ecosystem could reveal these controls. An ecosystem that is highly perturbed by external influences may have its controls damaged or reduced such that the system becomes unstable. It should then be possible to impose external controls which could restore its stability. For example, these controls could be applied by external control of the population of certain of the components through cropping or importation, or through the import or export of specific nutrients from the system. However, such external controls must be used with the recognition that some of the internal controls may still exist; otherwise the external controls may be counterproductive.

It is not necessary to view the ecosystem as a 'super organism' capable of 'knowing' which internal control to exert at which time. Surviving stable ecosystems can be thought of as having evolved or succeeded to include the components whose controlling behavior is so essential.

Equation (6) can be made to respond stably by modifying r to include a feedforward (anticipatory) control (Hannon, 1985*b*, *c*) or a feedback (reactive) control (Wonham, 1984; Hannon, 1986). The feedback control can be a function of q, and/or the derivative or integral of q. Both feedback and feedforward controls require some sort of information transfer in the system. For a discussion of positive feedback in ecological systems, see Ulanowicz (Chapter 2) and DeAngelis *et al*., (1986) and DeAngelis and Post (Chapter 6).

Feedback and feedforward constitute the main control strategies for any system. If knowledge of the current output is used to modify the inputs to control the system, we have a feedback control situation. Feedforward control uses current knowledge of the disturbance (rather than output) as the basis for a corrective action (Takahashi *et al*., 1970).

Another class of controls exists that does not require information transfer. They are called vibrational controls (Meerkov, 1980; Bellman *et al*., 1986*a*, *b*; Bentsman, 1987) or periodic controls (Bentsman & Hannon, 1987). Basically, these controls are periodic variations (zero–mean) in the flows between components in an ecosystem or between the ecosystem and the surrounding environment. If the amplitudes and frequencies of these variations are within the appropriate range, the ecosystem, unstable without such variations, could theoretically be stabilized by their introduction.

9.2.4 *Control theory applied to ecosystems*

To illustrate briefly how the basic control approaches could be applied to a real ecosystem we have chosen the Dame & Patten (1981) description of a six component oyster reef ecosystem. We will demonstrate the use of the flow accounting system and the results of application of the feedback and periodic controls to an otherwise exponentially unstable ecosystem model.

In Fig. 9.3, the diagram of the energy flows and stocks for an oyster reef ecosystem is given. The data of the diagram are reduced to the accounting system format in Table 9.2.

The matrices B and G can now be formed and the differential eq. (6) can be stated. Now, suppose that we wished to increase the predator export (represented by the sixth entry of vector r) by 1.0 kcal/m^2·day. (This quantity is certainly unrealistically large but it helps to clearly reveal the principles involved.) Let us employ a feedback concept to demonstrate how we achieve a stable transition from the current to the desired values of predator export. For simplicity, imagine that there is a feedback loop around each process and that the elements of the feedback vector, r_f, are

Fig. 9.3. The steady state oyster-reef ecosystem. Flow units are kcal/m^2·day (stock units; kcal/m^2).

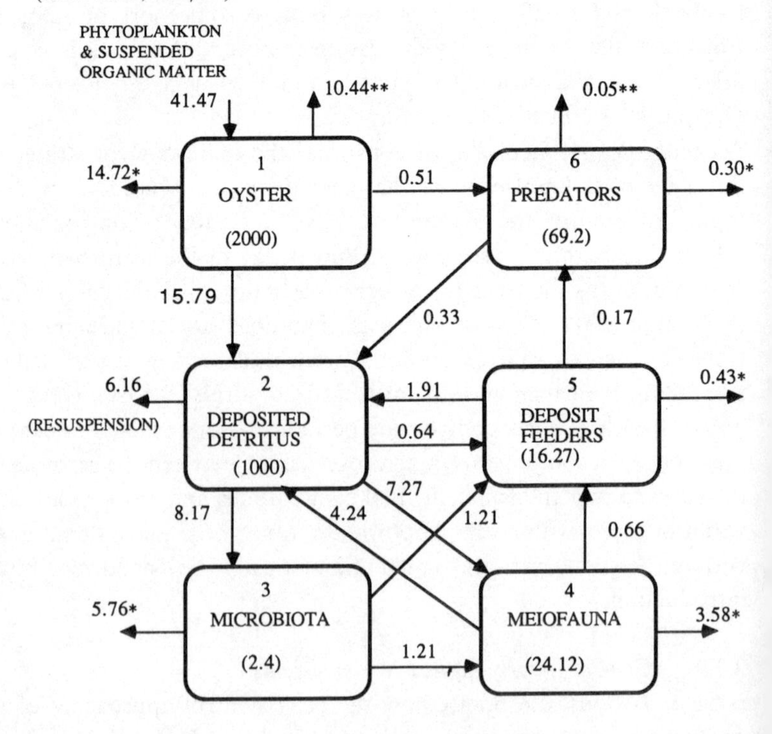

Table 2. *Oyster reef production–consumption matrix (P), along with vectors for exports – imports + basal metabolism (r), total output excluding waste heat (q), waste heat (w), and total output including waste heat (q′). (Basal metabolism requirements were not specifically provided for this system and they are therefore assumed zero for simplicity.) Flow is from the row components to those named in the columns.*

	P									
	(1)	(2)	(3)	(4)	(5)	(6)	r	q	w	q'
Oysters (1)	0	15.8	0	0	0	0.5	10.4	26.7	14.7	41.5
Detritus (2)	0	0	8.2	7.3	0.6	0	6.2	22.3	0	22.3
Microbiota (3)	0	0	0	1.2	1.2	0	0	2.4	5.6	8.2
Meiofauna (4)	0	4.2	0	0	0.7	0	0	4.9	3.6	8.4
Deposit feeders (5)	0	1.9	0	0	0	0.2	0	2.08	0.4	2.5
Predators (6)	0	0.3	0	0	0	0	0.1	0.4	0.3	0.7
Net input (e)	41.5	0	0	0	0	0				

proportional to the corresponding elements of q. Such an arrangement is called self-consumption. If the proportionality coefficients are sufficiently large (i.e. the self-consumption rate is high enough in the appropriate processes), the real parts of the eigenvalues of the new system matrix (say, N') become negative. This situation indicates that the model of the oyster reef ecosystem has been stabilized by feedback controls. A detailed discussion of this procedure is given in Hannon (1986) and is summarized in Appendix A9.1.

The response of the oyster reef ecosystem with the stabilizing feedback control r_f, is shown in Figs. 9.4 and 9.5. The changes in the total output vector q to stably accommodate the increase in predator export beginning at $t = 0$, is shown in Fig. 9.4. Total output values reach a new steady state in about 30 to 40 days with this particular control vector. The demand for input from the external environment (net input change) caused by the feedback vector is shown in Fig. 9.5. The initial demand for imports is very high for processes 1, 2 and 6, but these changes in the net inputs slowly diminish to zero (except of course the predator net output change which eventually reaches the desired 1.0 kcal/m$^2\cdot$day of export). Throughout this period of change, the assumption is that the net input e is sufficiently available to meet the new export demand. As stated earlier, the imposed demand for predator export is 100 times larger than practical harvesting techniques would allow. As a result of this overstated export, the demand for stabilizing inputs from the environment around the reef is also impractically large. But the principle is demonstrated. These inputs from the surrounding environment would be provided by the processes themselves if they possessed the feedback controls, or from man if the ecosystem could not supply them. The latter case is similar to the timely application of fertilizer in grain agriculture ecosystems.

Finally, let us briefly examine the nature of the periodic control when used with this system. The periodic control cannot be used unless the matrix N in system (6) has a negative trace (sum of the diagonal terms).

Fig. 9.4. The changes in the total output vector, q, in response to a 1.0 kcal/m^2·day increase in predator export with feedback controls.

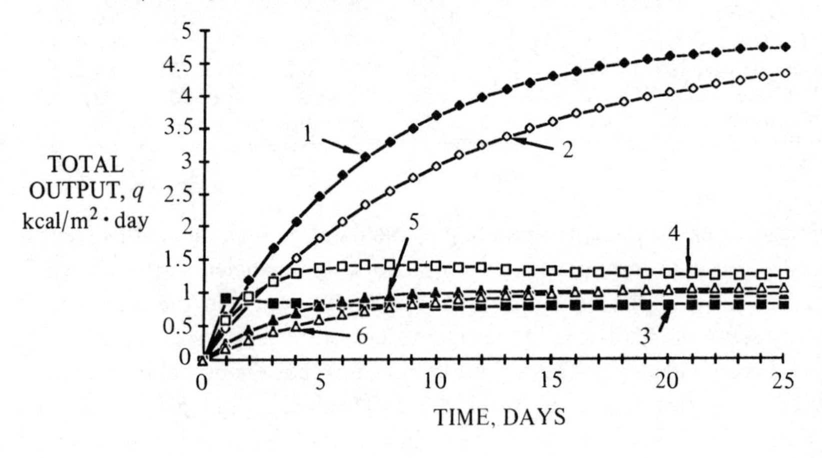

Fig. 9.5. The changes in the net output vector, r, in response to a 1.0 kcal/m^2·day increase in predator export with feedback controls.

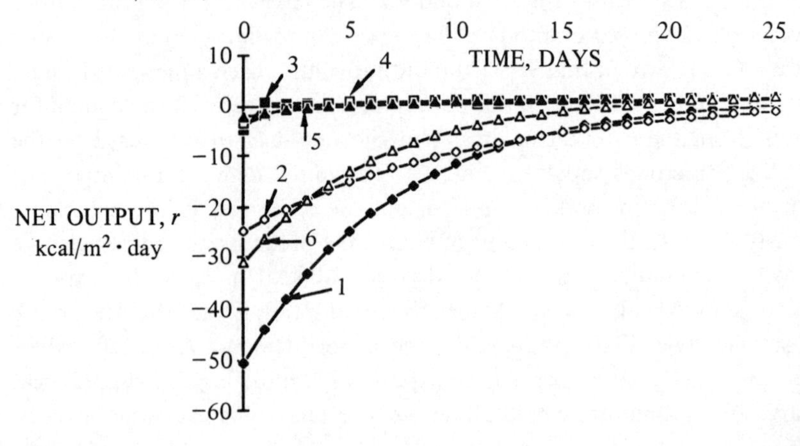

This will never be the case in the flow analysis approach, so feedback controls must be employed to modify at least one of the diagonal elements of N so that the trace of N be negative. Using again for simplicity a set of self-consumption feedbacks and employing the criterion of Bellman *et al.* (1985), we find that only one of the several possible flows (from the Deposit Feeders to the Microbiota) has a stabilizing potential. The stabilizing amplitudes and frequencies of this flow are highly constrained as was determined by repeated solution of (6), with cycles introduced in the (5,3) element of N. The region of the stabilizing frequencies and amplitudes is shown in Fig. 9.6. The analysis is discussed in detail in Bentsman & Hannon (1987) and is summarized in Appendix A9.2.

The data of Fig. 9.6 show, for example, that a periodic net input to the Deposit Feeders (which in turn allows them to cycle their feeding on the microbiota) can stabilize this ecosystem. With a cycle frequency of once in seven days, the stabilizing amplitude could be about 1.2 kcals/m^2·day, the average value of the flow from 3 to 5. These periodic flows can be viewed as cycles in the (5,3) element of the G matrix. Both donor and recipient must alternately cycle their non-basal flows (w) such that all the total outputs remain constant. Unfortunately, no data on the variation of flows in this oyster reef ecosystem were given (Dame, 1976, 1979; Dame & Patten, 1981). From Fig. 9.6, we see that smaller amplitudes are associated with lower frequencies.

Given the natural solar cycles and their clear effect on the producer, it

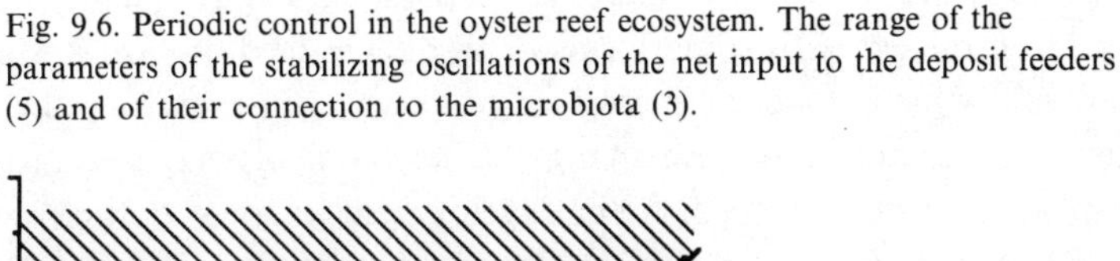

Fig. 9.6. Periodic control in the oyster reef ecosystem. The range of the parameters of the stabilizing oscillations of the net input to the deposit feeders (5) and of their connection to the microbiota (3).

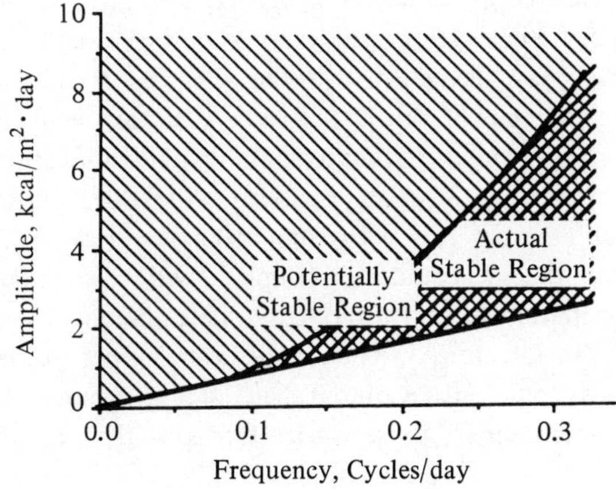

seems possible that some sort of stabilizing periodic flow could exist. This application to the oyster reef system is expected to convey a biological possibility of ecosystem stabilization by already existing or intentionally introduced oscillations. An example of an open loop ecosystem control is considered in Hannon (1985*b*, *c*).

Obviously, in each of the examples, the choice of the controls is arbitrary. Only two considerations constrain this arbitrariness: first, the controls must possess a biological reasonableness. For example, in many large scale ecosystem experiments, are there any indications, either directly by observation or indirectly by inference from flow data variability, of these types of controls? Through careful experimentation, it should be possible to determine where such indications of controls exist naturally and find some evidence of them. The second constraint on the arbitrariness of the control values is mathematical, as was the case with the periodic control example.

9.3 Conclusion

Large ecological systems might have compartments so physically separated that the existing feedback controls can only partially contribute to system stability; regardless of the magnitude of these feedback controls, system stability can then only be achieved by superimposing periodic (zero–mean) fluctuations on certain of the exchange flows in the system (Wang & Davison, 1973; Anderson & Moore, 1981; Trave *et al.*, 1985). Periodic controls may also be useful when the feedback controls are physically limited. We speculate that ecosystems can frequently be classed as decentralized with some forms of feedback, feedforward and periodic controls which undergo incessant adaptive adjustment in the face of disturbances and ecosystem changes. These controls may also be aimed in part at governing the ecosystem development in some optimal way.

While real ecosystems may deviate to varying degrees from the proposed paradigm, the richness of the phenomena that this setup is capable of generating on the one hand, and the availability of identification and testing procedures on the other, opens a possibility of development of models that describe and predict real ecosystem behavior with great accuracy. Thus, control theory could be coupled with ecosystem experiment to discern the mathematical form of either natural controls or man-made ones needed to be introduced to insure the desirable ecosystem functioning. The ecosystem could have acquired these stabilizing controls in the normal evolutionary manner: those earlier forms of the present stable ecosystem which did not develop the stabilizing controls simply ceased to exist.

9.4 Summary

The theory of feedback control as a possible stabilizing mechanism in ecosystem analysis has been summarized. One problem in the theory is the identification of the informational links by which such controls operate. Periodic controls, for example, zero–mean sine functions added to certain exchange flows in the system, might also contribute to system stability. Their advantage is that they operate without need for information from the rest of the system. The theory of ecosystem periodic control has been presented and applied to data from an oyster reef ecosystem.

Acknowledgement

Work supported in part by the Illinois Department of Energy and Natural Resources.

Appendix

A9.1 Feedback type controls

The general solution to the dynamic eq. (6) for a constant disturbance in the net output $r(0^+)$ is:

$$q = [I - e^{Nt}](I - G)^{-1} r(0^+) \quad t \geqslant 0 \tag{A1}$$

where $N = B^{-1}(I - G)$, which is unstable. Note that $(I-G)^{-1}r(0^+)$ is the final steady-state total output vector. For stability, the e^{Nt} must decay to zero. Here it does not because the trace of N is positive. This condition is guaranteed since B is a positive diagonal matrix and the diagonals of G are positive and less than one. Stabilizing equation (A1) requires changes in B and/or G that make all of the eigenvalues of N negative. This is done by adding terms to the control variables r, which in effect, simulate the control action of the natural system or which inform the ecosystem managers of their needed control action. These controls can be made functions of q or its derivative or integral. The example used in the text contains a set of feedback controls that are functions of q.

To stabilize the dynamic system the flow:

$$r = r(0^+) + r_0 - Qq \quad \text{(A2)}$$

where r_0 is a vector of control constants which allow one to choose the level of the outputs in the final steady state; Q is the flow control matrix chosen for the text example as a diagonal matrix for convenience. The elements of Q must be chosen with care. If they are too large the system response is too fast and the resulting r is too large.

Equation (A2) is a combination of the *change* in the net output due to the desired export $r(0^+)$ and the control net output $r_0 - Qq$. Equation (A2) can be rewritten in terms of the desired change in net output, $r(0^+)$, and the total output, as:

$$r = [I + Q(I-G)^{-1}]\, r(0^+) - Qq \quad \text{(A3)}$$

and the solution for the total output is:

$$q = [I - e^{Lt}](I-G)^{-1} r(0^+) \quad \text{(A4)}$$

where $L = B^{-1}(I - G + Q)$.

The elements of the control matrix Q are chosen to make the real parts of the eigenvalues of L negative. In developing these elements for the oyster reef, the Frobenius theorem (Varga, 1962) on negative eigenvalues is invoked. The diagonal elements used in the text example are: $\{-10.5, -5.2, -4.2, -1.1, -2.4, -30\}$. They were chosen to cause a reasonably uniform approach by all components to their respective steady states. The 'best' choice of Q requires an optimality theory (Hannon, 1986).

An enormous penalty must be paid, in the form of dramatically increased need for net input, if the controls are started with a delay, after the desired net output change has begun. In our example, the control is acting on *current* information on the system condition and so the situation is always theoretically stable in the long run, regardless of the lag time. In more realistic situations, the control would be acting on lagged *descriptions* of the system. In this case, the controls may be destabilizing, depending on how quickly the reaction to the controls takes effect.

A9.2 **Periodic control**

Not all real parts of the eigenvalues of N can be always made negative by the choice of Q, i.e. if the information gathering processes of the system are somehow limited, resulting in lack of 'on line' measurements necessary for the functioning of feedback controls. In such cases, however, the system can still be stabilized by oscillations of some of its parameters. Periodic changes in the parameters of an ecosystem that induce stable behavior can be viewed as an additional control mechanism, termed periodic control.

For the purposes of illustrating the mathematical periodic control formalism, let us now choose the vector r in system (6) as $r = -D(t)\,q$, where $D(t)$ is a periodic, zero–mean matrix. In this case eq. (4) becomes

$$\dot{q} = [N + B^{-1}D(t)]\,q. \tag{A5}$$

Because eq. (A5) is time-varying, eigenvalues can no longer describe its stability. It is possible, however, to associate stability properties of the oscillatory system (A5) with a certain constant matrix that describes its *average* behavior. The stabilizing action of periodic controls consists of converting the right-half plane eigenvalues of system (6) into 'left-half plane on-the-average' ones. In this case, stabilization is achieved without the need for additional information flows provided that the amplitudes and frequencies of the periodic controls are within a critical range.

Assume, for simplicity, that the *ij*th element of the periodic control matrix $D(t)$ is given by $d_{ij}(t) = c_{ij}\cos(\omega_{ij}\,t)$, where c_{ij} is the amplitude and ω_{ij} is the frequency of the oscillation.

In order to describe the average behavior of system (A5), we introduce the parameter ε as

$$\varepsilon \stackrel{\Delta}{=} \max_{ij}\,(1/\omega_{ij})$$

and define

$$\alpha_{ij} \stackrel{\Delta}{=} c_{ij}\,\varepsilon \text{ and } \beta_{ij} \stackrel{\Delta}{=} \omega_{ij}\,\varepsilon$$

so that the *ij*th element of $D(t)$ can be rewritten as $\mathrm{d}_{ij}(t) = (\alpha_{ij}/\varepsilon)\cos(\beta_{ij}\,t/\varepsilon)$.

With this notation, the periodic control matrix $D(t)$ takes the form

$$D(t) = \frac{1}{\varepsilon}\Delta'\left(\frac{t}{\varepsilon}\right),$$

and system (A5) becomes

$$\dot{q} = \left[N + \frac{1}{\varepsilon}B^{-1}D'\left(\frac{t}{\varepsilon}\right)\right]q. \tag{A6}$$

Thus, if the α_{ij} and β_{ij} are assumed to be constant, the amplitudes c_{ij} and the frequencies ω_{ij} of the zero-mean periodic terms $d_{ij}(t)$ are parameterized by a positive ε. It has been proven (Bellman *et al.*, 1985) that there exists an ε_0 = constant > 0, such that for any ε satisfying the inequality $0 < \varepsilon < \varepsilon_0$, the stability properties of system (A6) are defined by the eigenvalues of a constant matrix

$$M = \lim_{T\to\infty} \frac{1}{T}\int_0^T \Phi(\tau)^{-1} N \Phi(\tau)\,\mathrm{d}\tau, \tag{A7}$$

where $\Phi(t)$ is the state transition matrix of

$$\frac{\mathrm{d}q}{\mathrm{d}\tau} = B^{-1}D'(\tau)\,q, \tag{A8}$$

where $\tau = t/\varepsilon$.

Specifically, for sufficiently small ε, system (A6) is asymptotically stable if all the eigenvalues of M have negative real parts. As seen from this result, the elements of matrix M are defined in terms of the elements of matrices N, B^{-1}, 'amplitude/frequency' ratios α_{ij}, and 'frequency/frequency' ratios β_{ij}. Consequently, M provides a link between α_{ij}, β_{ij} and stability of (A5): If α_{ij} and β_{ij} are found which place all the eigenvalues of M in the left-half plane, then there exists an ε such that oscillations with amplitudes α_{ij}/ε and frequencies β_{ij}/ε guarantee asymptotic stability of system (A5). The matrix

$$M' \overset{\Delta}{=} M - N$$

can be thought of as a 'correction' of N induced by oscillations.

References

Anderson, B. & Moore, J. (1981). Time varying feedback laws for decentralized control. *IEEE Trans. Autom. Control*, *AC-26*, **5**, 1133–8.

Barber, M., Patten, B. & Finn, J. (1979). Review and evaluation of input–output analysis for ecological applications. In *Compartmental Analysis of Ecosystem Models*, vol. 10, Statistical Ecology, ed. J. Matis, B. Patten & G. White, pp. 43–72. Fairland, Md: International Cooperative Publishing House.

Bellman, R., Bentsman, J. & Meerkov, S. M. (1985). Stability of fast periodic systems. *IEEE Trans. Autom. Control*, *AC-30*, **3**, 289–91.

Bellman, R., Bentsman, J. & Meerkov, S. M. (1986*a*). Vibrational control of nonlinear systems, vibrational stabilizability. *IEEE Trans. Autom. Control*, *AC-31*, **8**, 710–16.

Bellman, R., Bentsman, J. & Meerkov, S. M. (1986*b*). Vibrational control of nonlinear systems, vibrational controllability and transient behavior. *IEEE Trans. Autom. Control*, *AC-31*, **8**, 717–24.

Bentsman, J. (1987). Vibrational control of a class of nonlinear systems by nonlinear multiplicative vibrations. *IEEE Trans. Autom. Control*, *AC-32*, **9**, 711–16.

Bentsman, J. & Hannon, B. (1987). Cyclic control in ecosystems. *Math. Biosci.*, **87**, 47–62.

Boling, R. & Van Sickle, J. (1975). Control theory in ecosystem management. In *Ecological Analysis and Prediction*, ed. S. Levin, pp. 219–29. Philadelphia: SIAM.

Costanza, R. & Hannon, B. (1990). Multicommodity ecosystem analysis: dealing with apples and oranges in flow and compartmental analysis. In *Complex Ecology: The Part-Whole Relation in Ecosystems*, ed. B. Patten & S. Jørgensen. The Hague: SPB Academic Publishing. (In press.)

Dame, R. (1976). Energy flow in an intertidal oyster population. *Estuarine and Coastal Mar. Sci.*, **4**, 243–53.

Dame, R. (1979). The abundance, diversity and biomass of macrobenthos on North Inlet, South Carolina, intertidal oyster reefs. *Proc. of the Natl Shellfisheries Assoc.*, **69**, 6–10.

Dame, R. & Patten, B. (1981). Analysis of energy flows in an intertidal oyster reef. *Mar. Ecol. Prog. Ser.*, **5**, 115–24.

DeAngelis, D., Post, W. & Travis, C. (1986). *Positive Feedback in Natural Systems.* Biomathematics 15. Berlin: Springer-Verlag.

Finn, J. (1976). Measures of ecosystem structure and function derived from analysis of flows. *J. Theor. Biol.*, **56**, 363–80.

Goh, B. (1979). The usefulness of optimal control theory to ecological problems. In *Theoretical Systems Ecology*, ed. E. Halfon, pp. 385–99. New York: Academic Press.

Hannon, B. (1973). The structure of ecosystems. *J. Theor. Biol.*, **41**, 535–46.

Hannon, B. (1979). Total energy costs in ecosystems. *J. Theor. Biol.*, **80**, 271–93.

Hannon, B. (1985*a*). Ecosystem flow analysis. In *Ecological Theory for Biological Oceanography*, ed. R. Ulanowicz & T. Platt. *Canadian J. of Fish. and Aq. Sci.*, **213**, 97–118.

Hannon, B. (1985*b*). Conditioning the ecosystem. *Math. Biosci.*, **75**, 23–42.

Hannon, B. (1985*c*). Linear dynamic ecosystems. *J. Theor. Biol.*, **116**, 89–98.

Hannon, B. (1986). Ecosystem control theory. *J. Theor. Biol.*, **121**, 417–37.

Herendeen, R. (1981). Energy intensities in ecological and economic systems. *J. Theor. Biol.*, **91**, 607–20.

Kercher, J. (1983). Closed-form solutions to sensitivity equations in the frequency and time domains for linear models of ecosystems. *Ecol. Mod.*, **18**, 209–21.

Levine, S. (1977). Exploitation interactions and the structure of ecosystems. *J. Theor. Biol.*, **69**, 345–55.

Levine, S. (1980). Several measures of trophic structure applicable to complex food webs. *J. Theor. Biol.*, **83**, 195–207.

Lowes, A. & Blackwell, C. (1975). Applications of modern control theory to ecological systems. In *Ecosystems Analysis and Prediction*, ed. S. Levin, pp. 299–305. Philadelphia: SIAM.

Luenberger, D. (1979). *Introduction to Dynamic Systems: Theory, Models, and Application.* New York: Wiley.

Meerkov, S. M. (1980). Principle of vibrational control: theory and applications. *IEEE Trans. Autom. Control*, *AM-25*, **4**, 755–62.

Mulholland, R. & Sims, C. (1976). Control theory and regulation of ecosystems. In *Systems Analysis and Simulation in Ecology*, ed. B. Patten, pp. 373–90. New York: Academic Press.

Olson, J. (1961). Analog computer models for movement of nuclides through ecosystems. In *Radioecology* (*Proc. Natl. Symp. Radioecol., 1st.*), ed. V. Schultz & A. Klement, pp. 121–5. Princeton: Van Nostrand–Reinhold.

O'Neill, R. (1979). A review of linear compartmental analysis in ecosystem science. In *Compartmental Analysis of Ecosystem Models*, ed. J. Matis, B. Patten & G. White, pp. 3–28. Fairland, MD: Internat. Cooperative Pub. House.

Patten, B. C., Bosserman, R., Finn, J. & Cale, W. (1976). Propagation of cause in ecosystems. In *Systems Analysis and Simulation*, vol. 4, ed. B. C. Patten, pp. 457–579. New York: Academic Press.

Takahashi, Y., Rabins, M. & Auslander, D. (1970). *Control and Dynamic Systems*. Reading: Addison-Wesley Publishing.

Trave, L., Tarras, A. & Titli, A. (1985). An application of vibrational control to cancel unstable decentralized fixed modes. *IEEE Trans. Autom. Control*, *AC-30*, **3**, 83–6.

Ulanowicz, R. (1986). *Growth and Development: Ecosystems Phenomenology*. New York: Springer-Verlag.

Varga, R. (1962). *Matrix Iterative Analysis*. Englewood Cliffs: Prentice-Hall.

Wang, S. & Davison, E. (1973). On the stabilization of decentralized control systems. *IEEE Trans. Autom. Control*, *AC-18*, **5**, 473–8.

Wonham, W. (1984). Regulation, feedback and internal models. In *Adaptive Control of Ill-Defined Systems*, ed. O. Selfridge & E. Rissland, pp. 75–88. New York: Plenum Press.

10

Do economics-like principles predict ecosystem behavior under changing resource constraints?

R. HERENDEEN

10.1 Introduction

This article is written in the flavor of my presentation at the symposium 'Search for Ecosystem Principles: A Network Perspective'. I came to that symposium interested in the connections between economics and ecology, and with the goal of making better predictions of how the stocks of specific organisms or groups of organisms in ecosystems respond over time to:

1 changing availability of energy and nutrients,
2 changing cropping or stocking,
3 changing exposure to toxic materials.

These conditions will hereafter be called DRP, for dynamic resource perturbations. The time scale of interest is from days to years for typical exploited ecosystems, as in silviculture or agriculture. I call this the 'managerial' time scale; it is the one used in typical economic planning. (In terms of the ecosystem compartments, the desired time scale is from the turnover time of the fastest compartment of interest to several times that of the slowest compartment of interest.)

My expectation, and hope, was that a whole-system approach, as exemplified by the network perspective and its explicit attention to interdependencies, would give more accurate predictions than approaches that do not incorporate that perspective. This ought to be especially true for a system with tight cycling, the kind favored by proponents of sustainability. Three proposed whole-system optimizing principles seemed applicable. On closer scrutiny, however, I noted two difficulties with the principles. First, in some cases they had not been developed to the point of specific, experimentally testable hypotheses. Second, there was a lack of uniformity regarding their temporal, spatial, and organizational ranges of applicability.

I will discuss the three principles in the context I have sketched above, and present preliminary computer-simulation testing of one of them. Throughout this article I will ask, explicitly and insistently, many more questions than I answer, in my search for usable predictive principles. I will also attempt to maintain the intent of this article's title: the use of 'economics-like' principles. To avoid false optimism, I will state early my conclusion: economics-like principles do not now perform better than others. However, because so little testing has occurred, further investigation is warranted.

10.2 What are economics-like principles?

There are a number of economic analogies used in connection with ecological systems (Rapport & Turner, 1977; Bernstein, 1981; Bloom *et al.*, 1985). A minimum requirement for their application is the existence of system-wide indicators and quantities. Price is an economic example. On the assumption of perfect information and the ability of the market to 'clear' (i.e. for sellers and buyers to agree on mutually acceptable prices, and to carry out transactions), price represents scarcity on a system-wide basis. It results from interdependent action by all system components. The ecological system-wide quantities that will be discussed in this article are exergy (Jørgensen, 1982; Jørgensen & Mejer, 1983), ascendency (Ulanowicz, 1980, 1986), and energy intensity (Hannon, 1979, 1985*a*; Hannon, *et al.*, 1986).

Two economics principles that have been proposed for ecological application are substitution and optimization.

Principle 1. Choice and substitution. System compartments may receive inputs from several others. The principle of substitution states that over some range one input may be substituted for another. In economics, the degree of substitution is determined by relative prices of the inputs and the relative ease with which the substitution can be made.

Principle 2. Optimization using system-wide quantities. System-wide optimization can be used in economics, for example, to determine how much output is appropriate from each industry to achieve maximum system-economic growth. If each industry were instead to maximize its own output, as dictated by its existing capital stock, the overall growth would be less (Dorfman *et al.*, 1958, ch. 11).

This article will be concerned almost exclusively with optimization because there is less background with substitution in whole systems.

Table 10.1. *Optimizing principles in ecology. 'Reflexive' refers to the extent to which whole-system indicators are used to predict behavior of individual compartments.*

Principle	Extent	Temporal scale	Reflexive?
Optimal foraging (Werner & Mittelbach, 1981)	Feeder and food organisms	Hours	No
Predictive principle on managerial time scale (desired goal of present author)	Ecosystem	Days	Yes
Succession (E. P. Odum, 1969)	Ecosystem	Decades	Slightly
Maximum power (H. T. Odum, 1983)	Ecosystem	Milennia	Uncertain

Optimal foraging theory (e.g. Werner & Mittelbach, 1981), which does investigate choice and substitution, applies only to a feeder organism and several food organisms, not to the many interacting compartments in an entire ecosystem.

Optimization has had great predictive success in other fields, notably physics: least energy, least action (in mechanics), maximum entropy (in thermodynamics), minimum time (in optics). Rosen (1967) discusses some of these in a biological context. In biology, natural selection is a prime optimizing principle, subject to careful definitions. Others are listed in Table 10.1. In Table 10.1, reflexivity refers to a possible contradiction: that emergent system-wide properties are used to predict specific compartment traits. The desired goal is a predictive principle applying over a period of days to years for a whole ecosystem. The three principles mentioned in the introduction are all optimizing principles, each involving one of the quantities mentioned above: exergy, ascendency, and energy intensity. These will be discussed in Section 10.2.2.

10.2.1 *Questions of consistency and applicability to ask of any principle*

Having already indicated reservations about the range of applicability of the several optimizing principles, I will emphasize at the outset several questions that must be addressed regarding the whole issue of prediction and the use of any principle. These questions cover system boundary, disaggregation, temporal resolution, and experimental testing. To some extent they are not separable; for example, spatial and temporal considerations often interact. This list is undoubtedly not complete, nor is it predominantly original. (Questions of system identification have been discussed in Halfon (1979) and Hirata & Ulanowicz (1985), for example.) Nonetheless, the issues deserve restatement and continuous attention.

(1) What is the system's organizational level (level of disaggregation)?

The general approach of ecosystem analysis using a network perspective is to break the whole into compartments and then study and exploit how these interact. I have seen little discussion of how many compartments the system should be broken into (but see Chapter 4 by Allen & O'Neill in this volume). Analysts seem to accept the compartmentalization used by the field workers. For example, the oyster reef ecosystem (Dame & Patten, 1981), which has received extensive theoretical treatment (Hannon, 1985*b*; Patten, 1985) and which is the suggested reference system for this volume, is divided into six biotic components based on the experimental difficulties of measurement. Does this constraint based on practical aspects of data acquisition correspond to functional characterization?

A related question is whether any abiotic compartments are included in the optimization, i.e. does optimization apply to a combined living system and its physical environment, as proposed in the Gaia hypothesis (Lovelock, 1979)? Jørgensen & Mejer (1983) and Jørgensen (personal communication, 1986) state that some abiotic components should be included, while, for example, Hannon makes no mention of this in any of his writings. If only biotic components are included, does that include dead biotic material, i.e. detritus? Given how important detritus often is to system structure and dynamics, there is a strong argument for its inclusion. (For example, 7 of the 12 direct energy flows in the 6-compartment oyster reef model of Dame & Patten (1981) involve detritus.)

(2) What are the proper units for measuring stocks and flows?

If the kernel of the network perspective is the interaction between components, what are they exchanging...energy (high or low thermodynamic quality)? nutrients? moisture? shelter? parasite removal? transportation? The pattern of interaction depends critically on the choice of flow variables. For example, a flow pattern measured in terms of energy tends to be pyramidal (chain-like) with few feedback loops, while in terms of nutrients the pattern is less pyramidal (more web-like) with more feedback (Finn, 1980). Metabolizable biomass energy has been the most frequent choice in analyses till now, but there is no proof that use of other flow variables, instead of, or in addition to, energy would not be superior. Ulanowicz (1986) acknowledges this problem, but does not resolve it.

A related question is that of weighting. This arises in two instances. First, stocks and flows for different compartments can be measured in physically different units. For example, one flow might be in kcal/yr and another in grams nitrogen/yr, and yet various sums and other manipulations may require commensurability. This is especially true for two of the three optimands discussed below: exergy requires that a summed system stock be calculated, and ascendency requires summed system output (throughflow).

Second, even if the units are commensurate, there may be valid reasons for weighting some flows more than others, such as to account for indirect effects. For example, for the consumption of 1 kcal of mouse by a hawk to be possible requires a far greater amount of primary production than 1 kcal.

(3) Over what temporal scale does the principle apply?

Given my desired managerial time scale, I would like the principle to apply to a range from as short as the turnover time of the fastest component to as long as several times the turnover time of the slowest. Does the proposed principle cover this range? The greatest difficulty will arise for the short time scale. If not constrained by biological factors such as length of reproductive period or seasonally varying feeding habits, the minimum time interval is finally limited by physical processes such as diffusion. In practice the turnover times will be the result of the aggregation level. Jørgensen & Mejer (1983, p. 23) hypothesize that exergy optimization should apply at '...all temporal...levels of resolution...', while Hannon does not state an explicit time scale. Ulanowicz (personal communication) leans strongly towards evolutionary time scales.

(4) What spatial extent is covered?

Spatial extent covers two aspects. First, it is an indirect indicator of the actual number of organisms and hence expected random statistical variation in measurements. Second, it raises the question of whether spatial variation, transit time effects, etc., are important mechanisms in the dynamics of the system.

(5) How 'evolved' is the system?

It would seem that system-wide principles would be more likely to apply to ecosystems that are older in evolutionary time, and less likely in younger or even arbitrary assemblages.

(6) How stable was the environment assumed to be during the system's evolutionary history?

How do the dynamic resource perturbations compare with those under which the system evolved?

(7) How closed is/was the system to genetic change and to the introduction of new species?

The managerial time scale may be consistent with no genetic change in some applications, and not in others (e.g. insects in agricultural systems).

(8) What reference levels are assumed for a non-living system?

Reference levels are especially important in the use of exergy as an optimand. Roughly stated, exergy is a measure of the extent to which the living system concentrates and orders material relative to the absence of life. It is therefore necessary to define concentrations in the absence of life, the reference levels.

(9) What is the mechanistic description of how the 'message' of optimization is transmitted to the individual compartments?

If optimization leads to good predictions, one hopes that there will be a physical/biological explanation for how the individual compartments, acting more or less independently, satisfy a system-wide optimizing principle. It is possible that identifying the mechanism may lag far behind recognition of the principle's predictive abilities. For example, the existence of the electron was inferred from atomic spectra long before it was 'seen' as a track on an emulsion.

(10) What are specific hypotheses following from the principles, and what are feasible experiments to test them?

The optimizing principles are to be tested against each other and against non-optimizing principles. In the end, validation through experiments on real living systems is necessary. There is also an important role for computer simulation experiments: to translate the principles into specific experimentally testable hypotheses. Ascendency, exergy, and other system-wide indicators cannot be measured directly; the hypotheses must therefore apply to quantities that can. Questions of the latter type include the following: (i) Do the optimizing principles predict different steady-state results (e.g. compartment stock levels) for a perturbed ecosystem, or do predictions differ only in the transition period between steady states? (ii) Which is expected to give the more easily measured result: introducing scarcity through reduced light input or through increased cropping?

For actual field or laboratory experiments, the number of experiments needed to quantify the parameters of the appropriate models should be estimated, along with their expected difficulty. For example, in the input–output (recipient-controlled) framework discussed below, it is necessary to measure both stocks in each compartment and flows between them during calibration, and to measure stocks during the validation phase. Furthermore, because it is likely that validation will involve transient behavior, the measurements must be made frequently.

10.2.2 *Proposed optimands or quantities used in optimands*

Three proposed optimands will be discussed. At the beginning, I stress that optimization principles are stated in the context of the assumed dynamics of the system. Without a sensible physical/biological model, optimization leads to absurdities. For example, if a living system maximizes negentropy (Schrodinger, 1967), why doesn't it produce diamonds, which are perfectly ordered carbon crystals? The answer is that it maximizes negentropy within constraints. A correct statement of an optimizing principle is exemplified by Ulanowicz (1980) in proposing ascendency as an optimand:

> A self-organizing community flow network behaves over an adequate interval of time so as to optimize its ascendency subject to hierarchical, thermodynamic and environmental constraints.

The choice of model, which incorporates constraints, is thus at least as important as that of optimizing principle. In the example presented below, a nonlinear, discrete input–output model is used to determine the *maximum* output[1] allowed by the existing stocks, and hence the maximum potential growth, of each compartment. The problem is that maximum output from all compartments can lead to extinction: the cows, by producing as much bovine biomass as they can, require maximum input and eat all the grass. This may be only a momentary occurrence, but it can destroy the system. If it could be avoided, even briefly, system demise might be avoided. This problem, which is typical of a broad class of models, can occur during times of increased as well as decreased resource availability. I call it the structural imbalance problem. In the model, output less than the maximum is admissable. Optimization may then

[1] 'Output' as used in this article is the same as 'throughflow' as used by Patten (1985) and Ulanowicz (1986). Output is used because of its economics overtones and because it can also be used for a non-conservative flow system, though that is not done here.

choose an appropriate outut level that avoids structural imbalance. When structural imbalance ought to be avoided, and when it should be accepted as reality, depends ultimately on real experimental results.

Fig. 10.1 shows the interaction of optimizing principle and a dynamic model for growth and decline. It is assumed that in the absence of optimization the individual compartments will have maximum output, which can lead to structural imbalance. If the optimization depends on a calculated quantity such as energy intensity, it must be calculated at each step and then used as an input to the optimization process at the next step.

In Table 10.2 are listed two proposed optimands and one quantity (energy intensity) proposed for inclusion in an optimand. These incorporate one or more of these aspects of ecosystem structure:

Fig. 10.1. Example of optimization applied to a dynamic growth model. Optimization may dictate that output can be less than its potential maximum. At steady state it is the maximum. The optimand is calculated at each step.

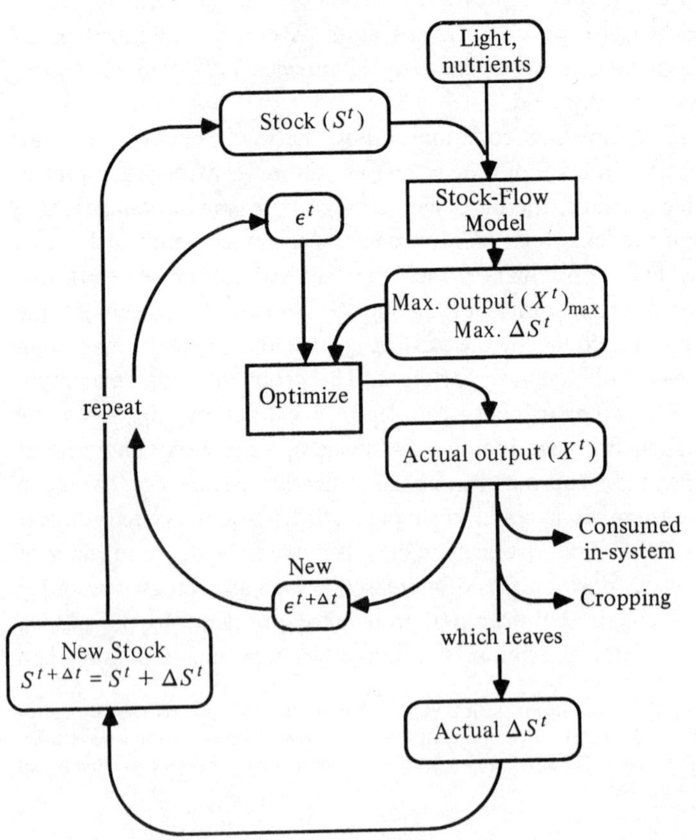

1 measures of system 'orderedness',
2 measures of indirect linkages, such as the dependence of top carnivores on primary producers.

The optimands differ in their degree of explicit dependence on stock, flow, and time. It is useful to compare the relative contributions from overall system size and from internal structure. The structural part of the optimand is related to the structural imbalance problem, which is at the heart of most instabilities. If the structural part is dominated by the part dependent on total size, the optimand will be insensitive to this common cause of instability.

Table 10.2. *Proposed optimands in an* n-*compartment ecosystem. (Energy intensity is not itself an optimand, but is a component of one.)*

	Explicit dependence on: *Stocks*	*Flows*	*Time*
Exergy (Jørgensen, 1982) $EX = R\theta \sum_{i=1}^{n} C_i \ln \frac{C_i}{C_{i,\mathrm{ref}}}$	Yes	No	No
Ascendency (Ulanowicz, 1980, 1986) $A = T \sum_{k=0}^{n+2} \sum_{j=0}^{n+2} \frac{X_{kj}}{T} \ln \frac{X_{kj}/X_k}{\sum_{i=0}^{n+2} X_{ij}/T}$	No	Yes	No
Energy intensity (Hannon, 1973; present author) $\varepsilon^{t+\Delta t} = (E^t\Delta t + \varepsilon^t \hat{S}^t)(\hat{X}^t\Delta t - X^t\Delta t + \hat{S}^t)^{-1}$	Yes	Yes	Yes

For exergy:
R = gas constant; θ = absolute temperature; C_i = concentration of stock in compartment i; $C_{i,\mathrm{ref}}$ = reference concentration of stock in compartment i.
For energy intensity:
X_{ij} = flow from compartment i to j; E = vector of energy inputs; X = matrix of flows; $\hat{X}$ = diagonal matrix of total outputs ($X_i = \sum_{j=1}^{n} X_{ij} + Y_i$, where Y_i = net output of compartment i); t = time; Δt = time step; $\hat{S}$ = diagonal matrix of stocks.
For ascendency:
In the sums from $j = 0$ to $n+2$, for $j = 1$ to n the X's are flows, defined as for energy intensity. $j = n+1$, $n+2$ refer to respiration and exports and hence $\sum_{j=n+1}^{n+2} X_{kj} = Y_j$. For $i = 1$ through n, X_j has the same meaning as for calculating energy intensity. $j = 0$ refers to system inputs (e.g. energy). T is then $\sum_{j=1}^{n} X_j$ + (system inputs).

The first optimand, *exergy*, proposed by Jørgensen (1982, 1986) and Jørgensen & Mejer (1983), depends only on compartment stocks. It is a measure of the orderedness, or concentration, of the stocks relative to a life-free reference level. The expression for exergy can be obtained by both thermodynamic and information-theoretic arguments. Exergy per liter (EX) as given by Jørgensen & Mejer (1983) is

$$EX = R\theta \sum_{i=1}^{n} C_i \ln (C_i/C_{i,\mathrm{ref}}), \tag{1}$$

where

R = the gas constant (units = kcal/mole-deg)

θ = absolute temperature of the ecosystem[2]
C_i = concentration of compartment i's stock (units = mole/l)
$C_{i,\mathrm{ref}}$ = reference concentration of compartment i's stock

n = number of compartments

Because of the logarithmic form of eq. (1), exergy can easily be written as the sum of a size term representing the overall concentration of the system's stock above that of the reference level, and a structural term representing the distribution of stocks among the compartments as compared with the distribution of the reference level:

$$EX = R\theta C_o[\ln (C_o/C_{o,\mathrm{ref}}) + \sum_{i=1}^{n} x_i \ln (x_i/x_{i,\mathrm{ref}})], \tag{2}$$

where

C_o = summed concentrations of compartments

$x_i = C_i/C_o$

$x_{i,\mathrm{ref}} = C_{i,\mathrm{ref}}/C_{o,\mathrm{ref}}$

The choice of reference levels is important to the relative contributions of the size and structural components.

A potential criticism of exergy is that because it is based only on stocks explicitly, it ignores the richness of interdependency exhibited by flows. It appears that no communication is required between compartments at all:

[2] Assume an aquatic ecosystem; the temperature is uniform throughout.

the system could be dead! This is not a serious objection, however, because the necessary accompanying dynamic model must incorporate stock-flow connections (it will also incorporate time dependence). For example, it must incorporate constraints on the minimum number of compartments. Otherwise it would seem that a system could achieve greater orderedness by the simplifying action of allowing some compartments to go to extinction. The point is that the connection is implicit through the dynamic model but not explicit in the definition of the optimand. Exergy is defined at one instant in time.

The second optimand, *ascendency*, (Ulanowicz, 1980) is also a measure of orderedness, but unlike exergy, it is defined exclusively in terms of flows. It uses direct flows only; there is no explicit accounting of indirect effects, such as the connection of autotrophs and carnivores via herbivores. Ulanowicz (1980) states that ascendency:

> ...perhaps...is best described as the coherence of the flow network, i.e. an indicator of the degree to which the flow system differs from either a homogeneous network or a collection of totally independent parts...[It] can also be interpreted as the average degree of unambiguity with which an arbitrary compartment communicates with any other compartment in the system.

The expression for ascendency is in Table 10.2. Indirectness is accounted for implicitly through the dynamic model. In contrast to exergy, however, the size and structural terms are not added, but multiplied. Ascendency is also defined at a single instant of time.

The third quantity, *energy intensity*, explicitly combines both stocks and flows. It does not explicitly include orderedness, but instead traces indirect effects. The expression for energy intensity is in Table 10.2. The energy intensity of the output of compartment j is the total amount of energy that must be fixed (in primary production) by the ecosystem to allow production of one unit of that output (Hannon, 1973, 1979; Bullard & Herendeen, 1975, Herendeen, 1981). As can be seen in Table 10.2, energy intensity explicitly incorporates time; energy intensity at any time depends on its value one time step before. Energy intensity is a normalized quantity, incorporating structure but not overall size. The proposed optimand introduces size by multiplying energy intensity by system *net output* (Hannon, 1979, 1985*a*).

10.3 Experiments, real and computer

My previous (unpublished) work has indicated that at steady state, output levels will be at the maximum allowed by the existing stock. Optimization will give results that differ from those of non-optimizing approaches only during transient response to DRP. Experiments must therefore cover dynamic situations.

Assuming that the dynamic model is calibrated (a formidable experimental task), we look for differences in stock levels over time for the two or more competing predictive models. Such differences will likely be very difficult to measure. If, on the other hand, one theory predicts system collapse while the other does not, the difference is easier to measure.

10.3.1 *What has been done so far to test these optimizing principles?*

To date there has been no explicit experimental test of these optimizing principles. One optimand has been investigated at a preliminary level in computer simulation. It is (Hannon, 1979, 1985*a*):

$$\sum_{i=1}^{n} \varepsilon_i^t Y_i^t,$$

where

ε_i^t = energy intensity of compartment i at time t,

Y_i^t = net output of compartment i at time t.

Net output is the sum of cropping and stock change (growth or decline)[3]. Equivalently, it is the surplus remaining after consumption by other compartments and self-use have been subtracted from the output of a compartment.

The rationale for using the above optimand can be sketched in four steps. The steps exploit analogies, so that this is not a logical proof, but a plausibility argument.

1. Energy intensity is system-wide (i.e. it depends on stocks and flows in all compartments) and varies significantly among compartments and over time in response to DRP.
2. Energy intensity has strong and suggestive analogies to price in economic systems.
3. Maximization of the sum of (price) × (net output) is a useful predictive tool in certain economic models.

[3] Under the assumption that cropping is not anticipated, use of this optimand is equivalent to maximizing growth weighted by energy intensity. The degree to which past experience should be anticipated is discussed by Rosen (1985).

4 Therefore, maximization of the sum of (energy intensity) × (net output) is useful in certain ecological models, such as discussed below.

10.3.1i *Dynamic energy intensity*

In this section I derive an expression for dynamic energy intensity, discuss its analogy to price, and demonstrate its variability under DRP. The starting point is the familiar balance diagram for embodied energy in a steady state system, Fig. 10.2 (Bullard & Herendeen, 1975). At steady state there is no reference to stock; energy intensity is totally flow-based. While the flows are often measured in terms of energy, they need not be. They can even be expressed in different units for different compartments. To emphasize this point flows are expressed in units of 'gloof'/day. Fig. 10.2 represents a balance of embodied energy for each compartment, in which the energy embodied in the output is equal to that embodied in the inputs plus the actual energy input, if there is any (as for autotrophs). Energy embodied in a flow is not necessarily actual consumable energy; it is the amount of energy that must be consumed in the entire system for that flow to occur. The balance for compartment j is expressed mathematically as:

$$\sum_{i=1}^{n} \varepsilon_i X_{ij} + E_j = \varepsilon_j X_j, \tag{3}$$

where

X_{ij} = flow from compartment i to compartment j (units = gloof/day)

X_j = total output (or throughflow) of compartment j (units = gloof/day)

E_j = energy input to compartment j (units = kcal/day),

Fig. 10.2. Steady state embodied energy balance.

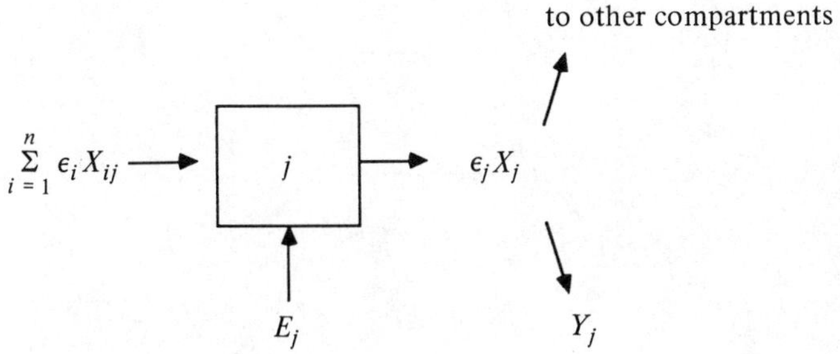

$\varepsilon_j =$ energy intensity of output of compartment j (units = kcal/gloof),

$Y_j =$ net output of compartment $j = X_j - \sum_{i=1}^{n} X_{ji}$ (units = gloof/day).

The solution to the resulting set of n balance equations, in vector notation, is

$$\varepsilon = \mathbf{E}(\hat{X} - X)^{-1}, \tag{4}$$

where

$X =$ matrix of flows,

$\hat{X} =$ diagonal matrix of total outputs,

$\mathbf{E} =$ vector of energy inputs.

By summing all energy into the entire system we arrive at an overall balance:

$$E = \varepsilon \mathbf{Y}, \tag{5}$$

where

$$E = \sum_{i=1}^{n} E_i \text{ (units = kcal/day).}$$

Eq. (5) states that at steady state, all energy input to the system is embodied in the flows to net output.

In the dynamic case, stock must appear explicitly in the energy balance equation, Fig. 10.3. The existing stock is considered an input to the production process and an output of that process, and its energy intensity is updated each time step. This is required for consistency: the stock 'next

Fig. 10.3. Dynamic embodied energy balance.

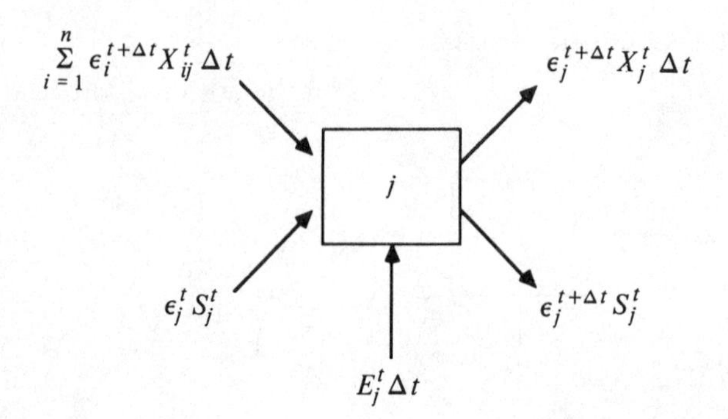

time' is the sum of the stock 'now' and the change in stock. Stock change is contained in output (it is what is left over after consumption by other compartments and by cropping) and therefore has the energy intensity of that output. If the energy intensity of the existing stock did not change, it would have an intensity different from that of the change in stock. Soon different portions of the stock in compartment j would have different energy intensities, which violates the assumption of homogeneity within each compartment. (The compartments are aggregated so as to average out age class effects. If these effects are important, additional compartments should be used.) The presence of the discrete time step Δt is necessary to make flows and stocks dimensionally commensurate[4].

Solving the balance equation pictured in Fig. 10.3 gives for the dynamic energy intensity:

$$\varepsilon^{t+\Delta t} = (\mathbf{E}^t \Delta t + \varepsilon^t \hat{S}^t)(\hat{X}^t \Delta t - X^t \Delta t + \hat{S}^t)^{-1}, \quad (6)$$

where, in addition to the quantities already defined, the superscript refers to the time, and $\hat{S}^t$ = diagonal matrix of stocks at time t (units = gloof).

Equation (6) explicitly shows the time-dynamic nature of energy intensity: its present value depends on its previous value. At steady state, eq. (6) reduces to eq. (4). Note that eq. (6) is not in normalized form; the stocks and flows appear separately, rather than merely as ratios. Normalizing is avoided to include the possibility that the output may be less than its possible maximum, as can occur as a consequence of some optimizing approaches. Even in this case Fig. 10.3 and eq. (6) still apply.

The overall system energy balance equation is now

$$E^t \Delta t = \varepsilon^{t+\Delta t} \mathbf{Y}^t \Delta t + (\varepsilon^{t+\Delta t} - \varepsilon^t) \mathbf{S}^t. \quad (7)$$

Now the energy into the system is allocated to net output plus an adjustment to the energy intensity of the existing stock. At steady state eq. (7) reduces to eq. (5). If net output is assumed to be the sum of basal metabolism plus cropping (exports) and change of stock, eq. (7) can be rewritten as

$$E^t \Delta t = \varepsilon^{t+\Delta t}(\mathbf{BASAL}^t + \mathbf{CROPPING}^t)\Delta t + \varepsilon^{t+\Delta t}\mathbf{S}^{t+\Delta t} - \varepsilon^t \mathbf{S}^t.$$

[4] Discrete rather than continuous time analysis is used for two reasons: (i) A non-zero minimum time interval is consistent with the idea of a compartmentalized ecosystem. It is assumed that certain age class effects, etc., can be averaged, which implies that there is a minimum time interval below which we must not go in analysis and interpretation. (ii) Even if continuous analysis is used, in practice a discrete approach will almost surely be needed to solve the resulting differential equations. Hannon (1985*b*) uses a continuous approach.

Table 10.3. *Energy flows and stocks in a 4-compartment model (Logofet & Alexandrov, 1983). Units: flows,* $kcal \cdot m^{-2} \cdot yr^{-1}$; *stocks,* $kcal \cdot m^{-2}$. *Small corrections have been made to balance the table. In the initial steady state shown here, dissipation is arbitrarily split* 1:1 *between search energy (dependent on output and on relative scarcity of inputs) and basal metabolism (proportional to stock). Flow is from row to column; for example, animals ingest* 38.05 $kcal \cdot m^{-2} \cdot yr^{-1}$ *of plants.*

	Plants	Animals	Decomp.	Detrit.	Dissipation (Search/Basal)	Cropping	Stock Change	λ Output	Stock
Plants	0	38.05	0	337.4	304/304	4.1	0	987.55	8490
Animals	0	0	0	58.21	8.37/8.37	0	0	74.95	1.25
Decomposers	0	0	0	305	134.95/134.95	0	0	584.9	35
Detritus	0	36.9	584.9	0	0/0	78.81	0	700.61	8836
Energy input	987.55	0	0	0					

Now the energy is allocated to system basal metabolism and cropping plus the difference in embodied energy of the initial stock (at the present energy intensity) and the new stock (at the new energy intensity).

To demonstrate how energy intensity varies under DRP, we study a specific system, the 4-level bog of Logofet & Alexandrov (1983). This work presents steady state stock and flow data, in terms of metabolizable biomass (Table 10.3). Logofet & Alexandrov's steady state data form a starting point for an assumed dynamic model based on recipient control. Linear recipient control, with constant coefficients relating inputs to outputs, is inexorably unstable. There are several proposed approaches to impart some degree of stability (not complete stability under *all* DRP,

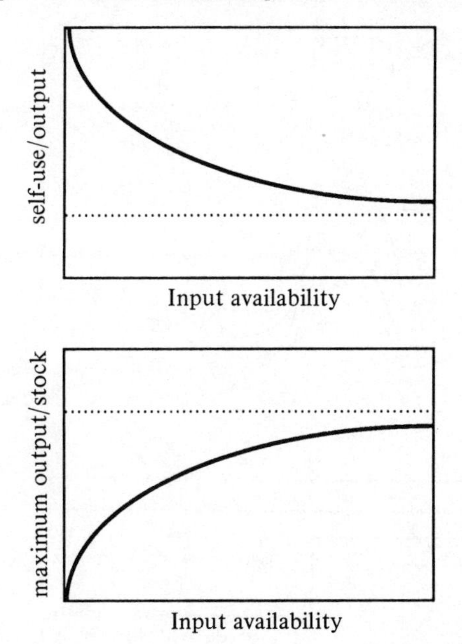

Fig. 10.4. Nonlinear relationships used to achieve stability.

An example of input availability for compartment j:

$$\left[\frac{S_1/S_j}{S(0)_1/S(0)_j} \bullet \frac{S_2/S_j}{S(0)_2/S(0)_j} \bullet \cdots\cdots\cdots \frac{L/S_j}{L(0)/S(0)_j}\right]^{\frac{1}{n}}$$

L = light level (kcal/day); '(0)' refers to stock and light levels at an initial steady state, at which availability = 1.0; n = number of inputs, including light if appropriate; the light term is included only if j = autotroph.

which is just as undesirable as complete instability). I will use two nonlinear relationships that are biologically reasonable and that stabilize the system adequately for the DRP used here. Both incorporate a density dependent response of individual compartments to scarcity of their inputs (Fig. 10.4); energy intensity will then express overall scarcity.

The first relationship expresses the necessity of an organism to dissipate more energy as a fraction of total output in obtaining inputs when they are scarcer. (This 'search' energy is in addition to a maintenance (basal)

Fig. 10.5*a*. For caption see opposite.

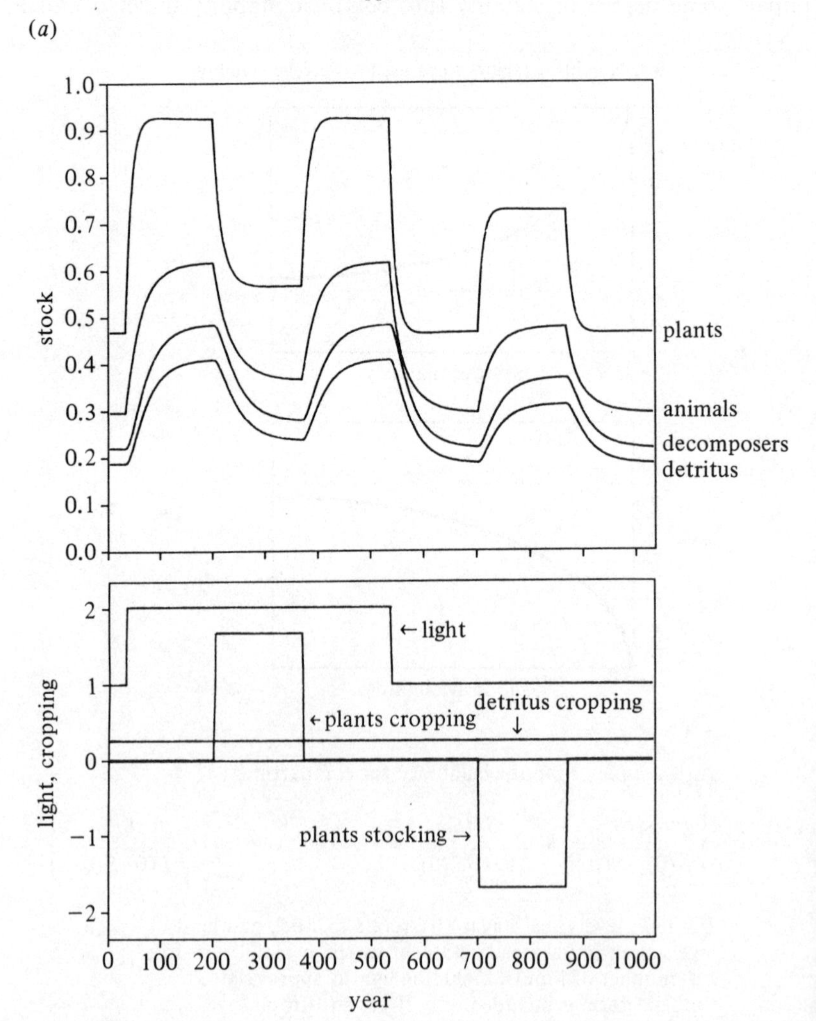

Fig. 10.5. Stocks and energy intensities in dynamic model ecosystem. Response of 4-compartment bog system to changes in light availability and changes in cropping/stocking of plants. System is initially at steady state as given in Table 10.1. Light is doubled at year 33 and returned to initial level at year 533. Cropping of plants is increased to 500 kcal/m²·day at year 200, returned to original level at year 367, reduced to −500 kcal/m²·day (i.e. system is stocked) at year 700, and finally returned to original level at year 867. (*a*) Response of stocks. (*b*) Response of energy intensities. All graphs are arbitrarily normalized for appearance. Actual initial stocks are given in Table 10.3. Initial energy intensities: plants, 1.45; animals, 1.94; decomposers 2.63; detritus, 2.00 (all in kcal/kcal). Time step used in simulation: 0.033 yr.

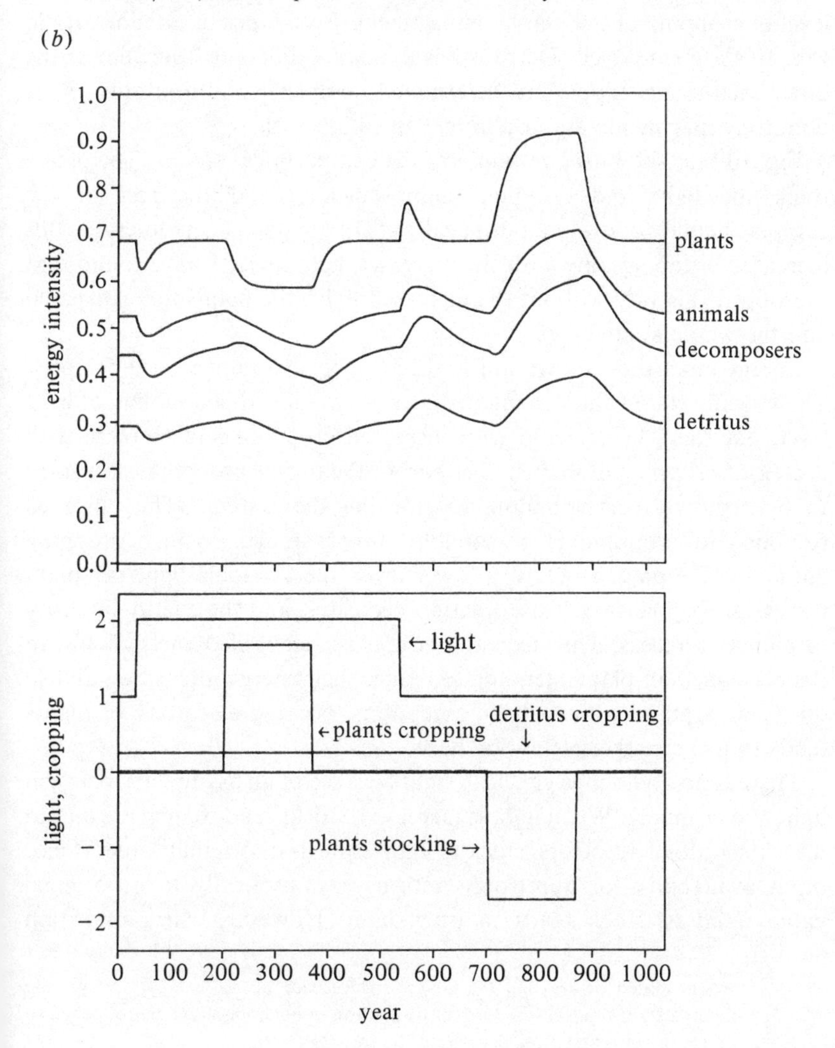

energy requirement, which is assumed proportional to stock, independent of output.) The second expresses the assumption that when inputs are scarce a compartment stock will not be able to produce as much output, and hence consume as much input, independent of how much of that output is self-use[5].

Energy intensity is a derived quantity, an abstraction imposed by us, and without optimization the dynamics of the system are not explicitly dependent on it, by assumption. Fig. 10.5 shows the time behavior of the stocks and energy intensities in the 4-level bog system in response to a DRP regime that includes step-function changes in light intensity and level of cropping of the plant compartment. Both types of nonlinearity in Fig. 10.4 are employed. Detritus has dynamics different from that of the other sectors; its inputs are determined completely by mortality of its donor compartments and are unrelated to its stock.

Fig. 10.5(*a*) shows the response of the compartment stocks. The plants track the light and cropping monotonically, and the rest of the compartments follow monotonically. The system can tolerate this increased cropping only with the increased light level; I have found that if cropped this way without the increased light, the plants are extirpated and the whole system dies.

Energy intensities, shown in Fig. 10.5(*b*), behave somewhat differently. Their steady state values change little or not at all with a doubling of light level, but they do decrease with increased cropping and increase with decreased cropping of plants. (For part of the regime cropping is assumed to be negative, corresponding to stocking the system.) The observed response to cropping is reasonable: for example, positive cropping maintains the plant stock at a lower level, the available light per plant increases, the plants' self-use fraction decreases, and the energy intensity of plants decreases. This decrease reduces the embodied energy input to the consumers of plants, tending to reduce their energy intensities, and so on. (This argument could have exceptions, because a scarcity of plants tends to increase the self-use by herbivores.)

There is no analogous effect for light because of an asymmetry between light and cropping. When light is increased, stocks tend to increase and to drive the ratio of autotroph stock to light towards its original value. Hence input availability for autotrophs returns asymptotically to its original value. That is, stock tends to track light. However, when autotroph

[5] It was stated before that the flow variables need not be energy, and yet the identification of a self-use term as dissipation in obtaining food seems locked to an energy interpretation. This seems inescapable.

cropping is increased, autotroph numbers are depressed but the light does not follow, and input availability remains greater.

On the other hand, the steady state energy intensity scarcely changes with the light level, which follows from the fact that the original cropping is a small fraction of the total output, so that cropping plays a minor role in the final balance of production and consumptive uses. In fact, for a materially closed system (no cropping or stocking) represented by the nonlinear model used here, all steady states will have the same energy intensities.

What is most interesting is the *transient* behavior of the energy intensities when the light level is abruptly changed; the energy intensity undergoes a large change in a direction opposed to its final change at the new steady state. These transients occur because of a delay in the above-mentioned tendency of stocks to track the light. The delay is a consequence of the interplay of stock and output for autotrophs, and of all compartments. Suddenly doubling a compartment's output does not suddenly double its stock. For example, for plants in Table 10.3 the ratio of stock to flow is 8.6 years, so that a doubling of output would take at least that long to double the stock. Actually it would take much longer because not all of the doubled production would go to stock increase; much would be consumed by animals, etc., whose own stocks would also be increasing.

These 'contrary' transients are thus expected for light changes, but not for changes in cropping. Furthermore, the size of the transient can be rather large. Thus energy intensity shows variability that is qualitatively different from that shown by stocks at times of abrupt DRP.

10.3.1ii *Comparison of energy intensity and price*

I now consider the analogy between energy intensity in ecosystems and price in input–output economics (which is recipient controlled). First, price in input–output, as derived by Leontief (1970) and slightly modified by me after personal correspondence with Leontief, is given by:

$$\mathbf{p}^{t+\Delta t} = (\mathbf{VA}^t\Delta t + \mathbf{p}^t(1+r\Delta t)\,\hat{S}^t)\,(\hat{X}^t\Delta t - X^t\Delta t + \hat{S}^t)^{-1} \tag{8}$$

where the quantities not already defined are

$\mathbf{VA}^t$ = vector of added time t
(units = value/day),

$\mathbf{p}^t$ = vector of prices at time t (units = value/gloof),

r = interest rate (units = day^{-1}).

Price can be considered a weighting factor that allows the 'value added' in a producing sector, i.e. a compartment, to equal the difference between output weighted by its price and the sum of inputs each weighted by their prices. If the value added is specified, then the prices are determined, assuming that all sectors agree on what the prices should be, i.e. that the 'market clears'. Thus price follows from value. The interest rate r in eq. (8) is included to account for the revealed tendency of humans to expect interest payments for deferring consumption. In eq. (8) $r\Delta t$ is the fraction of the stock's (capital stock in economic terms) value that would be paid to interest in time step Δt if money were borrowed to purchase it. Alternatively, it can be considered the interest 'lost' if the firm (the economic analog of compartment) spent it on the capital stock instead of lending it out.

For comparison, I repeat the energy intensity expression:

$$\varepsilon^{t+\Delta t} = (\mathbf{E}^t \Delta t + \varepsilon^t \hat{S}^t)(\hat{X}^t \Delta t - X^t \Delta t \hat{S}^t)^{-1}. \tag{6}$$

Equations (6) and (8) are remarkably similar. Introduction of an interest rate into the energy intensity equation is discussed by Hannon (1982, 1985*b*). Interest is apparently related to dissipation, but there is no clear evidence, yet, of its significance. (Further investigation of the role of interest rate is desirable.) If value added is analogous to energy input, then energy intensity is analogous to price. Value added can also be defined in terms of 'non-produced input'. In economic systems, labor is sometimes defined as a non-produced input, and then prices are all in terms of person-hours per unit of output. In a materially closed ecosystem, the only non-produced input is free energy. If the energy is the value added, then the energy intensity is the price.

In constant coefficient (linear) input–output economics, maximization of

$$\sum_{i=1}^{n} p_i Y_i$$

leads to maximum stable growth. Without this maximization, the growth is unstable and eventually leads to destruction of one sector and hence the entire economy (Dorfman *et al.*, 1958)[6]. No mention is made of decline in this reference, but a similar conclusion should apply; that is, the use of

[6] In a typical economic input–output model, every producing sector must have at least one input from another producing sector and send some output to another producing sector; otherwise it is not an *input–output* sector at all. By the assumed linearity, a sector cannot produce if any input cannot be obtained.

such an optimand will stabilize growth and decline. Therefore the last step in the analogies is to conjecture that in an ecosystem maximization of

$$\sum_{i=1}^{n} \varepsilon_i Y_i \qquad (9)$$

tends to induce a desired stability for a modeled ecosystem under DRP. How much stability ought to be built in is an unsettled question, but *some* is necessary.

Equation (9) is one optimand, originally proposed (1979) and extended (1985*a*) by Hannon. In words, it is the sum of the net outputs weighted by the energy intensities. At steady state this is, according to eq. (5), equal to the energy absorbed (and then dissipated or exported) by the system, which makes this optimization sound like H. T. Odum's Maximum Power Principle (Odum & Pinkerton, 1958; Odum, 1983), even though that principle is loosely stated (Costanza & Herendeen, 1984). However, in the dynamic case this equality does not hold, as some of the energy is being allocated to changing the energy intensity of the stock, as shown by eq. (7).

10.3.2 *Preliminary and limited simulation testing of an optimizing principle*

I have tested eq. (9) as a maximand (i.e. maximized optimand) for a linear and several nonlinear recipient control models of ecosystems containing from two to six compartments, with particular attention to the response to abrupt decreases in light level. Given that there is probably little agreement about whether increased or decreased stability is a desired result, the tentative criterion for judging desirability of this maximand is the demonstration of a consistent difference in stability between optimizing and non-optimizing approaches.

An example in which optimization increases stability is given in Fig. 10.6. There, without optimization, a simple hypothetical 2-compartment system dies when the light is abruptly reduced at day 400. With maximization of the expression given in eq. (9), the system survives. This is accomplished by the consumer compartment having less than maximum output for 19 time steps, as dictated by the optimizing principle, in the transition period, day 400–day 450. Optimization thus gives a different result for just these 19 time steps, but by so doing prevents a system collapse[7].

[7] This is a rather special case. First, only the first type of nonlinearity in Fig. 10.4 is used, that of making self-use dependent on input availability. Second, a non-

The result in Fig. 10.6 is not typical: sometimes optimization increases stability; sometimes it reduces it. At this point no clear trend has emerged, but it is too early to abandon the search. There remains the opportunity to do a coherent computer-experiment test of a factorial combination of optimands, models, DRP regimes, and criteria for predictive success.

This is not an invitation to an indiscriminate computer exercise. Rather, it is an invitation to a carefully prepared simulation test, one result of which can be an hypothesis couched in terms directly amenable to experimentation. I urge that proposers of the optimization principles and potential users measure their ideas and plans against questions of the type

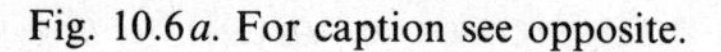
Fig. 10.6*a*. For caption see opposite.

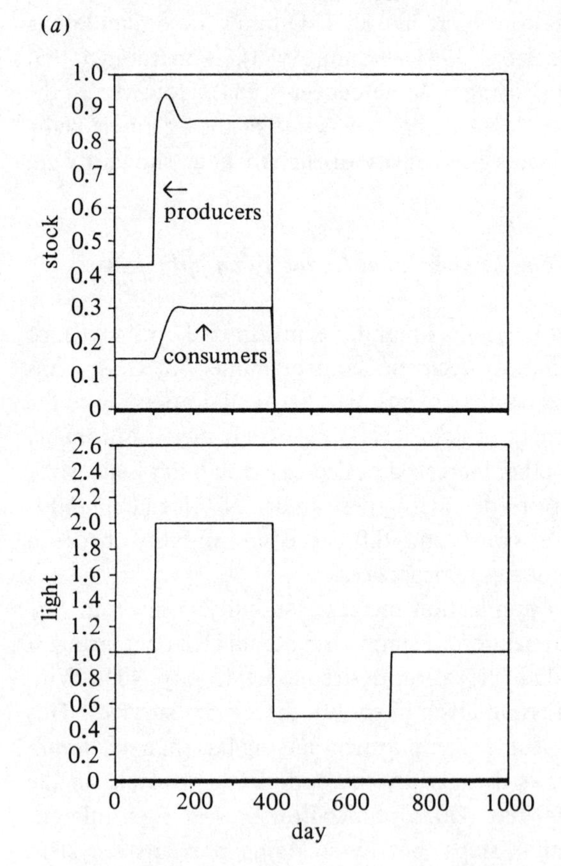

zero interest rate is used in the energy intensity because the difference between optimizing and non-optimizing does not occur for a rate of zero in this example. I have not tested the role of interest rate enough to comment further.

in Section 10.2.1. If the questions can be answered satisfactorily, then the testing can proceed in earnest.

10.4 Summary

Three optimizing principles that have been forwarded as useful in predicting ecosystem behavior over time have been compared for their

Fig. 10.6. One example of optimization increasing the stability of a hypothetical materially closed 2-compartment ecosystem. Light level is doubled at day 100, reduced to half its original value at day 400, and returned to its original value at day 700. (*a*) Non-optimization: system dies when light is reduced; consumers completely deplete producers. (*b*) Optimization: sum of net outputs weighted by energy intensities is maximized at each time step. System survives by one compartment, usually consumers, reducing output below the maximum possible with the existing stock. This occurs 19 times in the transition period following the reduction of light, days 400–450, and at no other time. Time step used in simulation: 1 day. Graphs are arbitrarily normalized for appearance.

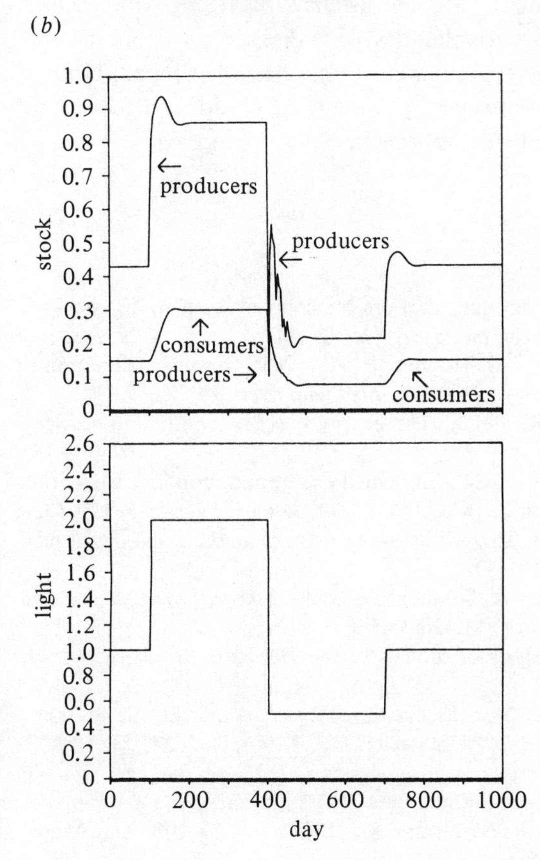

theoretical structure, range of spatial and temporal applicability, and level of verification. The principles, which all involve a system-wide quantity, i.e. one which depends on stocks and/or flows over all compartments, are:

1. maximization of system exergy, which depends explicitly on the orderedness of compartmental stocks,
2. maximization of system ascendency, which depends explicitly on the orderedness of intercompartmental flows,
3. maximization of net output weighted by energy intensities, the latter depending explicitly on both flows and stocks in dynamic systems.

Experimental verification of these principles is non-existent; even translation of the principles to testable hypotheses has in general not been done. The different principles are most likely to lead to different predictions in dynamic rather than steady state conditions. I have proposed simulations to compare predictions in the dynamic regime, after all principles are scrutinized for consistency regarding time scales, appropriate flow and stock variables, degree of disaggregation, and so on. An expression for dynamic energy intensity has been developed. Using it in principle 3 does not, however, yield any clear advantage of that principle over a non-optimizing approach.

References

Bernstein, B. J. (1981). Ecology and economics: complex systems in changing environments. *Ann. Rev. Ecol. and Syst.*, **12**, 309–30.

Bloom, A. J., Chapin, III, F. S. & Mooney, H. A. (1985). Resource limitation in plants – an economic analogy. *Ann. Rev. Ecol. and Syst.*, **16**, 363–92.

Bullard, C. & Herendeen, R. (1975). The energy costs of goods and services. *Energy Policy*, **3**, 268–78.

Costanza, R. & Herendeen, R. (1984). Embodied energy and economic evaluation in the United States economy: 1963, 1967, 1972. *Resour. Energy*, **6**, 129–63.

Dame, R. & Patten, B. (1981). Analysis of energy flows in an intertidal oyster reef. *Mar. Ecol. Prog. Ser.*, **5**, 363–80.

Dorfman, R., Samuelson, P. & Solow, R. (1958). *Linear Programming and Economic Analysis* New York: McGraw-Hill.

Finn, J. T. (1980). Flow analysis of models of the Hubbard Brook ecosystem. *Ecology*, **6**, 562–71.

Halfon, E. (1979). *Theoretical Systems Ecology*. New York: Academic Press.

Hannon, B. (1973). The structure of ecosystems. *J. Theor. Biol.*, **41**, 535–46.

Hannon, B. (1979). Total energy cost in ecosystems. *J. Theor. Biol.*, **80**, 271–93.

Hannon, B. (1982). Energy discounting. Pages 73–100. In *Energetics and Systems*, ed. W. Mitsch, R. Ragade, R. Bosserman & J. Dillon, pp. 73–100. Ann Arbor: Ann Arbor Science Publishers.

Hannon, B. (1985*a*). Ecosystem flow analysis. *Can. Bull. of Fisheries and Aq. Sci.*, **213**, 97–118.

Hannon, B. (1985*b*). Linear dynamic ecosystems. *J. Theor. Biol.*, **116**, 89–110.

Hannon, B., Costanza, R. & Herendeen, R. (1986). Measures of energy cost and value in ecosystems. *J. Environ. Econ. Manage.*, **13**, 391–401.

Herendeen, R. (1981). Energy intensity in ecological and economic systems. *J. Theor. Biol.*, **91**, 607–20.

Hirata, H. & Ulanowicz, R. (1985). Information theoretical analysis of the aggregation and hierarchical structural of ecological networks. *J. Theor. Biol.*, **116**, 321–41.

Jørgensen, S. E. (1982). Exergy and buffering capacity in ecological systems. In *Energetics and Systems*, ed. W. Mitsch, R. Ragade, R. Bosserman & J. Dillon, pp. 61–72. Ann Arbor: Ann Arbor Science.

Jørgensen, S. E. (1986). Structural dynamic model. *Ecol. Mod.*, **31**, 1–9.

Jørgensen, S. E. & Mejer, H. (1983). Trends in ecological modelling. *Analysis of Ecological Systems: State of the Art in Ecological Modelling*, **5**, ed. W. Lauenroth, G. Skogerboe & M. Flug, pp. 61–72. New York: Elsevier.

Leontief, W. (1970). The dynamic inverse. In *Contributions to Input–Output Analysis*, ed. A. Carter & A. Brody, pp. 17–46. New York: Elsevier.

Logofet, D. O. & Alexandrov, G. A. (1983). Modelling of matter cycle in a mesotrophic bog. I. Linear analysis of carbon environs. *Ecol. Mod.*, **21**, 247–58.

Lovelock, J. E. (1979). *Gaia, A New Look at Life on Earth.* Oxford: Oxford University Press.

Odum, E. P. (1969). The strategy of ecosystem development. *Science*, **164**, 262–70.

Odum, H. T. (1983). *Systems Ecology*. New York: Wiley.

Odum, H. T. & Pinkerton, R. C. (1958). Time's speed regulator: the optimum efficiency for maximum power. *Am. Sci.*, **43**, 331–343.

Patten, B. C. (1985). Energy cycling in the ecosystem. *Ecol. Mod.*, **28**, 1–71.

Rapport, D. & Turner, J. (1977). Economic models in ecology. *Science*, **195**, 367–73.

Rosen, R. (1967). *Optimality Principles in Biology*. London: Butterworths.

Rosen, R. (1985). *Anticipatory Systems*. Oxford: Pergamon Press.

Schrodinger, E. (1967). *What is Life*? Cambridge: Cambridge University Press.

Ulanowicz, R. E. (1980). An hypothesis on the development of natural communities. *Ecol. Mod.*, **85**, 223–45.

Ulanowicz, R. (1986). *Growth and Development: Ecosystems Phenomenology*. New York: Springer.

Werner, E. E. & Mittelbach, G. G. (1981). Optimal foraging: field tests of diet choice and habitat switching. *Am. Zool.*, **21**, 813–29.

Concluding Remarks

Network ecology: indirect determination of the life-environment relationship in ecosystems

BERNARD C. PATTEN

C.1 Introduction

Established traditions of thought in science fragment reality; things in the real world are objectified in the brain as autonomous, distinct entities, inherently separate and disconnected from other things outside which constitute their environment. Objects are in turn made up of smaller things within them, that science in its best (analytical) mode decomposes into the cells, compounds, atoms and elementary particles of biology, chemistry and physics. When the reductionist work is done, understanding, prediction and control are all supposed to result. But, as we know, the world doesn't necessarily work that way; usually, treating parts of reality in local isolation leads to unexpected, unexplained and often unwanted consequences in extended spheres; these are variously recognized as 'side', 'secondary' or 'indirect' effects, or when system synthesis is involved, 'emergent' properties. The general inability of science to build the properties of fragments into reasonable replicas of the properties of wholes is opening the entire reductionist–mechanistic paradigm to question, as recognition of the need to treat directly wholes, not parts, grows in various fields.

In physics, the wave-particle duality implies two fundamentally irreconcilable world-views, and both relativity and quantum mechanics are pointing presently to a model of 'unbroken wholeness of the universe, rather than analysis into independent parts' (Bohm, 1980, p. xv). Network ecology takes an intermediate ground; it does not abandon objects, with which all ecologists are forced to work. But it does take a step in the direction of unbroken wholeness by investigating the properties of collections of things complexly interconnected in webs of mutual dependency relationships. Out of this model of 'punctuated wholeness', part field and part particle, a more unitary picture of the object–environment relation emerges.

Many particulate ecologists would argue that this book about ecological networks is about nothing at all. Such networks like food webs, their archetype, lack a material status. They cannot be weighed or measured or held in the hand or put under a microscope for examination. They are immanent quantities with an existence only in the minds of a small group of scientists who try to encompass perhaps too much of the full complexity of the world all at once. James Lovelock (1979, 1982) does this conceptually in his *Gaia hypothesis* portraying the whole earth as a planetary control system auto-regulated by its collective life processes. I do it mathematically in my concept of *environs* (Patten, 1978), which have no beginning or end in time and yet have been drawn as 'elementary particles' for ecology (Patten, 1982*a*). In Gaia, 'the Earth's living matter, air, oceans, and land surface form a complex system which can be seen as a single organism and which has the capacity to keep our planet a fit place to live' (Lovelock op. cit., p. vii).

Such a Gaian planet is a Clementsian conception – 'superorganism', an early twentieth-century systems construct of ecology introduced when there was no language of systems by which ecologists could express their sense of wholeness and natural unity. Lovelock's cybernetic, almost teleological, thesis clashes with the dominant reductionist, mechanistic philosophy of ecology, which disdains superorganisms and eschews other forms of extraorganismal holism as found in Darwin, Forbes, Clements, Shelford, Tansley, Thienemann, Lindeman, the Odums and many others in its 'continuing struggle to focus on material, observable entities rather than ideal constructs' (Simberloff, 1980, p. 13). Reductionism, 'the attribution of reality exclusively to the smallest constituents of the world and the tendency to interpret higher levels of organisation in terms of lower levels' (Thorpe, 1965), may not, in its orientation to the small and palpable, always accurately distinguish what is ontic. The time honored 'species' and 'population' categories of reductionist ecologists have in final analysis little more physical status than the 'ecosystems' or 'networks' these ecologists reject. The materialism in all these constructs, just as in life itself, ends abruptly at the boundaries of the individual member organisms they comprise. The rest is mental.

Without mechanisms, Lovelock's 'geophysiology' and the 'genons' and 'creaons' with which I have populated my own theory may never become orthodox in a non-systems ecology that fails to take account of 'the writings...of Whitehead, Woodger, Waddington, Piaget or of von

Bertalanffy's general systems theory,...an ecology...no longer a "science of communities" nor a "science of ecosystems", let alone a science concerned with "the structure and function of Gaia".... Odum is today on his own' (Goldsmith, 1988, p. 65). But, far from being true, the Eugene Odum of my own Georgia to whom this quotation refers is not the last living relic of ecological holism. Quite the contrary – he is a twentieth-century precursor of its flowering to come, a sentinel of the inevitable in humanity's endeavor to understand and come to terms with complex environmental wholeness. Odum's life-summarizing statement displayed on his bust in the foyer of our Institute of Ecology is perhaps best taken as a lament (a) for the need to mark time on holism in his era, (b) for not having much that was substantially new to offer, and (c) for the yawning gap between intuition and realization, that he and his contemporaries could do little to close in this century. It says, simply:

An ecosystem is greater than the sum of its parts.

This book is the brainchild and product of two youthful editors who represent successors to the present generation of holists, the vanguard of a new generation of next century ecologists who will not shrink from natural complexity because it is too difficult, or from the abstraction needed to simplify, quantify and generally make sense of it. Gaia and environs are serious, though perhaps impermanent, constructs of an early science of an environmental holism still not very tractable within current paradigms, but in the network ecology now slowly emerging as chronicled in this book, issues of whole earth and organism–environment unity are sure to come up for serious attention in acceptable scientific form. The mechanisms of indirect effects – what happens when A influences B through C and its subsequent ramifying consequences in the invisible networks of nature – will, if we are only patient and persistent enough, be found in the obscure recesses of unknown organizational properties of the abstractions we call 'networks'.

To deny the existence of ecological networks is an act of profound disregard of the obvious. Consider only one of them from our time. Human AIDS (Acquired Immune Deficiency Syndrome) is spreading, like the plague of former centuries. Each case of AIDS is a marker tracing out in time and space a reticulum of human sexual contact. This network is unseen; it is generally untraceable back through the 'input environs' of its victims to origins, but of its pernicious reality and readiness to ensnare the naive, undisciplined or unwary there is no doubt. Denial of this network because it is nonphysical or invisible may bring with it the gravest of physical consequences that can accrue to life – its cessation.

AIDS is only one of many multifarious webs of social discourse, intercourse and commerce embodied in customs, institutions and culture, by which humanity conducts its global civilization. All means of transportation and communication are part of this web. All the rules of law and convention that constrain function to be orderly and organized are part of it. Some of the subwebs are physical, like railroads, highways and telephone lines; others are abstract, like the family, speech communities, airline routes, financial networks, and so forth. 'Old boy' networks are part and parcel of the political process, the art of applied indirect effects in which one says A, means B, and achieves C. All humans are enmeshed in and conduct their daily lives filling roles within networks of many different categories, at a variety of levels of organization. We all possess instinctive or learned understanding of how to function in these diverse roles, as befits our status and relationship to different networks at different times and places. What is more, every other distinct life form in nature, be it plant, animal or microbe, primitive or advanced is likewise enmeshed in networks. All function as cogs in wheels (environs) that spin at different rates and over different spans in ecosystems, wheels that in the end add up to natural selection – life, death, change, evolution. Taken all together, the aggregate complexity of ecosystems that results is more than the human mind, a network itself of no small respectability, can comprehend.

It is no wonder, then, that the holistic treatment of complex nature as a set of networks, each a small world of interrelationships between some discrete living or nonliving entity, group or individual, and participated in by one or more others of the same or different kind or rank, has waited until now. The metaphysical themes of a totally unified nature, that in biology have been on scientific hold since Darwin unintentionally drew the organism in sharp relief to its background matrix, are now returning for serious development. Contemporary problems of wide scale holism, expressed in environmental pollution, toxification, degradation and destruction; the greenhouse effect and global change; the ozone hole; acid rain; Bopal; Valdez; Three-Mile Island and Chernobyl; the spread of radioactivity – and AIDS; and in the next century a new set of unwanted side effects from recombinant DNA biotechnology – all these serve as reminders that the biosphere is in fact one single interconnected unit wherein events at one locale may spread quickly, or slowly and insidiously – but inevitably – around the globe to affect all things, at all organizational levels, everywhere. The underlying mechanism for such proliferation, implicit or explicit, is the *network*. Necessity has made its serious study as a scientific object in its own right mandatory and urgent; the advent of modern computers has made it possible.

In this concluding chapter, I have been invited by the editors to give a personal perspective on the future of network approaches in ecology. I have chosen to do this by focusing on what seems the key manifestation of the operation of ecological networks, and what might not occur without them – *indirect effects*. Because these seem so ubiquitous and pervasive in different activities, places and levels of organization, and because we so little understand them, I predict that all ecology into a far future, commencing sometime in the next century, will become network oriented. The essence of collective existence, of the cells, organs and organ systems within organisms, as well as of organisms in the populations, communities and ecosystems of synecology, is captured in the network model. Ecology *is* networks, and after a quarter of a century of trying to bring them into focus for study of the organism–environment relationship, I hold to the same conclusion now that I arrived at a long time ago – 'to understand ecosystems ultimately will be to understand networks' (Patten & Witkamp, 1967, p. 824).

My work with Martin Witkamp was a turning point in my recognition of the power of networks to influence organisms. In laboratory microcosms, Witkamp demonstrated, using a radioactive tracer, that changes only in the ways organisms were interconnected changed their function, right down to physiological rates of feeding and excretion – organism properties, not community properties. I lectured during that period on what I called 'the network variable in ecology', trying to say what should have been an obvious corollary of everyone's definition of system as a partially interconnected set of component parts (e.g. Zadeh & Desoer, 1963): *change the network, change the system* – and the things and their processes within it. Nothing I have learned since has deflected me from this conclusion; all observation and study have reinforced it.

The key property of the network phenomenon is *influence* – cause at a distance, or indirect determination. This is a subject I have come to in trying to ascertain why a systems study of the world is necessary in the first place. As I now see it, if indirect effects are unimportant in general determination, then there is probably no need to study systems as wholes. Science can continue to dissect them down to component parts and then simply build its understanding of wholes by adding the parts together; the ways to do this still, of course, remain to be discovered. We scientists and all humanity would be better off if this were only true. Unfortunately, it appears not to be. My own work, and that and the intuition of many others, leads me to the unshakeable belief that not only is the whole world connected together in what is probably a coevolutionary biosphere at every level of organization, but the dominant form of determination is that of parts by wholes, *holistic determination* (Bunge, 1959), in which indirect causes and effects – those acting at a 'temporal' and 'procedural' distance (see later) – are dominant (Higashi & Patten, 1989).

This has become my working hypothesis about systems, and in recent years I have pursued the indirect effects phenomenon in both qualitative and quantitative dimensions. It is the main outlines of these pursuits, especially some new developments taking form right now, that I will present here by way of demonstrating the future potential of the network approach to give new perspective to old issues of ecology (such as the significance of competition and predation; see Section C.5.2), as well as pose new classes of questions (such as the indirect effects hypothesis itself) about the nature of reality and communal life in it. If I were to state a worldview based on my present understanding and intuition about network indirect effects, it would go something like this:

> Collective life in ecosystems, from individual binary interactions to the whole global biosphere, is a coevolutionary continuum of unbroken wholeness encompassing network mediated cybernetic mechanisms of distributed or diffuse control, implicit in dominant indirect effects, that holistically regulate the life-environment relation. Within any system, the significant determinants of change are global, not local.

One can readily see the implications; the problem of holistic science becomes one of system definition – specifically, determining a minimal universe sufficient to encompass the indirect effects that are relevant to a given scope of inquiry and no more. For now, with faith in the future, I will simply relegate this to a rather significant and intimidating class of unresolved 'modeling problems', and go on.

C.2 Empirical indirect effects

In the real world, 'side', 'secondary' or 'indirect' effects are manifested everywhere. A miracle pesticide is discovered that will rid the world of malaria and other insect-vectored human scourges, and some years later entire avian species become endangered; they cannot reproduce because their eggshells have been embrittled by the new compound. Radioactivity or toxic chemicals are disposed into the environment in 'safe' concentrations that with passage of time become bioaccumulated or biomagnified into hazardous levels at the tops of food chains. A 'green revolution' is fomented in food-science laboratories that will conquer starvation, but inevitably it falters for lack of ability in the Third World to acquire fertilizers, and if it did succeed it would only rachet the cycle of human misery to new potential heights by causing greater over-population. Gene splicing is promoted as the new biotechnology that will cure many of the world's ills and promote human welfare. From the lessons of the past, ecological cautions are in place, but as early direct benefits are realized, restraint will give way to 'progress' and postponable risks, leaving it to future generations to suffer consequences and redress the indirect human, biological and environmental fallout.

Hurlbert (1975), reviewing the ecological side effects of toxic chemicals, observed, 'If the direct, toxicological effects of pesticides on the growth, survival, or reproduction of organisms may be called primary, irrespective of the ecological or physiological methods of exposure...and irrespective of the time interval between exposure and ultimate effect..., then ecosystem changes that follow from these primary effects are conveniently termed secondary effects' (p. 82). He then documented numerous examples of *primary* vs. *secondary effects*, and (p. 136) concluded with an effort 'to sketch the "real world" context in which secondary effects must be considered and to describe some difficulties of conclusively demonstrating secondary effects, and of cause-and-effect relationships in general, in large-scale ecosystems.' His principal points were: (1) primary effects are important; 'they come first, are more limited in their variety, are more amenable to scientific study and documentation, and are more directly susceptible to corrective action than are secondary effects'; (2) nevertheless, 'secondary effects resulting from these might exceed them in magnitude'; (3) ecosystems 'are the tangled context in which the effects of pesticides most need to be understood.'

The above applied ecology lessons have been slow entering academic ecology, but this too has begun to discover in its data the inevitable expression of indirect effects. Size-selective feeding by fish is a known and well documented phenomenon (Brooks & Dodson, 1965; Brooks, 1968; Archibald, 1975; Bartell, 1981 for review). It results in modification of zooplankton length and abundance. This indirectly influences phyto-plankton production by changing rates of phosphorus release (Bartell, 1981). Sterner (1986) showed that increased reproductive rates of a natural algal community caused by nitrogen regeneration from an herbivorous zooplankter (*Daphnia pulex*) approximately equalled the zooplankton induced predatory mortality. The author wrote, 'This result demonstrates that nutrient regeneration by herbivores is at least sometimes a strong indirect effect in natural communities.' Seale (1980) presented evidence that larval salamanders both directly and indirectly affect nitrogen cycling, and thereby the rate of natural eutrophication in a pond. Eutrophication is a process usually associated with direct nutrient loading from external sources, not with endogenous activities of a vertebrate within the affected system. Brabrand *et al.* (1984) showed that defecation of iron by omnivorous fishes can increase plankton productivity. Zaret & Paine (1973) demonstrated revision of the entire trophic structure of Lake Gatun, Panama, by the successful invasion of a top predatory fish, *Cichla ocellaris*. The whole 'keystone species' concept of the latter author (Paine, 1966), in fact, presumes a capacity for widespread ecosystem change to follow upon removal of the keystone form. Montague (1980), working in a Georgia salt marsh, found that fiddler crabs (*Uca pugnax*) altered the productivity and growth of *Spartina* grass by their burrowing activities. In a terrestrial ecosystem, phytophagous insects were shown to regulate forest primary productivity (Mattson & Addy, 1975). And, in a series of papers, the importance of indirect interactions between granivorous insects, birds and mammals, and annual plants in determining floristic composition of two desert communities has been extensively documented (Inouye, 1981; Davidson, Inouye & Brown, 1984; Brown & Munger, 1985; Brown *et al.*, 1985). Finally, the recent book by Kerfoot & Sih (1987), describing direct and indirect effects of predation in aquatic communities, has done much to focus current interest on the capacity of simple, direct interactions like predation to induce complex indirect effects.

In the following sections, directions for both qualitative and quantitative theories of indirect effects are outlined, and an attempt made to unify these distinct aspects into a single comprehensive theory, emphasizing what is new, or has potential for newness, through the wider application of the network perspective.

C.3 Network indirect effects: qualitative aspects

Roberts (1976) has discussed some uses of signed digraphs in ecology. Levins' *loop analysis* (Levins, 1974; Puccia & Levins, 1985) is an example of the application of such methods to the qualitative analysis of interactive ecological systems. Each digraph arc (i,j) is signed $s_{ij} = +$ or $-$ according as the effect from its originating node is augmenting $(+)$ or inhibiting $(-)$ to its terminating node. Suppose, in the digraph example of Fig. C.1 *a*, j is a predator of k ($s_{jk} = -$), i symbiotically aids j ($s_{ij} = +$), and i and k are competitors ($s_{ik} = s_{ki} = -$). Cycles are facilitating or *deviation amplifying* if they contain an even number of minus signs, and inhibiting or *deviation damping* if they do not (Maruyama, 1963). The two cycles $j \to k \to i \to j$ and $k \to i \to k$ are both deviation amplifying, and therefore this system will tend to be unstable. Such observations follow from the simple (unextended) digraph structure, but what is the situation when extended structure is considered? Are multiple passages around cycles relevant to qualitative analysis? Components j and k are directly linked in the small

Fig. C.1. (*a*) Single-signed digraph of text example; j predates k, i aids j, and i and k compete. (*b*) Double-signed version of (*a*); direct (δ) predation is $s_{jk}{}^{\delta} = (+, -)$, commensalism $s_{ij}{}^{\delta} = (0, +)$ and competition $s_{ik}{}^{\delta} = s_{ki}{}^{\delta} = (-, -)$. (*c*) Steps in algebraic reduction of digraph (*b*) to reveal the graph-wide (υ) relationship, amensalism [$s_{jk}{}^{\upsilon} = (0, -)$], propagated directly and indirectly from j to k in the network; the direct interaction is nihilism $s_{jk}{}^{\delta} = (+, -)$. (*d*) Digraph reduction, converting absence of a direct interaction from j to i [local neutralism, $s_{ji}{}^{\delta} = (0, 0)$] to commensalism, $s_{ji}{}^{\upsilon} = (0, +)$. (*e*) Both local and graph-wide relationships between k and i are the same, competition [$s_{ik}{}^{\delta} = s_{ik}{}^{\upsilon} = (-, -)$].

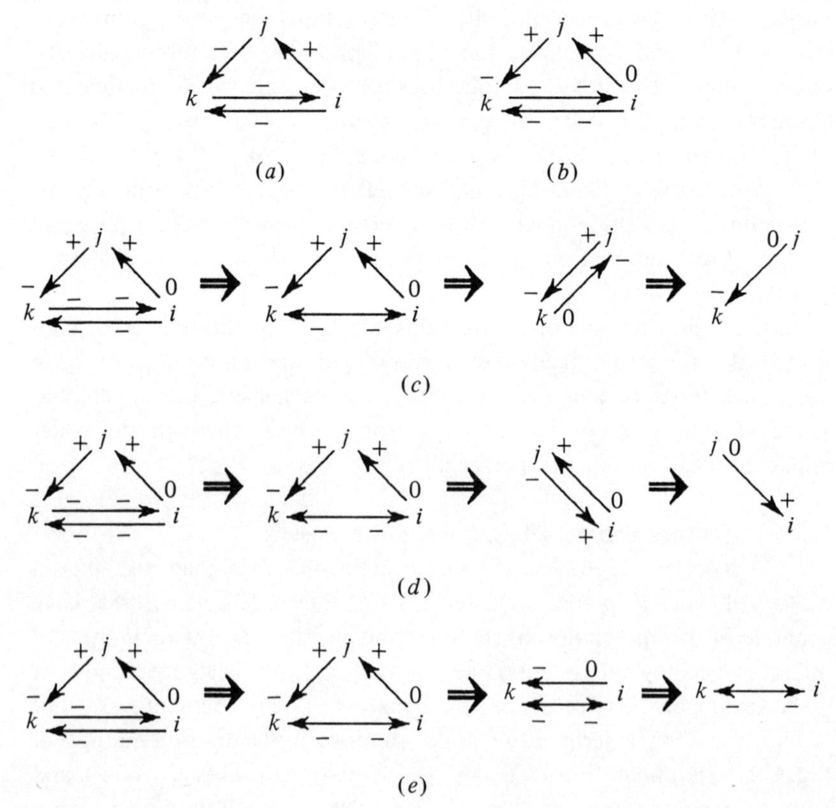

by a predation relation, but what is their true interactive status in the context of the whole system, globally, or in the large?

C.3.1 *Definition of qualitative interaction types*

There are only nine ordered pairwise combinations of the two signs + and −, plus a third element (0) to represent a neutral or null interaction. Each defines in qualitative terms a directed ecological relation. Based on the nine sign pairs, I will name and define these relations below as distinct interaction types. In doing this, I am aware that many exceptions, special cases, sources of incompleteness, changes of reference frame, prior meanings, etc. all compromise the terminology and open the theory to criticism before it is even started. Therefore, I hasten to assert that the names are intended for loose interpretation only as working terms serving to convey a general ecological sense of the interactions.

Let s_{ij}^{δ} denote a sign pair signifying a *direct* (δ), *local*, *immediate* or *micro interaction* oriented from i to j, to be distinguished later from *direct plus indirect*, *global*, *universal* or *ultimate* (υ), or *macro interactions*, s_{ij}^{υ}. As an example, the ordered sign pair $s_{ij}^{\delta} = (+,-)$ signifying a *direct effects arc* directed from i to j, $i^{+}\!\rightarrow^{-}j$, represents an interaction type that can be called *nihilism*; the archetypal ecological example of this type is *predation*, which produces an energy-matter flow f_{ij} against the grain of usual behavior (i to j), that is from j to i over a *direct flow arc*, $j \rightarrow i$. Note that while the substance flows in this case have a j to i orientation, corresponding effects are transposed, i to j. This is generally true for 'abundance' and 'behaviorally' mediated effects, but for 'chemically' mediated processes (e.g., allelopathy) flow and effects arcs often have the same orientation (see Section C.3.3, for description and examples of the three kinds of effects just mentioned). Sign pairs $s_{ij}^{\delta} = s_{ji}^{\delta} = (-,-) = i^{-}\!\leftrightarrow^{-}j$ and $s_{ij}^{\delta} = s_{ji}^{\delta} = (+,+) = i^{+}\!\leftrightarrow^{+}j$ denote *competition* and *mutualism*, respectively; these are symmetric relations, indicated by bidirectional arcs ($\leftrightarrow$). Other interactions include: $s_{ij}^{\delta} = (0,+) = i^{0}\!\rightarrow^{+}j$, *commensalism* directed from i to j; $s_{ij}^{\delta} = (0,-) = i^{0}\!\rightarrow^{-}j$, *amensalism*; $s_{ij}^{\delta} = (-,0) = i^{-}\!\rightarrow^{0}j$, *dissipation* or *catabolism* at i; $s_{ij}^{\delta} = (+,0) = i^{+}\!\rightarrow^{0}j$, *growth* or *anabolism* at i; $s_{ij}^{\delta} = (-,+) = i^{-}\!\rightarrow^{+}j$, *altruism*; and $s_{ij}^{\delta} = s_{ji}^{\delta} = (0,0) = i^{0}\!\leftrightarrow^{0}j$, *neutralism*, which is also symmetric. Growth and dissipation are generally considered unary (reflexive, or self referential) rather than binary interactions; usage here will be taken to imply a partner in the relation even though no effect (0) is transmitted to the latter. Neutralism also has some ambiguity associated with it. The sign pair (0,0) will be taken to mean either no (null) interaction between two entities, or an interaction with a zero (neutral) outcome.

C.3.2 *Digraph reduction algebra*

Legovic & Patten (1981, unpubl.) developed a digraph reduction algebra that resolves such doubly signed arcs representing direct pairwise micro-interactions, to determine graph-wide macro-relationships between any pair of components in a signed digraph. To illustrate how this works, consider the single-signed digraph of Fig. C.1 *a* double-signed as shown in Fig. C.1 (*b*). Components j and k are locally predator and prey, $s_{jk}{}^{\delta} = (+, -) = j^{+}\!\rightarrow^{-} k$, but is this their real relationship as expressed within the network taken as a whole? The answer is no; the graph reduction sequence of Fig. C.1 (*c*) shows their ultimate in-system relation as mediated by all the relationships in the entire network to be amensalism, $s_{jk}{}^{\upsilon} = (0, -) = j^{0}\!\rightarrow^{-} k$. Thus, local (i.e., direct, δ) nihilism, $s_{jk}{}^{\delta} = j^{+}\!\rightarrow^{-} k$, is transformed in this case to universal (direct plus indirect, υ) amensalism, $s_{jk}{}^{\upsilon} = j^{0}\!\rightarrow^{-} k$, wherein j suppresses k, but with no ultimate benefit or cost to itself. The graph-wide relationship between j and i is commensalism (Fig. C.1 *d*), but now j is the benefactor, $s_{ji}{}^{\upsilon} = (0, +) = j^{0}\!\rightarrow^{+} i$, rather than the beneficiary it appears to be in the local case, $s_{ij}{}^{\delta} = (0, +) = i^{0}\!\rightarrow^{+} j$. Finally, i and k remain competitors in the large as they are in the small, $s_{ik}{}^{\delta} = s_{ik}{}^{\upsilon} = (-, -) = i^{-}\!\leftrightarrow^{-} k$ (Fig. C.1 *e*).

From these examples, it is clear that the kind of biological interaction observed at a proximate level may become a qualitatively different type at a higher level of organization when all the consequences of context are taken into account. This principle of holism is another manifestation of hierarchical network organization, and Legovic & Patten (1981) observed in several digraph reductions they investigated, a tendency for locally negative interactions to become transformed into globally positive ones. They characterized this holistic synthesis of positives from the negatives of local 'struggles for existence' as a kind of 'serendipity of nature' inherent in network organization. We will observe this 'hedonic property' of networks expressed later in a quantitative approach to the assessment of global interaction types in network models.

The qualitative graph reduction algebra of Legovic & Patten was based on strictly unextended relationships. Development for the extended case from a quantitative methodology will be outlined in Section C.5. First, several real world examples from the kinds of 'complex ecological systems' reviewed in Section C.2 will prove instructive.

C.3.3 *Empirical examples*

Miller & Kerfoot (1987) recognized three different types of qualitative indirect effects:

1 *abundance indirect effects* – mediated by population sizes of the interacting species;
2 *behavioral indirect effects* – mediated by behaviors of the interacting organisms; and
3 *chemical indirect effects* – mediated by biogenic chemical agents.

They illustrated these, each in turn, with the following examples taken from the literature on 'complex' ecological interactions. While these examples are isolations of a few direct relationships from much more complex interactive networks, they serve to illustrate the distinction between direct and indirect interaction types.

Lubchenko (1978) studied the interactions between a snail (S; *Littorina littorea*) which grazes an alga (E; *Enteromorpha*) that is competitively dominant over a subordinate alga (C; *Chondrus crispis*). When *Littorina* is present, its grazing on *Enteromorpha* enables *Chondrus* to thrive, an abundance indirect effect ($a\rightarrow$) of S transmitted to C through E. With $s_{SE}{}^{\delta} = (+,-) = S^{+}\!\rightarrow^{-}E$ and $s_{EC}{}^{\delta} = (-,-) = E^{-}\!\leftrightarrow^{-}C$, the algebra involved in this interaction is $s_{SE}{}^{\delta} \times s_{EC}{}^{\delta} = (+,-)(-,-) = (-,+) = s_{SC}{}^{v} = S^{-}a\!\rightarrow^{+}C$. That is,

$$\underset{\text{(predation)}}{S^{+}\!\rightarrow^{-}E}\underset{\text{(competition)}}{^{-}\!\leftrightarrow^{-}C} \Rightarrow \underset{\text{(altruism)}}{S^{-}a\!\rightarrow^{+}C}.$$

By consuming C's competitor (E), S facilitates C; by competing with S's food source (E), C's effect on S is negative. Neither Lubchenko nor Miller & Kerfoot identified the fact that the resultant net indirect effect of the snail on *Chondrus* could be interpreted as altruistic, or discussed the complex interaction in such terms.

Werner *et al.* (1983) experimentally studied another three species interaction between two centarchid fishes, largemouth bass (LB; *Micropterus salmoides*) and a predator of the bluegill (BG; *Lepomis machrochirus*), plus the planktonic cladoceran *Daphnia pulex* (DP), a bluegill prey. Small BG prefer to feed in open water where DP is abundant. When LB is present, however, the BG forage in vegetation where the cladocerans are not as abundant as in open water. In ponds containing BG and DP only, the daphnids are grazed down and disappear within ten days. If LB are added, the cladoceran prey survive more than 20 days, even if BG predatory mortality from the bass is compensated for by replacement. The latter is a behavioral indirect effect ($b\rightarrow$) induced by the cover seeking habit of bluegills in response to the presence of the LB predator, which reduces their foraging effectiveness on DP. Without BG replacement, the bass and daphnids are indirect mutualists as a result of the abundance indirect effect of two sequential predation relations:

$$\underset{\text{(predation) (predation)}}{\mathrm{LB}\ {}^{+}\!\!\rightarrow^{-}\mathrm{BG}\ {}^{+}\!\!\rightarrow^{-}\mathrm{DP}} \Rightarrow \underset{\text{(mutualism)}}{\mathrm{LB}\ {}^{+}\!\!\leftarrow\! a \!\rightarrow^{+}\mathrm{DP}}.$$

The algebra of this interaction, resulting from $s_{\mathrm{LB,\,BG}}{}^{\delta} = (+,-) = \mathrm{LB}\ {}^{+}\!\!\rightarrow^{-}\mathrm{BG}$ and $s_{\mathrm{BG,\,DP}}{}^{\delta} = (+,-) = \mathrm{BG}\ {}^{+}\!\!\rightarrow^{-}\mathrm{DP}$, is $s_{\mathrm{LB,\,BG}}{}^{\delta} \times s_{\mathrm{BG,\,DP}}{}^{\delta} = (+,-) \times (+,-) = (+,+) = s_{\mathrm{LB,\,DP}}{}^{\upsilon} = \mathrm{LB}\ {}^{+}\!\!\leftarrow\! a \!\rightarrow^{+}\mathrm{DP}$. The replacement (*r*) of consumed BG, however, elevates the status of this prey of bass to that of a competitor with its predator. This produces a behaviorally mediated indirect altruism directed from the bass to the cladocerans:

$$\underset{\text{(competition)}}{\mathrm{LB}\ {}^{-}\!\!\leftarrow\! a \!\rightarrow^{-}} \underset{\text{(predation)}}{\mathrm{BG}^{(r)}\ {}^{+}a\!\rightarrow^{-}\mathrm{DP}} \Rightarrow \underset{\text{(altruism)}}{\mathrm{LB}\ {}^{-}b\!\rightarrow^{+}\mathrm{DP}}.$$

The algebra, with $s_{\mathrm{LB,\,BG}}{}^{\delta} = (-,-) = \mathrm{LB}\ {}^{-}\!\!\leftarrow\! a \!\rightarrow^{-}\mathrm{BG}$ and $s_{\mathrm{BG,\,DP}}{}^{\delta} = (+,-) = \mathrm{BG}\ {}^{+}a\!\rightarrow^{-}\mathrm{DP}$, is $s_{\mathrm{LB,\,BG}}{}^{\delta} \times s_{\mathrm{BG,\,DP}}{}^{\delta} = (-,-)(+,-) = (-,+) = s_{\mathrm{LB,\,DP}}{}^{\upsilon} = \mathrm{LB}\ {}^{-}b\!\rightarrow^{+}\mathrm{DP}$. The net qualitative effect of bass on *Daphnia* changes with bluegill replacement from an abundance mediated indirect mutualism to a behavior mediated indirect altruism.

The third study discussed by Miller & Kerfoot (1987) illustrates a chemical indirect effect. Smiley, Horn & Rank (1985) studied the relationships between willows (W; *Salix orestera*) which contain a toxin, salicin, used by beetles (B; *Chrysomela aenicollis*) to produce defensive secretions. Various arthropod predators of the beetles (particularly ants, A) are repelled by the secretions. The chrysomelids favor willows with high salicin content, for the abundance of this compound transforms them from susceptible to resistant prey of A through the following chemical indirect effect ($c\!\rightarrow$). In trees with low salicin, the willow–ant relationship is an abundance mediated indirect mutualism:

$$\underset{\text{(predation) (predation)}}{\mathrm{W}\ {}^{-}\!\!\leftarrow^{+}\mathrm{B}\ {}^{-}\!\!\leftarrow^{+}\mathrm{A}} \Rightarrow \underset{\text{(mutualism)}}{\mathrm{W}\ {}^{+}\!\!\leftarrow\! a \!\rightarrow^{+}\mathrm{A}}.$$

With $s_{\mathrm{BW}}{}^{\delta} = (+,-) = \mathrm{B}\ {}^{+}\!\!\rightarrow^{-}\mathrm{W}$ and $s_{\mathrm{AB}}{}^{\delta} = (+,-) = \mathrm{A}\ {}^{+}\!\!\rightarrow^{-}\mathrm{B}$, the two predations in sequence yield $s_{\mathrm{AB}}{}^{\delta} \times s_{\mathrm{BW}}{}^{\delta} = (+,-)(+,-) = (+,+) = s_{\mathrm{AW}}{}^{\upsilon} = \mathrm{A}\ {}^{+}\!\!\leftarrow\! a \!\rightarrow^{+}\mathrm{W}$. When salicin content of the willow is high, however, the beetles are aided through a direct altruistic relation that accrues from their plant feeding. Beetle foraging directed against the plants becomes transposed, $s_{\mathrm{BW}}{}^{\delta\mathrm{T}} = (+,-)^{\mathrm{T}}$, into an altruistic interaction represented by the plant ${}^{-}\!\!\rightarrow^{+}$ beetle transfer of salicin, $s_{\mathrm{WB}}{}^{\delta} = (-,+) = \mathrm{W}\ {}^{-}\!\!\rightarrow^{+}\mathrm{B}$. The salicin based repellent causes direct amensalistic suppression of the beetles' ant predators, $s_{BA}{}^{\delta} = (0, -) = \mathrm{B}\ {}^{0}\!\!\rightarrow^{-} \mathrm{A}$, and the chain of effects from plants to ants produces a chemically mediated indirect amensalism directed to the latter:

$$\underset{\text{(altruism)}}{\mathrm{W}\ {}^{-}\!\!\rightarrow^{+}\ \mathrm{B}^{(s)}}\ \underset{\text{(amensalism)}}{{}^{0}\!\!\rightarrow^{-} \mathrm{A}} \Rightarrow \underset{\text{(amensalism)}}{\mathrm{W}\ {}^{0}c\!\!\rightarrow^{-} \mathrm{A}}.$$

The algebra of this interaction, given $s_{WB}{}^{\delta} = (-, +) = \mathrm{W}\ {}^{-}\!\!\rightarrow^{+} \mathrm{B}$ and $s_{BA}{}^{\delta} = (0, -) = \mathrm{B}\ {}^{0}\!\!\leftrightarrow^{-} \mathrm{A}$, is $s_{WB}{}^{\delta} \times s_{BA}{}^{\delta} = (-, +)(0, -) = (0, -) = s_{WA}{}^{\upsilon} = \mathrm{W}\ {}^{0}c\!\!\rightarrow^{-} \mathrm{A}$.

Again, as with the prior examples, neither the original authors nor Miller & Kerfoot treated this complex interaction beyond an elementary description of the observed results. They did not couch their discussions in terms of the specific qualitative interaction types revealed by the methods of this section; had they been able to do so, their treatment of empirical observations would have been strengthened, and they might have come to the systems oriented postulate that local interactions (such as the competition and predation of high standing in organism–population ecology) are not, in general, very determining of behavior; immediate, direct interactions serve mainly to define networks that, once in place and operating, holistically determine the real, operational, ultimate or direct plus indirect relationships between entities involved in complex ecological systems. This postulate from holism is supported also from results on the quantitative aspects of network indirect effects, as shown in the next sections.

C.4 Network indirect effects: quantitative aspects

In addition to the qualitative consequences of extended path structure in networks, there are corresponding quantitative manifestations. Patten (1982*a*), for example, indicated that causal propagations associated with combinatorially increasing numbers of extended, $j \rightarrow \ldots \rightarrow i$, paths traced out as transitions proceed through networks, vastly dominate in magnitude the direct effects associated with direct transitions over paths of length 1, $j \rightarrow i$. Wiegert & Kozlowski (1984) criticized this result, stating that most of the numerical indirect effects were not associated with indirect paths, but were due to time delays at the nodes. To this, Higashi & Patten (1989) responded by pointing out errors that led the critics to misclassify most indirect effects into the direct category. They, Higashi & Patten, reclassified indirect paths into *route* or *procedurally indirect* and *time delayed* categories, *p*- and *t-indirect*, respectively, and hypothesized as a general network property that *p, t-indirect* > *p-direct*, *t-indirect* ≫ *p, t-direct* effects. Investigation of the magnitude dominance of these several classes of indirect effects is the objective of the quantitative theory.

The numerical results on dominance of indirect paths and effects, as reported by Patten (1982*a*) and elsewhere (Patten, 1982*b*, *c*, 1984, 1985; Patten, Richardson & Barber, 1982; Higashi & Patten, 1986), are expressions of general properties of network organization. Masahiko Higashi (unpubl.) has derived algebraic properties of networks that tend to confer dominance of indirect over direct effects as a general characteristic. These results, summarized in the subsections below, are expressed in terms of indirect/direct ratios being greater than, less than, or equal to one. Increases in system order, connectedness, intranodal transition or stasis (looping), cycling, and strength of direct interactions all tend to contribute to indirect (I)/direct (D) ratios > 1.

C.4.1 *System order and connectivity*

The number of nodes in a network defines system *order*; nodes are interconnected by simple (unextended) paths, whose number defines *connectivity*. The digraph of Fig. C.2(*a*) represents an acyclic flow network from node j to node i. Solid arrows depict direct arcs, and broken ones represent indirect subnetworks of varying complexity, which may or may not include path branching, convergence, cycling or feedback. Let f_{ij} be the *steady state direct flow* of conservative substance from node j to i in a digraph; the sum of such flows into or out of a node constitutes the *throughflow* at that node, $T_i = \sum_{k=0}^{n} f_{ik} = \sum_{k=0}^{n} f_{ki}$, where n is the system order (number of nodes), subscript zero denotes the extrasystem environment, and $k \neq i$. The flow from j to i can be expressed as a *nondimensional flow intensity* by normalizing it with respect to throughflow: $g(i,j) \equiv g_{ij} = f_{ij}/T_j$. The Greek letters in Fig. C.2 signify such nondimensional *transmittances* (for information) or *transferences* (for energy-matter), $g(\psi)$, for either direct or indirect paths, ψ, in the depicted networks. For example, in Fig. C.2*a* $\alpha = g(k_1, j)$, $\beta = g(i, k_1)$, and for the path $\psi = j \rightarrow k_1 \rightarrow i$ the intensity measure is $g(\psi) = g(k_1, j) \cdot g(i, k_1) = \alpha\beta$. For sets Ψ of paths, $\psi \in \Psi$, $g(\Psi)$ is the throughflow normalized measure of relative effects. As an example, for paths $\Psi = \{\psi = j \rightarrow \ldots \rightarrow k_m\}$, $g(\Psi) = \varepsilon$. In this and subsequent examples, the p,t-direct arc directly connecting compartment j to i will have measure g_{ij}, and this will consequently always appear in denominators of indirect (I)/ direct (D) ratios.

The indirect/direct effects ratio for the Fig. C.2(*a*) digraph is:

$$\mathrm{I/D} = g(\Psi)/g_{ij} = (\alpha\beta + \xi\gamma\beta + \xi\delta + \ldots + \varepsilon\eta)/g_{ij}.$$

It is apparent that, for a given strength of direct relationship g_{ij}, this ratio increases as the numerator increases, either through increased intensities of existing relations (stronger interactions), or by adding more nodes (increasing system order) or arcs (increasing connectivity). The first of these, which is conferred by stronger direct effects, will be treated more specifically in Section C.4.3 below.

C.4.2 *System looping and cycling*

The Fig. C.2(*a*) network contained no self transfers (loops) at or cycles (closed paths) through compartments j or i. Loops or cycles at other nodes may, however, have been implicit features of some of the

Fig. C.2. Contributing factors to indirect/direct ratios >1. (*a*) System order and connectivity; (*b*) looping (storage or state stasis) at and cycling through source node; (*c*) feedback and magnitude of direct effects.

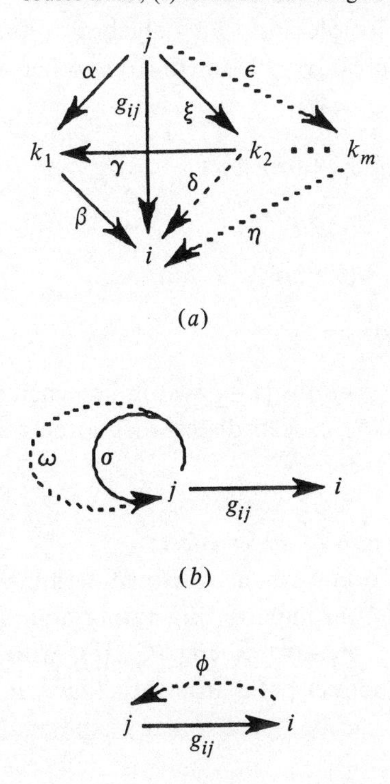

indirect paths shown only as broken arrows. In that case, the digraph would have been cyclic, and its cycles or loops would induce significant change in its microscopic dynamic characteristics, but would not change the I/D ratio with respect to j and i. In general, a loop signifies stasis or apparent nontransition of state from one time step to another. Cycles mean returning to an already visited node. The digraph of Fig. C.2(*b*) illustrates these features, with σ denoting the intensity of intranodal transfer (looping) through source compartment j, and ω that of cycling over all parallel indirect paths from j to itself without passing i along the way. The shortest indirect routes from j to i in this graph are those involving either the loop or cycle, with respective measures σg_{ij} or ωg_{ij}, and the possibilities are expressed as $(\sigma+\omega)\, g_{ij}$. But it is also possible to have two circuits around either the loop (with measure σ^2), or the cyclic subgraph (with measure ω^2), or once around each (with measure $\sigma\omega+\omega\sigma = 2\sigma\omega$); the combined possibilities are $\sigma^2+2\sigma\omega+\omega^2 = (\sigma+\omega)^2$. In general, many ($m$) such circuits are possible, each with combined magnitude given by $(\sigma+\omega)^m$ where $\lim_{m\to\infty}(\sigma+\omega)^m = 0$. These considerations allow an indirect relations measure to be formulated as $\sum_{m=0}^{\infty}(\sigma+\omega)^m g_{ij}$, whereupon [employing an algebraic theorem: $1+x+x^2+\ldots = (1-x)^{-1}$ if and only if $|x| < 1$, and diverges otherwise] the I/D ratio becomes:

$$\begin{aligned}\mathrm{I/D} &= [(\sigma+\omega)\, g_{ij}+(\sigma+\omega)^2\, g_{ij}+(\sigma+\omega)^3\, g_{ij}+\ldots]/g_{ij}\\ &= (\sigma+\omega)+(\sigma+\omega)^2+(\sigma+\omega)^3+\ldots\\ &= (\sigma+\omega)\,[1+(\sigma+\omega)+(\sigma+\omega)^2+(\sigma+\omega)^3+\ldots]\\ &= (\sigma+\omega)/[1-(\sigma+\omega)].\end{aligned}$$

This ratio exceeds one whenever $(\sigma+\omega) > [1-(\sigma+\omega)]$, i.e. when $(\sigma+\omega) > 1/2$. Thus, indirect effects tend to exceed direct ones in networks as looping or cycling, or both, increase.

C.4.3 *System feedback and strength of direct effects*

Special cases of cycling occur when a defined terminal compartment i recycles substance back through an originating node j. Such relationships are depicted in the digraph of Fig. C.2(*c*), with ϕ the feedback intensity. The shortest indirect paths from j to i have intensity measure $g_{ij}(\phi g_{ij})$, but these paths can be traveled many times, $g_{ij}(\phi g_{ij})^m$. The I/D ratio is therefore:

$$
\begin{aligned}
\mathrm{I/D}\ &[g_{ij}(\phi g_{ij}) + g_{ij}(\phi g_{ij})^2 + g_{ij}(\phi g_{ij})^3 + \ldots]/g_{ij} \\
&= (\phi g_{ij}) + (\phi g_{ij})^2 + (\phi g_{ij})^3 + \ldots \\
&= (\phi g_{ij})[1 + (\phi g_{ij}) + (\phi g_{ij})^2 + \ldots] \\
&= (\phi g_{ij})/[1 - (\phi g_{ij})],
\end{aligned}
$$

which exceeds one when $(\phi g_{ij}) > [1-(\phi g_{ij})]$, or $(\phi g_{ij}) > 1/2$. Thus, indirect to direct effects ratios in networks tend to increase both with increasing feedback and, counter-intuitively, with strength of *direct* effects as well.

Higashi (pers. comm., Appendix AC.1) has shown that the latter paradoxical increase in I/D with increase in D takes a parabolic form beyond a critical, relatively low value, $g_{ij} = g_{ij}^*$, of the direct effect. That is, with increasing strength of direct interactions, magnitudes of indirect effects increase at parabolically increasing rates that, above the critical point, exceed any increase in direct effects. Below the critical point, the I/D ratio decreases parabolically. This *parabolic rule* establishes that the tendency for indirect effects to dominate in networks cannot be overridden by increasing the strengths of direct interactions. Indirect effects will increase faster than any corresponding increase in a direct effects function.

With the dominance of indirect effects quantitatively guaranteed as a general property of complex ecological organization, produced by different combinations of nine elementary types of qualitative interactions between entities, we turn now to a consideration of the possibilities and directions for a unified qualitative–quantitative theory of network indirect effects.

C.5 Network indirect effects: unified theory

The nondimensional flow intensities g_{ij} pertain to output environs and are the basis for their analysis. For input environs, the corresponding measure $g_{ij}' = f_{ij}/T_i$ comes from input–output analysis (Leontief, 1966; Hannon, 1973), and looks backward to system history rather than forward to its future. Both coefficients, g_{ij} and g_{ij}', range in values between zero and one; together they lend themselves to an approach to quantifying single-signed, qualitative digraphs as recently suggested by R. E. Ulanowicz in a draft paper with C. Puccia (Ulanowicz & Puccia, 1988, unpubl.). Let $G = (g_{ij})$ and $G' = (g_{ij}')$ be two $n \times n$ transference matrices. Then, the difference between the direct effect of j on i in the input environ of the latter and the direct effect of i on j in the output environ of the former is given by $D = (d_{ij}) = G' - G^T$. Since $-1 \leqslant d_{ij} \leqslant +1$, the elements of D, representing the *net direct flow* from each j to each i in the network, can be interpreted as nondimensionally measuring the direct utility (positive or negative) experienced by i through its direct interactions with j. D may be termed a *relative* (or *nondimensional*) *direct, local, immediate* or *micro utility* (or *value*) *matrix*. Its dimensionless elements in effect quantify the single-signed digraphs of loop analysis discussed in the introduction to Section C.3 above. From D may be computed other utility measures; an *absolute* (or *dimensional*) counterpart of D, $\Delta = (\delta_{ij})$; a *relative universal, global, direct plus indirect, ultimate* or *macro utility matrix*, $U = (u_{ij})$; and an absolute counterpart of the latter, $\Upsilon = (\upsilon_{ij})$. These measures are developed below.

The concept of 'utility' just introduced, is in the spirit of the same term as originated by J. von Neumann and O. Morgenstern (1944, 1947) in mathematical game theory to describe the outcome or payoff of a game (Bohnert, 1954). All the measures D, Δ, U and Υ are contextual – *change the network, change the utilities* à la Patten & Witkamp (1967). D and U will be expressed in units of nondimensional, and Δ and Υ in dimensional, '*utiles*'. '"Utile" is a popular word for a "unit" on the utility scale established by any given utility function...' (Davis, 1954, p. 13). For further entries into the literature on utility theory as developed and employed in operations research and economics, see Luce & Raiffa (1957, Chapter 2) and Fishburn (1970).

K. Kawasaki (pers. comm., Appendix AC.2) has provided a proof that the sum of powers of D converge to a *transitive closure* matrix U,

$$I+D+D^2+\ldots+D^m+\ldots = (I-D)^{-1} = U,$$

accounting for all causality transmitted over all paths of all lengths in a network (Patten *et al.*, 1976, p. 521), if and only if all absolute values of the eigenvalues of D are less than one. The elements of U range $-\infty < u_{ij} < \infty$, and represent both the magnitudes and signs of integral effects propagated from j to i directly and indirectly over the system network. U thus provides a nondimensional measure of direct plus indirect utility, and from it (see below) a dimensional measure Υ can also be computed. In addition to single-signed digraph applications, Kawasaki's convergence theorem can also be employed to implement a quantitative version of the double-signed qualitative indirect effects analysis of Section C.3.2. These developments will be presented in the following sections using several hypothetical ecological examples and compartment models for illustration. The first of these will be described more didactically than the others.

C.5.1 ***Hypothetical examples***

C.5.1i *Example* 1: *Two predators, one prey*

The digraphs of Figs. C.3(*a*) represent two predators (compartments 2 and 3) competing for unequal shares of a single prey (1). The numbers in Fig. C.3(*a*.1) are hypothetical, but denote fluxes (dimensioned, say, $ML^{-k}T^{-1}$ in the mass–length–time dimensional system, where $k = 2$

Fig. C.3. (*a*.1) Flow or transition digraph for two predator, one prey Example 1; the flows (transitions) are hypothetical. (*a*.2) Corresponding direct (local) and (*a*.3) direct plus indirect (global) effects digraph for (*a*.1). (*b*.1) Flow (transition), (*b*.2) direct effects and (*b*.3) direct plus indirect effects digraphs for a hypothetical two prey, one predator system, Example 2. (*c*.1) Flow (transition) digraph for a value-infeasible, three element food chain. Example 3a. (*c*.2) Flow (transition), (*c*.3) local effects and (*c*.4) global effects digraphs for a value-feasible food chain, Example 3b.

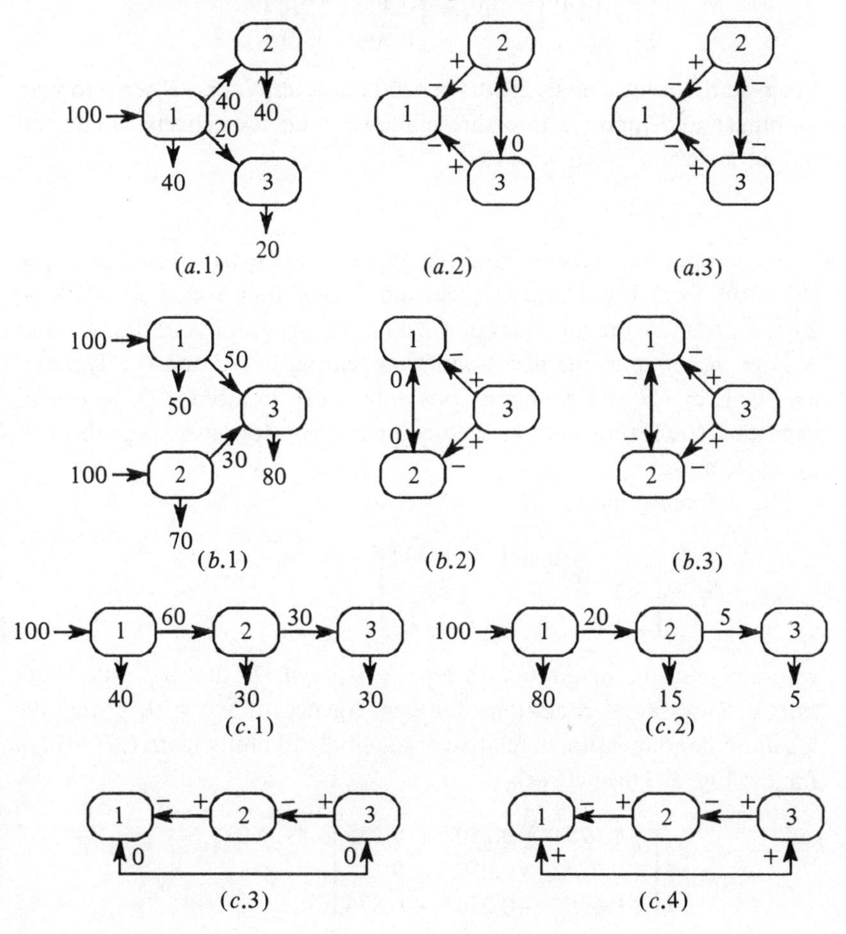

for area and $k = 3$ for volume) of energy or matter during some time period. Input and output environ flow intensities for this system are:

$$G_1' = \begin{bmatrix} 0 & 0 & 0 \\ 1 & 0 & 0 \\ 1 & 0 & 0 \end{bmatrix} \quad G_1 = \begin{bmatrix} 0 & 0 & 0 \\ 0.400 & 0 & 0 \\ 0.200 & 0 & 0 \end{bmatrix},$$

where the subscripts in G_1' and G_1 refer to this first example. Interpreting, in G_1 the two predators derive unequal relative (reflecting absolute; Fig. C.3*a*) allocations from inputs to the prey. In G_1', one unit of output of both predators requires a unit of inflow ($g_{21}' = g_{31}' = 1$) from the prey. Time and path integrated flow intensities over all paths of all lengths are represented by the following $(I-G_1')^{-1}$ and $(I-G_1)^{-1}$ matrices:

$$N_1' = \begin{bmatrix} 1 & 0 & 0 \\ 1 & 1 & 0 \\ 1 & 0 & 1 \end{bmatrix} \quad N_1 = \begin{bmatrix} 1 & 0 & 0 \\ 0.400 & 1 & 0 \\ 0.200 & 0 & 1 \end{bmatrix}.$$

From static environ analysis, the general matrixes N' and N serve to map outputs y and inputs z into throughflows T in, respectively, input and output environs. That is,

$$T = (N')^T y = Nz.$$

These computations for N_1' and N_1, applied to $y_1 = [40\ 40\ 20]^T$ and $z_1 = [100\ 0\ 0]^T$ from Fig. C.3(*a*.1), yield the throughflow vector $T_1 = [100\ 40\ 20]^T$. Since there are no cycles in the digraph, integral intensities, N_1' and N_1, are the same as the direct values appearing in G_1' and G_1. The unit diagonals in N_1' and N_1 signify one entry of introduced (G_1), or exit of departing (G_1'), substance at each compartment during passage through the system.

The difference matrix, $D_1 = G_1' - G_1^T$, is:

$$D_1 = \begin{bmatrix} 0 & -0.400 & -0.200 \\ +1 & 0 & 0 \\ +1 & 0 & 0 \end{bmatrix},$$

whose eigenvalue magnitudes are $|\lambda_1| = |\lambda_2| = 0.775$ and $|\lambda_3| = 0$. These satisfy Kawasaki's conditions for convergence of $\sum_{m=0}^{\infty} D_1^m$, and the resultant nondimensional, relative or unit integral utility matrix, $U = (u_{ij})$, for the Fig. C.3(*a*) network is:

$$U_1 = \begin{bmatrix} +0.625 & -0.250 & -0.125 \\ +0.625 & +0.750 & -0.125 \\ +0.625 & -0.250 & +0.875 \end{bmatrix}.$$

In D_1, a unit input into the prey compartment (column 1) generates positive direct relative utilities ($d_{21} = d_{31} = +1.000$) at compartments 2 and 3, whereas unit inflows to 2 and 3 cause direct relative disutilities ($d_{12} = -0.400$ and $d_{13} = -0.200$) to be experienced at compartment 1. In the U_1 matrix, reading columnwise, a unit input into the prey compartment generates, after all gains and losses are accounted for, equal integral net benefits ($u_{11} = u_{21} = u_{31} = +0.625$) to all three compartments. In column 2, a unit of inflow in Predator 2 produces $u_{22} = +0.750$ units of positive utility to that compartment at a net integral cost of $u_{12} = u_{32} = -0.250$ units of disutility experienced by both Prey 1 and Predator 3. In column 3, one unit of input into Predator 3 generates $u_{33} = +0.875$ units of utility for that compartment, while Prey 1 and Predator 2 suffer net disutilities of $u_{13} = u_{23} = -0.125$ units.

In absolute terms, since throughflows at each of the nodes in the digraph differ, actual utilities and disutilities [dimensioned $ML^{-k}T^{-1}$] will vary from the relative relationships just described. With the given throughflow vector T_1, the absolute direct benefits gained by Predators 2 and 3 from consuming Prey 1 in the quantities indicated in Fig. C.3(*a*) are $\delta_{21} = +1.000 \times 40 = +40.0$ for Predator 2, and $\delta_{31} = +1.000 \times 20 = +20.0$ for Predator 3. The prey in these interchanges sustains direct absolute disutilities of $\delta_{12} = -0.400 \times 100 = -40.0$ due to consumption by Predator 2, and $\delta_{13} = -0.200 \times 100 = -20.0$ due to Predator 3. The total absolute effects, direct plus indirect, are as follows. The +0.625 relative benefit experienced by each compartment per unit input to Prey 1 translates, for each 100 units of prey throughflow, into absolute universal benefits of $\upsilon_{11} = +0.625 \times 100 = +62.5$ for the prey, $\upsilon_{21} = +0.625 \times 40 = +25.0$ for Predator 2, and $\upsilon_{31} = +0.625 \times 20 = +12.5$ for Predator 3. Input into Predator 2 produces an absolute positive utility of $\upsilon_{22} = +0.750 \times 40 = +30.0$ for that compartment, and negative utilities of $\upsilon_{12} = -0.250 \times 100 = -25.0$ for Prey 1 and $\upsilon_{32} = -0.250 \times 20 = -5.0$ for Predator 3. Input to Predator 3 results in $\upsilon_{33} = +0.875 \times 20 = +17.5$ units of positive utility for that compartment, and $\upsilon_{13} = -0.125 \times 100 = -12.5$ and $\upsilon_{23} = -0.125 \times 40 = -5.0$ units of disutility for Prey 1 and Predator 2, respectively. In this paragraph, the elements of dimensional, direct (local) and integral (direct plus indirect, or global) absolute utility or value matrices, Δ and Υ respectively, have been described.

The diagonal entries of U and Υ require interpretation. In U_1 and two rows of Υ_1 these elements are observed to be sums of absolute values of nondiagonal row entries. Rows represent incoming influence from other nodes, which is widely distributed throughout the network due to the transitive closure process. Thus, diagonal entries may reflect a utility outcome of direct and indirect adaptive linkage of each compartment into the rest of the system – how well it conforms to or articulates with the niche and extended niche (environ) context afforded by the network. For niche perspectives on environs see Patten & Auble (1980, 1981) and Patten (1981). Levine's (1977) concept of 'extended niche' is equivalent to an input environ. The term 'adaptation' already has diverse ecological and evolutionary connotations, and is best avoided here. To indicate in-system 'fitness' both as contribution to a coherent whole and success in receiving benefits sufficient to complete life cycles, the terms 'affordance' or 'input fittedness' and 'effectance' or 'output fittedness' may be employed. *Affordance* is the sum total of opportunities (direct and indirect interrelationships, with associated utilities and disutilities) a network provides to its constituents. Its reciprocal match interpreted in terms of characteristics or traits at the compartment level is *effectance*, the adjustment of components to the affordances that bind them into a system.

The niche-like concept of affordance was introduced in the study of animal vision by Gibson (1977), and has been extended by ecological psychologists to mean the many properties of environment which permit organism requirements to be fulfilled (Turvey & Shaw, 1978). This corresponds in ecology to the original 'habitat niche' of Grinnell (1917) which is, as Patten & Auble (1981, p. 916) have indicated, an 'input niche' restriction of input environs. The reciprocal of affordance, what the organism affords to its environment, Gibson called 'effectivity' and Patten (1982*a*) 'effectance'. Effectance has the reverse orientation from affordance, and may be taken to correspond to Elton's (1927) 'role' or 'function niche', which Patten & Auble (op. cit., p. 916) pointed out was an 'output niche' restriction of output environs. Fittedness, with its input and output modifiers, substitutes for both these terms, and is selected for its etymological link to fitness.

Input and output fittedness are therefore reciprocal matched pairs, network oriented systems concepts like input and output environs. Input fittedness is an offer of both macro and micro utilities and disutilities from the system into which a fit is to be made. Output fittedness is an expression, implicit in genetic or other fitness at the micro level, of acceptance by the organism (or other unit) and offer back of macro utilities and disutilities to the system at large and its other constituents. Thus, fitness is a local concept (and in many applications, is totally divorced from any broader systems framework) and fittedness a global, contextual one with two perspectives. Input and output fittedness are two perspectives on the same thing; the first is the direct and indirect afferent system properties incoming to components (affordance or utility from input environs), and the second efferent traits projected outward from components to the system (effectance or utility to output environs). These concepts measure in utility or value terms the degree of articulation or fit of parts and wholes. Burns (1989) has recently provided a more structural measure of this articulation; he equates the average of component fittedness to the 'redundancy' portion of Ulanowicz' (1986) measure of network *ascendency*.

Here, tentatively, the diagonal elements of U and Υ will be taken as utility oriented measures of the input–output fittedness pair. Accordingly, from U_1, Prey 1 may be interpreted as fitting least into the Fig. C.3 (*a*) system ($u_{11} = 0.625$ utiles), Predator 3 most ($u_{33} = 0.875$), and Predator 2 in between ($u_{22} = 0.750$). In terms of values actually experienced, prey are best articulated ($\upsilon_{11} = +62.5$ $\mathrm{ML}^{-k}\mathrm{T}^{-1}$–utiles), Predator 2 next ($\upsilon_{22} = +30.0$), and Predator 3 least ($\upsilon_{33} = +17.5$).

It is obvious from the foregoing that, just as the N' and N transitive closure matrices serve to map outputs and inputs, respectively, into throughflows T, the nondimensional direct and integral utility matrices, D and U, also reflecting the transitive closure property, map these throughflows into absolute (dimensional) positive and negative utilities, δ_{ij} and υ_{ij}. The formulations that achieve this mapping in each case are:

$$\begin{aligned}\Delta &= [\mathrm{diag}\, T]\, D \\ &= [\mathrm{diag}\, (N')^T y]\, D = [\mathrm{diag}\, Nz]\, D,\end{aligned}$$

and

$$\begin{aligned}\Upsilon &= [\mathrm{diag}\, T]\, U \\ &= [\mathrm{diag}\, (N')^T y]\, U = [\mathrm{diag}\, Nz]\, U,\end{aligned}$$

where [diag·] denotes diagonal matrices of the indicated vectors. The elements of Δ and Υ denote absolute utilities accruing explicitly and implicitly, and directly and indirectly, to all compartments from all others within the defined network. For the Fig. C.3 (*a*.1) system, these matrices are:

$$\Delta_1 = \begin{bmatrix} 0 & -40 & -20 \\ +40 & 0 & 0 \\ +20 & 0 & 0 \end{bmatrix} \qquad \Upsilon_1 = \begin{bmatrix} +62.5 & -25.0 & -12.5 \\ +25.0 & +30.0 & -5.0 \\ +12.5 & -5.0 & +17.5 \end{bmatrix}$$

The above expressions for Δ and Υ appear to provide a link between materialistic flows typically measured by process and ecosystem ecologists and benefits or costs to be derived from such flows, which is the kind of information that is meaningful in organismal and population ecology. In other words, these formulations provide a potential connection between the two sides of ecology (system vs. organism oriented) that have long been separated by thinking and methodology.

Only H. T. Odum among contemporary ecologists has steadfastly pursued a relationship between physical energy and issues of value. He quantifies 'energy quality' in terms of *embodied energy*, what he names '*emergy*' – 'the energy of one type required in transformations to generate a flow or storage...a measure of value in the sense of what has been contributed', and *transformity* – 'the amount of energy of one type [or quality] required to generate a unit of energy of another type' (Odum, 1988, p. 1135). 'Flows of energy develop hierarchical webs in which inflowing energies interact and are transformed...into energy forms of higher quality.... [These] new forms carry the embodiment of larger amounts of lower-quality energy used in the transformation process. Tracing of embodied energy through webs enables flows and products to be related to energy sources. Higher quality flows require more embodied energy and have greater amplifier effects when they feed back. Consequently, embodied energy is a measure of value...' (Odum, 1983, p. 251).

Emergy is calculated by a traceback procedure through a historical sequence of energy transformations that corresponds, in a limited sense (acyclic chain vs. arbitrarily complex cyclic network), to an energy input environ for the object or process at the top. '[I]nput environs..., which encompass the entire past history of a substance's movements within a system, are the subunits of systems that define and quantify embodiment' (Patten, 1985, p. 56). The catenary thinking inherent in the 'hierarchical webs' of Odum's traceback approach derives from classical Lindeman–Hutchinson trophic dynamics (Lindeman, 1942), in which bioenergy steps along the discrete trophic levels of simple food chains as a result of feeding events. In deriving emergy, Odum intuitively converts his acknowledged energy networks into embodying causal chains by an analogous, if informal, mental mapping to the macrochains achieved formally by the *network unfolding* methodology of environ analysis (Higashi *et al.*, 1990; Chapter 5 in this volume). The concept of emergy, while restricted to the energy realm, is nevertheless in the same spirit as the utility measures Δ and Υ, which appear to provide a much more general method of calculating 'embodiment' – the essential historical property of networks spanned by input environs. Network flows do not need to be, although they can be, expressed in energy terms for the Δ and Υ measures to apply. The above expressions for these measures take system outputs y or inputs z of any description, map them first into throughflows T at the component nodes of the network, and thence into positive (benefits) or negative (costs) values Υ, expressed at various points within the system.

To illustrate further, we can consider competition, which is central in population biology but virtually absent in considerations of ecosystems. The two hypothetical predators of Fig. C.3(*a*.1) are competitors for the common prey in the depicted situation. No flow arc directly interconnects these two compartments in either direction, hence qualitatively their local interaction is neutralism, ${s_{23}}^{\delta} = {s_{32}}^{\delta} = (0, 0)$, as depicted in Fig. C.3(a.2). But an implicit *exploitation competition* in which a physical resource is shared exists by virtue of the predators' uses of the same prey. The nonmaterialistic interaction of competition does not appear explicitly in either the flow or direct effects diagram, but it does emerge from its implicit status in the network, defined in relative (nondimensional) terms, utiles, in matrix U, and in absolute (dimensional) terms, $ML^{-k}T^{-1}$-utiles, in matrix Υ. In U, Predator 2 is shown to be twice as strong a competitor as 3 per unit prey received ($u_{32} = -0.250$ vs. $u_{23} = -0.125$). But, because the absolute feeding flows differ, both predators exert equal actual negative effects upon the other ($\upsilon_{32} = \upsilon_{23} = -5.0$). Just as these entries in both U and Υ define the implicit competition quantitatively, their signs define it qualitatively: ${s_{23}}^{\upsilon} = {s_{32}}^{\upsilon} = (su_{23}, su_{32}) = (s\upsilon_{23}, s\upsilon_{32}) = (-, -)$. An explicit effects arc carrying these binegative influences is depicted qualitatively (signs only) in Fig. C.3(*a*.3). I will return later to the further use of the U and Υ matrices for qualitative double-sign analysis to determine globally manifested types of interactions that exist by virtue of context between every node pair, directly interactive or not, within a network.

One final point from this example. If D is the simple difference between direct transition intensities, $G' - G^T$, $(I - D)^{-1} = U$ is not correspondingly such a difference between integral inflow and outflow intensities of each compartment, i.e. $(I - G')^{-1} - (I - G^T)^{-1} = N' - N^T$. This latter formulation, whose entries may like those of U be positive, negative or zero, is a *difference of integrals*, and has the interpretation of net integral direct and indirect flow intensities resulting from the entire network interconnection from j to i and i to j. U, on the other hand, is an *integral of differences*; and while it has a similar interpretation to $N' - N^T$, its values are different and the signs of these values turn out to be correct, whereas those of $N'-N^T$ are not, for a valid qualitative analysis of indirect effects. This is taken as evidence that U, and not $N'-N^T$, is the proper nondimensional measure of integral utilities in networks.

C.5.1ii *Example 2: Two prey, one predator*

The flow digraph of Fig. C.3(*b*.1) depicts results for a single predator (3) feeding unequally on two alternative prey items (1 and 2). Input, output and throughflow vectors are $z_2 = [100\ 100\ 0]^T$, $y_2 = [50\ 70\ 80]^T$ and $T_2 = [100\ 100\ 80]^T$. Flow intensity matrices for this system are:

$$G_2' = \begin{bmatrix} 0 & 0 & 0 \\ 0 & 0 & 0 \\ 0.625 & 0.375 & 0 \end{bmatrix} \quad G_2 = \begin{bmatrix} 0 & 0 & 0 \\ 0 & 0 & 0 \\ 0.500 & 0.300 & 0 \end{bmatrix},$$

and integral flow intensities:

$$N_2' = \begin{bmatrix} 1 & 0 & 0 \\ 0 & 1 & 0 \\ 0.625 & 0.375 & 1 \end{bmatrix} \quad N_2 = \begin{bmatrix} 1 & 0 & 0 \\ 0 & 1 & 0 \\ 0.500 & 0.300 & 1 \end{bmatrix}.$$

Difference matrices are:

$$D_2 = \begin{bmatrix} 0 & 0 & -0.500 \\ 0 & 0 & -0.300 \\ +0.625 & +0.375 & 0 \end{bmatrix} \quad \Delta_2 = \begin{bmatrix} 0 & 0 & -50 \\ 0 & 0 & -30 \\ +50 & +30 & 0 \end{bmatrix}.$$

D_2 has eigenvalue magnitudes $|\lambda_1| = |\lambda_2| = 0.652$ and $|\lambda_3| = 0$, which satisfy the convergence conditions for $\sum_{m=0}^{\infty} D_2{}^m$, and the nondimensional global utility matrix U that results, and its dimensional counterpart Υ, are:

$$U_2 = \begin{bmatrix} +0.781 & -0.132 & -0.351 \\ -0.132 & +0.921 & -0.211 \\ +0.439 & +0.263 & +0.702 \end{bmatrix} \quad \Upsilon_2 = \begin{bmatrix} +78.1 & -13.2 & -35.1 \\ -13.2 & +92.1 & -21.1 \\ +35.1 & +21.1 & +56.1 \end{bmatrix}.$$

The U_2 matrix shows that the two prey experience an *interference competition* of equal relative magnitude ($u_{12} = u_{21} = -0.132$ utiles) as a result of the predation they suffer (see Holt, 1977, on '*apparent competition*'). No flow arc connects these two compartments in Fig. C.3(*b*.1), and there is also no direct effects arc, denoting $s_{21}{}^{\delta} = s_{12}{}^{\delta} = (sd_{21}, sd_{12}) = (0, 0)$ (where d_{21} and d_{12} are elements of D_2) in Fig. C.3(*b*.2). But, a total (direct plus indirect) effects arc does connect them in Fig. C.3(*b*.3) as a result of their two outflows, f_{31} and f_{32}, to the predator. This indicated competition, $s_{21}{}^{\upsilon} = s_{12}{}^{\upsilon} = (su_{21}, su_{12}) = (-, -)$, is not of the exploitation type because Prey 1 and 2 jointly form the diet of the predator. The two prey do not compete for a joint resource; they compete implicitly, by avoidance and other means, not to become a resource. One unit of input into 1 and 2 generates $u_{31} = +0.439$ and $u_{32} = +0.263$ units of positive utility to Predator 3, at costs of -0.351 and -0.211 negative utiles to the respective source organisms. Throughflow-weighted value relationships, with dimensions $ML^{-k}T^{-1}$, appear in matrix Υ_2.

Based on both relative and absolute values of diagonal elements, fittedness ordering is Prey 2 best ($u_{22} = +0.921$, $\upsilon_{22} = +92.1$), Prey 1 next ($u_{11} = +0.781$, $\upsilon_{11} = +78.1$) and Predator 3 last ($u_{33} = +0.702$, $\upsilon_{33} = +56.1$). These utilities are generally higher than those of Example 1, in which input and output are half the present level (100 $ML^{-k}T^{-1}$). In Example 1 fittedness values were ($u_{11} = +0.625$, $\upsilon_{11} = +62.5$) for Prey 1, ($u_{22} = +0.750$, $\upsilon_{22} = +30.0$) for Predator 2, and ($u_{33} = +0.875$, $\upsilon_{33} = +17.5$) for Predator 3. The Example 2 predator, enjoying double the resource throughflow (200 $ML^{-k}T^{-1}$), experiences lower relative fittedness (+0.702) than the two competing Example 1 predators (+0.875 and +0.750), but higher absolute fittedness (+56.1, vs. +30.0 and +17.5). All three prey in these two examples have 100 $ML^{-k}T^{-1}$ of input and throughflow. Yet, compared to the single prey of Example 1 both members of the competing pair in the present case have high fittednesses both relatively (+0.921 and +0.781, vs. +0.625) and absolutely (+92.1 and +78.1, vs. +62.5).

C.5.1iii *Example* 3: *Simple food chains*

Consider the three level feeding cascade shown in Fig. C.3(*c*.1). This will be referred to as Example 3a. Input, output and throughflow vectors are $z_{3a} = [100\,0\,0]^T$, $y_{3a} = [40\,30\,20]^T$ and $T_{3a} = [100\,60\,30]^T$, and the relevant environ analysis matrices:

$$G_{3a}' = \begin{bmatrix} 0 & 0 & 0 \\ 1 & 0 & 0 \\ 0 & 1 & 0 \end{bmatrix} \quad G_{3a} = \begin{bmatrix} 0 & 0 & 0 \\ 0.600 & 0 & 0 \\ 0 & 0.500 & 0 \end{bmatrix}$$

$$N_{3a}' = \begin{bmatrix} 1 & 0 & 0 \\ 1 & 1 & 0 \\ 1 & 1 & 1 \end{bmatrix} \quad N_{3a} = \begin{bmatrix} 1 & 0 & 0 \\ 0.600 & 1 & 0 \\ 0.300 & 0.500 & 1 \end{bmatrix}$$

$$D_{3a} = \begin{bmatrix} 0 & -0.600 & 0 \\ +1 & 0 & -0.500 \\ 0 & +1 & 0 \end{bmatrix}.$$

The absolute values of the eigenvalues of D_{3a} are $|\lambda_1| = |\lambda_2| = 1.049$ and $|\lambda_3| = 0$. These do not satisfy the convergence conditions for $\sum_{m=0}^{\infty} D_{3a}{}^{m}$, precluding further analysis.

The inability to assess utilities in this case, I will conjecture, implies that Example 3a, with the given flow values assigned to it, is an infeasible configuration for a food chain. No value can accrue to the members of such a chain, which has high *progressive* (Lindeman, 1942) or *ecological* (Kozlovsky, 1968) *efficiencies* as denoted by the transition coefficients $g_{21} = 0.600$ and $g_{32} = 0.500$. These efficiencies are much greater than the mean values typically observed in empirical data, which vary greatly but are low enough to have motivated a '10% rule' (Slobodkin, 1960) as a generally observed level. Reasons usually stated for this are thermodynamic; physiological and biochemical processes of energy conversion are accompanied by second-law dissipation, which precludes high transformation efficiencies. The Fig. C.3(*c*.1) system suggests another kind of explanation that is not thermodynamic (ontic), but a purely mathematical (epistemic) expression of network organization: if a food chain cannot yield value to its members, then it will not have evolved and will not be found in nature. Network constraints can be added to thermodynamic constraints, in the present case, to achieve a dual system of explanation in ontic and epistemic realms. Explanations sufficient in one frame of reference may not be necessary, or may mask other explanations that are also sufficient and interesting in other frames.

Exploration of admissible transfer efficiencies along trophic sequences remains for future research, but one utility-feasible food chain, which is isomorphic to that of Example 3a, but with lower progressive or ecological efficiencies, is shown in Fig. C.3(*c*.2). This will be referred to as Example 3*b*; its low progressive efficiencies, more typical of empirical observations, yield computable utilities to its member compartments. Input, output and throughflow vectors are $z_{3b} = [100\ 0\ 0]^T$, $y_{3b} = [80\ 15\ 5]^T$ and $T_{3b} = [100\ 20\ 5]^T$, and the analysis matrices:

$$G_{3b}' = \begin{bmatrix} 0 & 0 & 0 \\ 1 & 0 & 0 \\ 0 & 1 & 0 \end{bmatrix} \qquad G_{3b} = \begin{bmatrix} 0 & 0 & 0 \\ 0.200 & 0 & 0 \\ 0 & 0.250 & 0 \end{bmatrix}$$

$$N_{3b}' = \begin{bmatrix} 1 & 0 & 0 \\ 1 & 1 & 0 \\ 1 & 1 & 1 \end{bmatrix} \qquad N_{3b} = \begin{bmatrix} 1 & 0 & 0 \\ 0.200 & 1 & 0 \\ 0.050 & 0.250 & 1 \end{bmatrix}$$

$$D_{3b}' = \begin{bmatrix} 0 & -0.200 & 0 \\ +1 & 0 & -0.250 \\ 0 & +1 & 0 \end{bmatrix} \qquad \Delta_{3b} = \begin{bmatrix} 0 & -20 & 0 \\ +20 & 0 & -5 \\ 0 & +5 & 0 \end{bmatrix}.$$

The eigenvalue magnitudes for D_{3b} are $|\lambda_1| = |\lambda_2| = 0.652$ and $|\lambda_3| = 0$, which satisfy Kawasaki's conditions for convergence of $\sum_{m=0}^{\infty} D_{3b}{}^m$. The convergent unit utility matrix U, and its dimensional counterpart Υ, are:

$$U_{3b} = \begin{bmatrix} +0.862 & -0.138 & +0.035 \\ +0.690 & +0.690 & -0.172 \\ +0.690 & +0.690 & +0.828 \end{bmatrix} \quad \Upsilon_{3b} = \begin{bmatrix} +86.2 & -13.8 & +3.5 \\ +13.8 & +13.8 & -3.5 \\ +3.5 & +3.5 & +4.1 \end{bmatrix}.$$

Let the three compartments be a Producer (1), Herbivore (2) and Carnivore (3). As in previous examples, the principal diagonals of both matrices are positive. In U_{3b}, for each unit of base resource or input consumed, the Herbivore gains $u_{21} = +0.690$ utiles and imposes $u_{12} = -0.138$ units of disutility on the Producer. The top predator derives the same amount of utility ($u_{32} = +0.690$) from each consumed unit of the Herbivore compartment, exacting a cost from the latter of ($u_{23} = -0.172$). Reading columnwise, in column 1 both Herbivore and Carnivore obtain equal benefits ($u_{21} = u_{31} = +0.690$) from the Producer. In column 2, each unit of input to the Herbivore is accompanied by $u_{12} = -0.138$ disutility units at the basal compartment. In column 3, a unit of inflow into the Carnivore slightly benefits the Producer ($u_{13} = +0.035$ utiles) and suppresses the Herbivore ($u_{23} = -0.172$). Corresponding absolute utilities, in dimensional $ML^{-k}T^{-1}$-utiles, appear in Υ_{3b}. Regarding affordance/fittedness, these measures are relatively highest for Producer ($u_{11} = +0.862$) and Carnivore ($u_{33} = +0.828$), and lower for the intermediate Herbivore ($u_{22} = +0.690$). Absolute affordance/fittednesses decrease along the food chain, however ($\upsilon_{11} = +86.2$, $\upsilon_{22} = +13.8$ and $\upsilon_{33} = +4.1$). As in prior examples, entries in matrix Υ_{3b} are magnitude symmetric across the principal diagonal, $|\upsilon_{ij}| = |\upsilon_{ji}|$. This reflects the *identity constraints* that can always be written to signify interactive linkage between any two compartments. That is, in the output environ orientation, if a flow or transition f_{ij} can be interpreted as the component of throughflow at compartment j that exits from j destined for i, f_{ij}'', or alternatively, in the input environ orientation, as the amount of throughflow at i derived from j, f_{ij}', then $f_{ij}'' \equiv f_{ij}'$. It is through this identity that the two compartments form an interactive system; before the joining they were isolated, unrelated entities. Identity constraints, specifying coupling, are the origin of systems.

In this example, the Producer (1) benefits indirectly as the top consumer (3) suppresses the intermediate Herbivore (2) through indirect consumption of original basal production. The Producer and top predator both contribute positively to each other through indirect interactions with an interconnecting compartment even though their direct flow interaction is null. In Fig. C.3(*c*.2) there is observed to be no flow arc connecting Producer to Carnivore 3, $f_{31} = g_{31} = 0$. This denotes a neutral direct or micro-interaction between the two compartments in the effects domain, i.e. $s_{31}{}^{\delta} = (sd_{31}, sd_{13}) = (0, 0)$ (Fig. C.3(*c*.3)). As indicated, the 0 signs in the ordered pair and digraph segment $1\,{}^{0}\!\leftrightarrow^{0}\,3$ of the figure represent the signs of elements d_{31} and d_{13} of D_{3b}, namely $sd_{31} = sd_{13} = 0$, ordered to take into account the transposition change from flows to effects. The integral, in-system macro relationship between host and top predator is not this direct effect, neutralism, however; global integration of direct and indirect effects over the entire interconnecting network results in a different relation, which is *apparent* or *sequential mutualism*. This is determined from the corresponding signs in the utility matrices, namely $su_{31} = su_{13} = +$, $sv_{31} = sv_{13} = +$, from which $s_{31}{}^{v} = (su_{31}, su_{13}) = (sv_{31}, sv_{13}) = (+, +)$ and the direct plus indirect effects digraph segment $1\,{}^{+}\!\leftrightarrow^{+}\,3$ illustrated in Fig. C.3(*c*.4) both follow.

The sign pairs required to identify direct and global interactions between every pair of compartments in a network, from the set of nine qualitative interaction types, can be read directly from signs-only versions of the D or Δ and U or Υ matrices. These sign matrices derived from D_1, D_2 and D_3, or Δ_1, Δ_2 and Δ_3, of the three previous examples are:

$$sD_1 = s\Delta_1 = \begin{bmatrix} 0 & - & - \\ + & 0 & 0 \\ + & 0 & 0 \end{bmatrix} \qquad sD_2 = s\Delta_2 = \begin{bmatrix} 0 & 0 & - \\ 0 & 0 & - \\ + & + & 0 \end{bmatrix}$$

$$sD_3 = s\Delta_3 = \begin{bmatrix} 0 & - & 0 \\ + & 0 & - \\ 0 & + & 0 \end{bmatrix}.$$

In each of these, the interaction types $s_{ij}{}^{\delta}$, directed from any row element i to any column element j, responsible for generating flows $f_{ij} \geqslant 0$ in the reverse directions from j to i in the respective flow digraphs (Figs. C.3*a*.1, *b*.1, and *c*.2), can be determined from the ordered pairs of signs (sd_{ij}, sd_{ji}). To illustrate, in Examples 1 and 2 flows $f_{31} > 0$ over flow arcs $1 \rightarrow 3$ are both predation generated since, from both $sD_1 = s\Delta_1$ and $sD_2 = s\Delta_2$, $s_{31}{}^{\delta} = (sd_{31}, sd_{13}) = (s\delta_{31}, s\delta_{13}) = (+, -)$, signifying the effects arc $3\,{}^{+}\!\rightarrow^{-}\,1$. Similarly, in Examples 1 and 3, flows $f_{21} > 0$ over flow arcs $1 \rightarrow 2$ are also produced by predator–prey interactions as, in $sD_1 = s\Delta_1$ and $sD_3 = s\Delta_3$, $s_{21}{}^{\delta} = (sd_{21}, sd_{12}) = (s\delta_{21}, s\delta_{12}) = (+, -)$, indicating $2\,{}^{+}\!\rightarrow^{-}\,1$. The double-signed effects arcs have reverse orientations to their corresponding flow arcs because flows to predators from prey result from feeding acts directed toward the prey from the predators. In Examples 1, 2 and 3, respectively, $f_{32} = 0, f_{21} = 0$ and $f_{31} = 0$; no flows interlink these compartments. From the sign matrices above, the direct effects in each case are seen to be neutralisms, $i^0 \leftrightarrow {}^0 j$: $s_{32}{}^{\delta} = (sd_{32}, sd_{23}) = (s\delta_{32}, s\delta_{23}) = (0,0)$, $s_{21}{}^{\delta} = (sd_{21}, sd_{12}) = (s\delta_{21}, s\delta_{12}) = (0,0)$, and $s_{31}{}^{\delta} = (sd_{31}, sd_{13}) = (s\delta_{31}, s\delta_{13}) = (0,0)$.

For total effects, the sum of direct and indirect, this same procedure is followed employing either U or Υ matrices. The sign matrices corresponding to these for the three examples are:

$$sU_1 = s\Upsilon_1 = \begin{bmatrix} + & - & - \\ + & + & - \\ + & - & + \end{bmatrix} \qquad sU_2 = s\Upsilon_2 = \begin{bmatrix} + & - & - \\ - & + & - \\ + & + & + \end{bmatrix}$$

$$sU_3 = s\Upsilon_3 = \begin{bmatrix} + & - & + \\ + & + & - \\ + & + & + \end{bmatrix}.$$

As the networks in question are all acyclic, the direct predations in each case are maintained also as the globally mediated interactions. But, as previously observed, direct neutralisms become indirect competitions in Examples 1 and 2, and indirect mutualism in Example 3. In Examples 1 and 2, net interactions between compartments 1 and 3 are both nihilisms as, from both $sU_1 = s\Upsilon_1$ and $sU_2 = s\Upsilon_2$, $s_{31}{}^{\upsilon} = (su_{31}, su_{13}) = (s\upsilon_{31}, s\upsilon_{13}) = (+,-)$. In Examples 1 and 3, net interactions between compartments 1 and 2 are also nihilistic; for both cases, $s_{21}{}^{\upsilon} = (su_{21}, su_{12}) = (s\upsilon_{21}, s\upsilon_{12}) = (+,-)$. The direct neutralisms observed between compartments 2 and 3 in Example 1, and 1 and 2 in Example 2, are global competitions: $s_{32}{}^{\upsilon} = (su_{32}, su_{23}) = (s\upsilon_{32}, s\upsilon_{23}) = (-,-)$ from $sU_1 = s\Upsilon_1$, and $s_{21}{}^{\upsilon} = (su_{21}, su_{12}) = (s\upsilon_{21}, s\upsilon_{12}) = (-,-)$ from $sU_2 = s\Upsilon_2$. In Example 3, the direct neutralism between compartments 1 and 3 becomes mutualism with a shift in observational scale to the global level: $s_{31}{}^{\upsilon} = (su_{31}, su_{13}) = (s\upsilon_{31}, s\upsilon_{13}) = (+,+)$.

The above set of hypothetical examples demonstrates conclusively that information derived from flows in networks exists for the construction of double-signed effects arcs that can serve to identify both local (through D or Δ, whose signs are the same) and global (in U or Υ), network mediated, interaction types in the qualitative sense. We now proceed to similar analyses applied to compartmental ecological models.

C.5.2 *Flow-storage model examples*

C.5.2i *Example 4: Oyster reef community*

Figure I.8 shows an energy flow model of an intertidal oyster reef ecosystem, as described by Dame & Patten (1981). This is the standard model for this book and is analyzed in several other chapters; it is described in greater detail in Appendix AI.3 of the Introduction. Flows are in kcal m^{-2} day^{-1}, and storages kcal m^{-2}. Input, output and throughflow vectors are $z_4 = [41.470\ 0\ 0\ 0\ 0\ 0]^T$, $y_4 = [25.16\ 6.176\ 8.172\ 3.579\ 0.430\ 0.359]^T$ and $T_4 = [41.470\ 22.244\ 8.172\ 8.481\ 2.510\ 0.686]^T$. The following computations were done by S. J. Whipple and K. Kawasaki. Direct and integral flow intensity and difference matrices are:

$$G_4' = \begin{bmatrix} 0 & 0 & 0 & 0 & 0 & 0 \\ 0.709 & 0 & 0 & 0.190 & 0.086 & 0.015 \\ 0 & 1 & 0 & 0 & 0 & 0 \\ 0 & 0.858 & 0.142 & 0 & 0 & 0 \\ 0 & 0.256 & 0.481 & 0.263 & 0 & 0 \\ 0.749 & 0 & 0 & 0 & 0.251 & 0 \end{bmatrix}$$

$$G_4 = \begin{bmatrix} 0 & 0 & 0 & 0 & 0 & 0 \\ 0.381 & 0 & 0 & 0.500 & 0.760 & 0.475 \\ 0 & 0.367 & 0 & 0 & 0 & 0 \\ 0 & 0.327 & 0.148 & 0 & 0 & 0 \\ 0 & 0.029 & 0.148 & 0.078 & 0 & 0 \\ 0.012 & 0 & 0 & 0 & 0.069 & 0 \end{bmatrix}$$

$$N_4' = \begin{bmatrix} 1 & 0 & 0 & 0 & 0 & 0 \\ 1 & 1.389 & 0.102 & 0.297 & 0.124 & 0.020 \\ 1 & 1.389 & 1.102 & 0.297 & 0.124 & 0.020 \\ 1 & 1.389 & 0.244 & 1.297 & 0.124 & 0.020 \\ 1 & 1.389 & 0.620 & 0.561 & 1.124 & 0.020 \\ 1 & 0.348 & 0.155 & 0.141 & 0.282 & 1.005 \end{bmatrix}$$

$$N_4 = \begin{bmatrix} 1 & 0 & 0 & 0 & 0 & 0 \\ 0.537 & 1.389 & 0.278 & 0.780 & 1.101 & 0.660 \\ 0.197 & 0.510 & 1.102 & 0.286 & 0.404 & 0.242 \\ 0.205 & 0.529 & 0.253 & 1.297 & 0.419 & 0.251 \\ 0.061 & 0.157 & 0.190 & 0.166 & 1.124 & 0.074 \\ 0.017 & 0.011 & 0.013 & 0.011 & 0.077 & 1.005 \end{bmatrix}$$

$$D_4 = \begin{bmatrix} 0 & -0.381 & 0 & 0 & 0 & -0.012 \\ +0.709 & 0 & -0.360 & -0.136 & +0.057 & +0.015 \\ 0 & +1.000 & 0 & -0.148 & -0.481 & 0 \\ 0 & +0.358 & +0.142 & +0.142 & -0.078 & 0 \\ 0 & -0.504 & +0.481 & +0.481 & 0 & -0.069 \\ +0.749 & -0.475 & 0 & 0 & +0.251 & 0 \end{bmatrix}$$

$$\Delta_4 = \begin{bmatrix} 0 & -15.79 & 0 & 0 & 0 & -0.514 \\ +15.75 & 0 & -8.172 & -3.035 & +1.465 & +0.326 \\ 0 & +8.172 & 0 & -1.206 & -1.206 & 0 \\ 0 & +3.035 & +1.206 & 0 & -0.661 & 0 \\ 0 & -1.465 & +1.206 & +0.661 & 0 & -0.172 \\ +0.514 & -0.326 & 0 & 0 & +0.172 & 0 \end{bmatrix}.$$

Eigenvalue magnitudes for D_4 are $|\lambda_1| = |\lambda_2| = 0.899$, $|\lambda_3| = |\lambda_4| = 0.222$, and $|\lambda_5| = |\lambda_6| = 0.052$, which satisfy the conditions for convergence of $\sum_{m=0}^{\infty} D_4{}^m$. The convergent universal utility matrix U_4 and its absolute value counterpart Υ, obtained by premultiplying U by the diagonal matrix of throughflows T_4, are:

$$U_4 = \begin{bmatrix} +0.833 & -0.223 & +0.071 & +0.013 & -0.027 & -0.012 \\ +0.424 & +0.599 & -0.194 & -0.036 & +0.065 & -0.001 \\ +0.394 & +0.547 & +0.741 & -0.200 & -0.061 & +0.007 \\ +0.208 & +0.287 & +0.002 & +0.946 & -0.056 & +0.005 \\ +0.001 & +0.066 & +0.437 & +0.166 & +0.911 & -0.061 \\ +0.423 & -0.435 & +0.255 & +0.068 & +0.177 & +0.976 \end{bmatrix}$$

$$\Upsilon_4 = \begin{bmatrix} +34.553 & -9.240 & +2.928 & +0.531 & -1.197 & -0.485 \\ +9.440 & +13.333 & -4.311 & -0.799 & +1.450 & -0.022 \\ +3.127 & +4.472 & +6.059 & -1.635 & -0.498 & +0.060 \\ +1.762 & +2.435 & +0.017 & +8.021 & -0.478 & +0.047 \\ +0.003 & +0.116 & +1.097 & +0.418 & +2.286 & -0.154 \\ +0.290 & -0.299 & +0.175 & +0.047 & +0.122 & +0.670 \end{bmatrix}.$$

The base of this oyster community food web model is twofold: inputs of plankton and suspended particulate matter through filter feeding of the pelecypods and other members of compartment 1, and the pool of accumulated detritus within the system, compartment 2, which comprises a secondary, time lagged or pooled energy source as the center of all cycling and related properties of the network. Inputs to compartment 1 must pass through 2 before distribution to other compartments, except the Predators (6) which directly feed upon 1 (Fig. I.8). The columns of U_4 provide relative information about the compartments as sources of value or disutility to others within the fixed network configuration; they denote the distribution within the system of time and path integrated net utilities derived from input or throughflows of energy at each of the components listed at the heads of the columns. Rows indicate in relative terms the function of compartments as receivers of positive or negative macro values from others; they quantify the within system contributions of integral utility to the indexed compartments derived from energy exchanges within the system. Columnwise interpretation is futuristic, that is, it has an output orientation. Interpretation by rows is retrospective, as appropriate to input environs. The Υ_4 matrix gives the same information as U_4, but expressed in absolute terms.

Columns 1 and 2 of U_4 quantify the relative utilities derived from the two energy source compartments. The first column shows that all compartments obtain positive benefits from compartment 1. The detrital pool (compartment 2) and top Predators (6) realize the greatest values from each unit of input or throughflow of the Filter Feeders (1), $u_{21} = +0.424$ and $u_{61} = +0.423$ utiles. Microbiota (3) also receive a relatively high degree of benefit, $u_{31} = +0.394$. The Meiofauna (4) receive only half the utility of these other compartments, $u_{41} = +0.208$. The Deposit Feeders (5) benefit the least, $u_{51} = +0.001$, although their position in the network and the origins, destinations and values of their inflows and outflows, and their standing stocks, in comparison with these same characteristics of compartments 4 and 6, give no prior indication why this should be so. In column 2, only the direct detritus feeding compartments 3, 4 and 5 register positive benefits from the detrital pool (2). Microbiota (3) are the greatest beneficiaries, $u_{32} = +0.547$, whereas Deposit Feeders (5) do not appear to live up to their name, $u_{52} = +0.066$. Both Filter Feeders (1) and Predators (6) suffer disutilities, $u_{21} = -0.223$ and $u_{61} = -0.435$ utiles, respectively, presumably because they contribute direct egestive and excretory losses of energy to the Detritus (2), uncompensated for by corresponding direct gains. Column 3 shows that all compartments, except the detritus pool substrate upon which the Microbes (3) draw, receive positive global value from the microbial community. Deposit Feeders (5) and, counter-intuitively, Predators (6) are the major beneficiaries: $u_{53} = +0.437$ and $u_{63} = +0.255$ utiles, respectively, per unit inflow or standing stock at 3. In column 4, compartments 1, 5 and 6 derive positive benefit, and 2 and 3 negative influence, from the Meiofauna (4). Column 5 shows that compartments 1, 3 and 4 are all disadvantaged by Deposit Feeders (5), whereas 2 and 6 are positively influenced. Predators (column 6) affect compartments 1, 2 and 5 negatively, and 3 and 4 positively, but all at relatively low levels of influence.

Each row of U_4 indicates integral relative utilities experienced by the indexed compartment per unit throughflow of the other compartments. The first row shows that Filter Feeders (1) experience negative ultimate effects from compartments 2, 5 and 6, but positive ones from compartments 3 and 4. In all cases except the relationship to Detritus (2), the utilities or disutilities are weak since no energy feedback to compartment 1 from the other compartments occurs within the network (Fig. I.8). The relatively strong disutility experienced from the relationship to Detritus (2), $u_{12} = -0.223$ utiles, must derive from the direct flow $f_{21} =$ 15.78 kcal m^{-2} d^{-1} in the model since integral direct plus indirect flows are all zero (first rows of the N_4' and N_4 matrices above). Row 2 shows that the detrital pool derives strongest integral benefit from the Filter Feeders (1), $u_{21} = +0.424$, and suffers moderate disutility through the activities of Microbiota (3), $u_{23} = -0.194$. Other macro effects on 2 are all small, positive from the Deposit Feeders (5) and negative from Meiofauna (4) and Predators (6). The third row indicates that the microbial community (3) receives strongest direct and indirect integral benefits from its detrital substrate (2) and the filter feeding antecedents (1) of this organic matter pool: $u_{32} = +0.547$ and $u_{31} = +0.394$ utiles. Moderate disutility is suffered by bacteria from the Meiofauna (4), $u_{34} = -0.200$. Effects from Deposit Feeders (5) are weak and negative, and those from Predators ten-fold weaker but positive. Row 4 shows strongest support for the Meiofauna coming from the detrital pool (2), $u_{42} = +0.287$, and Filter Feeders, $u_{41} = +0.208$. The remaining effects propagated to compartment 4 are weak, positive from Deposit Feeders (5) and negative from Predators (6). Row 5 shows Microbes (3) to provide the overwhelming support of the Deposit Feeders (5), $u_{53} = +0.437$, with Meiofauna (4) second, $u_{54} = +0.166$. Finally, the sixth row of U_4 indicates that all compartments except Detritus (2) contribute positively to the Predators (6), in the order $1 > 3 > 5 > 4$. The negative impact of the detrital pool is the strongest of all the relative direct plus indirect influences on the predator compartment, $u_{62} = -0.435$ utiles.

With the input, interflows, and outputs at the levels shown in Fig. I.8, the absolute global utilities and disutilities distributed within the oyster model system are as given in kcal m^{-2} d^{-1}-utiles in matrix Υ_4. Exclusive of the diagonal elements, positive values range from $\upsilon_{51} = +0.003$ for the effect of Filter Feeders (1) on Deposit Feeders (5), to $\upsilon_{21} = +9.440$ for the Filter Feeders' direct and indirect influence on Detritus (2). The absolute impact of the detrital pool on the filter feeders is very negative, $\upsilon_{12} = -9.240$. The Microbiota (3) too have a pronounced negative impact on the detrital pool (2), $\upsilon_{23} = -4.311$ kcal m^{-2} d^{-1}-utiles. Columns and rows 4, 5 and 6 indicate that these compartments generally propagate and receive smallish positive and negative absolute utilities compared to the values associated with columns 1, 2 and 3. As a group, the Predators (6) both distribute and receive the smallest levels of absolute positive and negative macro values within the system.

In the previous three examples, the nondiagonal elements of Υ matrices were observed to be magnitude symmetric around the principal diagonals, reflecting the identity constraints associated with joining one compartment to another in chains or trees (acyclic networks). In Υ_4, this property is lost. The identity constraints of intercompartmental coupling still hold, of course; this is a universal property of general dynamical interaction. But in Example 4 these constraints are masked as a result of the energy cycling and circular causal attributes of the flow and effects networks, respectively. The degree of departure from strict magnitude identity across the diagonal apparently reflects the strength of cycling processes in the utility propagating relations of a network, probably affording a future means to quantify the significance of cycling and feedback in the value substructures of complex ecological networks.

Fittedness relationships represented by the diagonal elements of U_4 and Υ_4 are as follows. In relative terms the nonliving detrital pool conforms least with the rest of the system, $u_{22} = +0.599$ utiles and the Predators most, $u_{66} = +0.976$. In ascending order, relative affordances/fittednesses are Detritus < Microbiota < Filter Feeders < Deposit Feeders < Meiofauna < Predators. In absolute terms these relationships are reversed. Predators derive least utility from the system as a whole, $\upsilon_{66} = +0.670$ kcal m^{-2} d^{-1}-utiles, and Filter Feeders and Detritus most, $\upsilon_{11} = +34.553$ and $\upsilon_{22} = +13.333$. Ascending inequality relationships are Predators < Deposit Feeders < Microbiota < Meiofauna < Detritus < Filter Feeders.

Finally, in connection with this example, the corresponding sign matrices defining local and global qualitative interaction types, $sD = s\Delta$ and $sU = s\Upsilon$, respectively, are as follows:

$$sD_4 = s\Delta_4 = \begin{bmatrix} 0 & - & 0 & 0 & 0 & - \\ + & 0 & - & - & + & + \\ 0 & + & 0 & - & - & 0 \\ 0 & + & + & 0 & - & 0 \\ 0 & - & + & + & 0 & - \\ + & - & 0 & 0 & + & 0 \end{bmatrix} \Rightarrow \begin{bmatrix} N & & & & & \\ P & N & & & & \\ N & P & N & & & \\ N & P & P & N & & \\ N & L & P & P & N & \\ P & L & N & N & P & N \end{bmatrix}$$

$$sU_4 = s\Upsilon_4 = \begin{bmatrix} + & - & + & + & - & - \\ + & + & - & - & + & - \\ + & + & + & - & - & + \\ + & + & + & + & - & + \\ + & + & + & + & + & - \\ + & - & + & + & + & + \end{bmatrix} \Rightarrow \begin{bmatrix} G & & & & & \\ P & G & & & & \\ M & P & G & & & \\ M & P & P & G & & \\ P & M & P & P & G & \\ P & K & M & M & P & G \end{bmatrix}.$$

To the right, sign pairs, $s_{ij}{}^{\delta} = (sd_{ij}, sd_{ji}) = (s\delta_{ij}, s\delta_{ji})$ in $sD_4 = s\Delta_4$ and $s_{ij}{}^{\upsilon} = (su_{ij}, su_{ji}) = (s\upsilon_{ij}, s\upsilon_{ji})$ in $sU_4 = s\Upsilon_4$ on the left, are represented as letters denoting the different interaction types, nonsymmetric $i\,{}^{a}{\rightarrow}{}^{b}\,j$ or symmetric $i\,{}^{a}{\leftrightarrow}{}^{a}\,j$, oriented from rows to columns (where a and b are 0, + or −): N = neutralism (0, 0), P = predation or nihilism (+, −), L = altruism (−, +), G = growth (+, 0), M = mutualism (+, +), and K = competition (−, −). Comparison of these direct and integral interaction-type matrices indicate the following. All local neutralisms $s_{ii}{}^{\delta}$ = N along the diagonals become growth interactions, $s_{ii}{}^{\upsilon}$ = G, at the holistic level. That is, each compartment under a scale shift to the entire holistic network becomes self reinforcing. Four of the five nondiagonal local neutralisms (N) in $sD_4 = s\Delta_4$ become globally expressed in $sU_4 = s\Upsilon_4$ as indirect mutualisms (M); the remaining one, $s_{51}{}^{\delta}$ = N, becomes an indirect nihilism, $s_{51}{}^{\upsilon}$ = P. All eight direct nihilisms (P) of $sD_4 = s\Delta_4$ are expressed globally as nihilisms also in $sU_4 = s\Upsilon_4$ suggesting local to global invariance of the trophic relations. The next model will demonstrate that this is not general, however; in it, none of the immediate feeding interactions will remain as nihilisms at the global level. Finally, of the two local altruisms, $s_{52}{}^{\delta} = s_{62}{}^{\delta}$ = L, one becomes mutualism, $s_{52}{}^{\upsilon}$ = M, and the other competition, $s_{62}{}^{\upsilon}$ = K, in global expression.

In summary, considering that there are 15 nondiagonal pairwise interactions in this oyster reef network, none of these are mutualisms locally (nor can they be; see later), but five such mutually beneficial interactions emerge in the whole-system context. In addition, all the local diagonal neutralisms are transformed into growth interactions in the universal case. In only one instance does competition emerge from the network organization. A count of individual signs in these interactions reveals the following. In the local case, there are 10 positive, 10 negative, and 16 neutral direct interactions. Holistically, the result is 25 positive, 11 negative and none zero. Thus, this network has the property, conferred by the process of path and time integration of initial effects derived from direct feeding interactions which themselves carry both positive (for eaters) and negative (for eaten) utilities, of transforming direct energy exchanges between its constituents into positively signed ultimate utilities in a ratio of $25/11 = 2.27$.

The magnitudes of these latter utilities are also distinctly positive, as follows. In D_4, the sum of all positive micro utilities (d_{ij}^+) is $\sum_i \sum_j d_{ij}^+ = +4.025$, and that of all direct disutilities $\sum_i \sum_j d_{ij}^- = -2.311$. The resultant positive/negative ratio of absolute values is 1.74. In absolute terms (Δ_4), positive and negative direct utilities balance out, $\sum_i \sum_j \delta_{ij}^+ = \sum_i \sum_j \delta_{ij}^- = \pm 32.547$. Globally ($\Upsilon_4$), absolute positive utilities sum to $\sum_i \sum_j \upsilon_{ij}^+ = +28.537$ and negative utilities to $\sum_i \sum_j \upsilon_{ij}^- = -19.118$, yielding a positive/negative magnitude ratio of 1.49 (vs. 1.00 locally). Integral relative (U_4) positive values sum to $\sum_i \sum_j u_{ij}^+ = +3.616$ and negative to $\sum_i \sum_j u_{ij}^- = -1.306$, for a positive/negative absolute value ratio of 2.77 (vs. 1.74 locally). These data do not, perhaps, overwhelmingly confirm Legovic & Patten's (1981; Section C.3.2) conjecture about synthesis of positives as a general holistic property of ecological networks. Still, overall integral value relationships within this model are substantially on the positive side. The next example, on a larger scale, shows a stronger result with a global positive/negative ratio of 4.58 for the count of signs, and 2.60 and 4.15 for absolute values of the nondimensional and dimensional macro utilities, respectively.

C.5.2ii *Example 5: Okefenokee marsh food web*

Table C.1 shows a matrix F_5 of feeding flows between $i, j = 1, 2, \ldots,$ 24 compartments interchanging dry biomass (g m^{-2} y^{-1}) in the food web of an Okefenokee Swamp aquatic macrophyte marsh ecosystem, Little Cooter Prairie. The negative throughflows normally on the principal diagonal of F matrices are set equal to zero for the present purposes. From this information, the other environ analysis matrices needed to compare direct vs. direct plus indirect utilities and qualitative interaction types in this system can be derived. These matrices have been computed by Stuart Whipple; they are too extensive for presentation and discussion here, but the results from their calculation pertaining to the identification of local

Table C.1. *Matrix of flows, F_5 but with negative throughflows normally on the diagonal set equal to zero, for an Okefenokee Swamp marsh food web model authored by A. Caudle-Smith, T. Savisky, J. Sheldon & S. J. Whipple. Flows are in units of grams dry biomass per square meter per year* ($g\ m^{-2}\ y^{-1}$). *Compartments, with absolute fittednesses expressed parenthetically in scientific notation* (*e.g.* $\upsilon_{11} = 3.8+2 = 3.8 \times 10^2$), *are*:

1	Superficial peat (3.8+2)	13	Algae eating macroinvertebrates (5.2+1)
2	Non-peat detritus (3.4+3)	14	Macrophyte eating macroinvertebrates (4.8+1)
3	Nutrients (1.2+2)	15	Saprovorous macroinvertebrates (3.4+3)
4	Peat decomposers (4.6+1)	16	Microinvertebrate eating macroinvertebrates (1.7+1)
5	Non-peat detritus decomposers (6.3+2)	17	Macroinvertebrate eating macroinvertebrates (1.3+1)
6	N fixing and nitrifying monera (8.1+2)	18	Algae eating vertebrates (2.1)
7	Peat forming macrophytes (5.2+2)	19	Macrophyte eating vertebrates (9.3−2)
8	Non-peat forming macrophytes (1.6+3)	20	Saprovorous vertebrates (2.0)
9	Algae (1.5+2)	21	Microinvertebrate eating vertebrates (5.5)
10	Herbivorous microinvertebrates (3.6+1)	22	Macroinvertebrate eating vertebrates (1.6+1)
11	Saprovorous microinvertebrates (7.5+1)	23	Vertebrate eating vertebrates (3.2)
12	Predaceous microinvertebrates (1.4+1)	24	Vertebrate eating macroinvertebrates (4.3)

Table C.1. (*cont.*)

	1	2	3	4	5	6	7	8	9	10	11	12
1	0	0	40.00	0	0	0	469.5	0	0	0	0	0
2	0	0	1.000	7.000	137.1	2.600	0	688.9	38.00	32.67	69.00	12.50
3	59.51	11.00	0	1.824	0.340	61.58	0.559	0.373	13.40	0	0	0
4	14.08	0	45.00	0	0	0	0	0	0	0	0	0
5	0	694.4	30.00	0	0	0	0	0	0	0	0	0
6	0	0	3.000	1.000	20.00	0	254.3	473.0	0.271	0.006	0.014	0.005
7	0	0	0.559	0	0	0	0	970.0	0	0	0	0
8	0	0	9.628	0	0	0	0	0	0	0.001	0.001	0.001
9	0	0	19.40	0	0	0	0	0	0	0	0	0
10	0	0	0	0	0	3.832	0	0	38.32	0	0	0
11	0	46.40	0	23.20	23.20	0	0	0	0	0	0	0
12	0	0	0	0	0	0	0	0	0	4.606	11.55	0
13	0	0	0	0	0	6.350	0	0	57.15	0	0	0
14	0	0	0	0	0	0	0	48.76	0	0	0	0
15	0	3259.0	0	16.12	335.6	0	0	0	0	0	0	0
16	0	0	0	0	0	0	0	0	0	4.070	10.22	3.031
17	0	0	0	0	0	0	0	0	0	0	0	0
18	0	0	0	0	0	0	0	0	2.058	0	0	0
19	0	0	0	0	0	0	0.009	0.082	0	0	0	0
20	0	1.780	0	0	0.198	0	0	0	0	0	0	0
21	0	0	0	0	0	0	0	0	0	0.794	1.993	0.591
22	0	0	0	0	0	0	0	0	0	0	0	0
23	0	0	0	0	0	0	0	0	0	0	0	0
24	0	0	0	0	0	0	0	0	0	0	0	0

Table C.1. (*cont.*)

	13	14	15	16	17	18	19	20	21	22	23	24
1	0	0	0	0	0	0	0.009	0	0	0	0	0
2	46.03	35.35	2931	0.866	0.695	1.866	0.033	1.640	1.014	6.499	0.932	0.061
3	0	0	0	0	0	0	0	0	0	0	0	0
4	0	0	0	0	0	0	0	0	0	0	0	0
5	0	0	0	0	0	0	0	0	0	0	0	0
6	6.670	5.120	112.8	0.823	0.660	0.105	0.019	0.179	2.035	3.042	1.137	0.058
7	0	0	0	0	0	0	0	0	0	0	0	0
8	0	0	0	0	0	0	0	0	0	0	0	0
8	0	0	0	0	0	0	0	0	0	0	0	0
10	0	0	0	0	0	0	0	0	0	0	0	0
11	0	0	0	0	0	0	0	0	0	0	0	0
12	0	0	0	0	0	0	0	0	0	0	0	0
13	0	0	0	0	0	0	0	0	0	0	0	0
14	0	0	0	0	0	0	0	0	0	0	0	0
15	0	0	0	0	0	0	0	0	0	0	0	0
16	0	0	0	0	2.000	0	0	0	0	0	0	0
17	0.630	0.340	12.52	2.000	0	0	0	0	0	0	0	0.400
18	0	0	0	0	0	0	0	0	0	0.018	0	0
19	0	0	0	0	0	0	0	0	0	0	0	0
20	0	0	0	0	0	0	0	0	0	0.018	0	0
21	0	0	0	0	0	0	0	0	0	2.382	0	0
22	0.640	0.640	5.120	2.910	8.740	0.036	0	0.062	0.715	0	0	3.610
23	0	0	0	0	0	0.001	0.005	0.002	0.530	3.170	0	0
24	0	0	0	0	0	0.010	0	0.018	0.028	3.441	1.038	0

vs. universal interaction types (from sD_5 and sU_5) will be described later. To give an idea of whole–part conformation, the diagonal elements of Υ_5 expressing absolute fittednesses are listed parenthetically alongside the compartment names in Table C.1. Non-Peat Detritus and Saprovorous Macroinvertebrates ($\upsilon_{22} = \upsilon_{15\,15} = +3400$ g m^{-2} y^{-1}-utiles), and Non-Peat Forming Macrophytes ($\upsilon_{88} = +1600$), have the highest fittedness values. Vertebrates as a group and their predators (compartments 18–24) have the

Table C.2. *Matrix of differences in local (direct) vs. global (direct plus indirect) interaction types in the Okefenokee Swamp marsh food web model, based on sign pairs* $s_{ij}^{\delta} = (sd_{ij}, sd_{ji}) = (s\delta_{ij}, s\delta_{ji})$ *in the difference matrices* sD_5 *and* $s\Delta_5$, *and* $s_{ij}^{\upsilon} = (su_{ij}, su_{ji}) = (s\upsilon_{ij}, s\upsilon_{ji})$ *in the utility sign matrices* $sU_5 = s\Upsilon_5$, *derived from environ analysis based on the Table C.1 flow matrix.* $N = (0, 0)$, *neutralism*; $G = (+, 0)$, *growth*; $P = (+, -)$, *nihilism*; $K = (-, -)$, *competition*; $L = (-, +)$, *altruism*; *and* $M = (+, +)$, *mutualism. The left member of each entry represents direct interaction types, and the right member direct plus indirect types manifested through transitive closure of the interconnection network.*

	1	2	3	4	5	6	7	8	9	10	11	12
1	NG											
2	NP	NG										
3	PP	PK	NG									
4	PP	LL	PP	NG								
5	NM	PP	PP	NK	NG							
6	NK	LK	LL	PM	PP	NG						
7	LL	NK	NM	NL	NP	LL	NG					
8	NM	LL	PM	NL	NM	LL	LL	NG				
9	NP	LM	PP	NK	NL	LK	NK	NP	NG			
10	NK	LL	NK	NL	NM	PP	NM	LK	PP	NG		
11	NM	LM	NM	PP	PP	LP	NP	LP	NL	NK	NG	
12	NP	LL	NP	NM	NM	LM	NP	LK	NM	PP	PP	NG
13	NK	LL	NM	NL	NM	LK	NP	NK	PP	NK	NM	NP
14	NP	LL	NP	NK	NM	LK	NK	PP	NL	NK	NK	NK
15	NP	PP	NM	PP	PP	LL	NK	NP	NP	NM	NK	NP
16	NP	LK	NP	NM	NM	LM	NP	NL	NM	PP	PP	PP
17	NP	LM	NM	NK	NM	LL	NK	NP	NM	NL	NL	NL
18	NK	LL	NM	NL	NM	LL	NP	NK	PP	NK	NM	NP
19	LK	LK	NM	NK	NM	LL	PK	PP	NL	NK	NL	NK
20	NP	LM	NM	NL	PP	LL	NK	NK	NP	NM	NK	NM
21	NP	LL	NP	NM	NM	LL	NP	NK	NM	PP	PP	PP
22	NP	LL	NM	NP	NM	LL	NK	NK	NM	NM	NM	NM
23	NP	PM	NM	NP	NM	LL	NK	NK	NP	NM	NM	NM
24	NM	LK	NP	NP	NM	LK	NM	NL	NM	NM	NM	NM

lowest, and Macrophyte Eating Vertebrates appear to be the least integrated category within the system ($\upsilon_{19\,19} = +0.093$ g m^{-2} y^{-1}-utiles). Nutrients (3), microbial guilds (4–6) and producers others than compartment 8 (7 and 9) have generally high fittedness (ranging between $\upsilon_{44} = +46$ to $\upsilon_{66} = +810$); micro- and macroinvertebrates (compartments 10–17 and 24) have intermediate levels (range $\upsilon_{24\,24} = +4.3$ to $\upsilon_{11\,11} = +75$). These relationships are dominated by throughflow, and change when the data are normalized by throughflow.

Regarding qualitative interaction types, Table C.2 indicates differences in the local vs. universal types based on sign pairs $s_{ij}^{\delta} = (sd_{ij}, sd_{ji})$ in sD_5, and $s_{ij}^{\upsilon} = (su_{ij}, su_{ji})$ in sU_5. The left member of each entry represents direct interaction types, and the right member direct plus indirect types manifested through transitive closure of the interconnection network. Results are summarized in the following table where, as before, K = competition, P = predation or nihilism, N = neutralism, L = altruism, G = growth or anabolism, and M = mutualism:

Table C.2. (*cont.*)

	13	14	15	16	17	18	19	20	21	22	23	24
1												
2												
3												
4												
5												
6												
7												
8												
9												
10												
11												
12												
13	NG											
14	NK	NG										
15	NM	NP	NG									
16	NL	NK	NL	NG								
17	PP	PP	PP	NK	NG							
18	NK	NK	NM	NP	NK	NG						
19	NK	NK	NL	NL	NL	NK	NG					
20	NP	NP	NK	NK	NK	NK	NK	NG				
21	NL	NM	NP	NK	NM	NL	NK	NM	NG			
22	PP	PP	PM	PP	PP	PP	NK	PP	LL	NG		
23	NM	NM	NP	NM	NM	PK	PP	PP	PP	PP	NG	
24	NP	NP	NP	NP	LL	PP	NM	PP	PM	LM	PP	NG

↓	K	P	N	L	G	M	Direct + indirect
K	0	3	51	10	0	0	64
P	0	43	45	2	0	0	90
N	0	0	0	0	0	0	0
L	0	0	23	26	0	0	49
G	0	0	24	0	0	0	24
M	0	5	62	6	0	0	73
Direct	0	51	205	44	0	0	300

The total number of binary interactions is 300 [((24 × 24) − 24)/2 nondiagonal + 24 diagonal]. The column sums in this table indicate that 68 percent (205 of the 300, including the 24 principal diagonal entries) of the direct interactions are null, that is, neutralisms. The relative sparseness of interconnections responsible for this can be inferred from Table C.1, which is isomorphic to the system's adjacency matrix. The remaining direct interactions are almost equally split between predations and altruisms. There are no competitions, anabolisms or mutualisms, or for that matter any of the other three possible interactions (commensalism, amensalism, and dissipation). These exclusions result from inherent inequality constraints on possible direct relationships in conservative flow based models, as described below.

Element pairs d_{ij} and d_{ji} of D matrices are defined by:

$$d_{ij} = g_{ij}' - g_{ji} = f_{ij}/T_i - f_{ji}/T_i$$

$$d_{ji} = g_{ji}' - g_{ij} = f_{ji}/T_j - f_{ij}/T_j.$$

If $f_{ij} \geqslant f_{ji}$, then $d_{ij} \geqslant 0$ and $d_{ji} \leqslant 0$, and conversely; similarly, $f_{ij} \leqslant f_{ji}$ implies and is implied by $d_{ij} \leqslant 0$ and $d_{ji} \geqslant 0$. Contradictions that arise in pairings of these inequalities preclude certain interaction types from occurring as direct relations. Specifically, the nine qualitative interaction categories result from the following relationships between d_{ij} and d_{ji}:

$d_{ij} > 0, d_{ji} < 0 \Rightarrow$ predation $\quad d_{ij} < 0, d_{ji} > 0 \Rightarrow$ altruism

$d_{ij} < 0, d_{ji} < 0 \Rightarrow$ competition $\quad d_{ij} > 0, d_{ji} > 0 \Rightarrow$ mutualism

$d_{ij} = 0, d_{ji} > 0 \Rightarrow$ commensalism $\quad d_{ij} = 0, d_{ji} < 0 \Rightarrow$ amensalism

$d_{ij} > 0, d_{ji} = 0 \Rightarrow$ growth $\quad d_{ij} < 0, d_{ji} = 0 \Rightarrow$ dissipation

$d_{ij} = 0, d_{ji} = 0 \Rightarrow$ neutralism.

Positiveness or negativeness of d_{ij} and d_{ji} are given in turn by the original flows; from the above definitions of d_{ij} and d_{ji}, $d_{ij} > 0 \Leftrightarrow f_{ij} > f_{ji}$, $d_{ij} < 0 \Leftrightarrow f_{ij} < f_{ji}$, and $d_{ij} = 0 \Leftrightarrow f_{ij} = f_{ji}$. As a result, some pairings of d_{ij} and d_{ji} are feasible while others are not.

Consider the general flow digraph segment,

$$j \xrightarrow{f_{ij}} i \xrightarrow{f_{ji}} j,$$

interrelating two compartments by direct interactions. The corresponding digraph of effects, with the form $i\,^a\!\!\rightarrow^b j$ for asymmetric relations and $i\,^a\!\!\leftrightarrow^a j$ for symmetric, is to be determined. Local predation, $i\,^+\!\!\rightarrow^- j$, occurs when, from the above list of inequality pairs, $d_{ij} > 0$ and $d_{ji} < 0$, or from the definitions of d_{ij} and d_{ji}, when the flow from j to i exceeds that from i to j, $f_{ij} > f_{ji}$. Altruism exists whenever $d_{ij} < 0$ and $d_{ji} > 0$, denoting the flow inequality $f_{ij} < f_{ji}$. If $d_{ij} = d_{ji} = 0$, implying $f_{ij} = f_{ji}$, then the interaction relation is neutralism, $i\,^0\!\!\leftrightarrow^0 j$; the most common case, as mentioned above, is that when no flow exchange occurs between the two compartments, $f_{ij} = f_{ji} = 0$.

In flow-storage models of the kind represented in all five examples of this section, predation, altruism and neutralism are the only three types admissible as direct interactions. This is apparent in the above table where only P, L and N have non-zero columns. The inequality relations that prevent the other interaction types from occurring as direct effects are as follows. Local competition, $i\,^-\!\!\leftrightarrow^- j$, requires that both $d_{ij} < 0$ and $d_{ji} < 0$. However, $d_{ij} < 0$ implies $f_{ij} < f_{ji}$, whereas $d_{ji} < 0$ results from $f_{ji} < f_{ij}$; therefore, both of these flow inequalities cannot simultaneously be true. Mutualism, $i\,^+\!\!\leftrightarrow^+ j$, exists if $d_{ij} > 0$ and $d_{ji} > 0$, denoting in the first instance $f_{ij} > f_{ji}$, and in the second $f_{ji} > f_{ij}$; this is also a contradiction. Commensalism, $i\,^0\!\!\rightarrow^+ j$, is defined when $d_{ij} = 0$ and $d_{ji} > 0$ deriving, respectively, from $f_{ij} = f_{ji}$ and $f_{ji} > f_{ij}$; this equality and the inequality cannot be jointly satisfied. The relation amensalism, $i\,^0\!\!\rightarrow^- j$, occurs when $d_{ij} = 0$ and $d_{ji} < 0$, requiring the contradictory relations $f_{ij} = f_{ji}$ and $f_{ji} < f_{ij}$. Growth, $i\,^+\!\!\rightarrow^0 j$, results from $d_{ij} > 0$ and $d_{ji} = 0$, implying both $f_{ij} > f_{ji}$ and $f_{ji} = f_{ij}$. Finally, local dissipation, $i\,^-\!\!\rightarrow^0 j$, is defined when $d_{ij} < 0$ and $d_{ji} = 0$, implying impossibly that both $f_{ij} < f_{ji}$ and $f_{ji} = f_{ij}$.

Thus, within the reference frame of network analysis of conservative flow–storage models, only three of the nine possible basic interaction types (nihilism, altruism and neutralism – those with the non-zero columns in the above table) are realizable as direct interactions. Due to sign constraints, the others, including such putatively direct relationships as growth, dissipation and competition, can only arise as indirect effects across network distances. Does this mean that such fundamental biological phenomena as anabolism and catabolism must always be investigated and interpreted as consequences of network indirect effects? These are known, certainly, as manifestations of networks at lower levels of organization – biochemical pathways.

Perhaps the restriction of admissible direct interaction types indicates no more than an inherent limitation in the flow based approach to a qualitative interaction theory. Competition is of special interest ecologically because it has received so much attention as an organizer of communities. That it can only arise as a consequence of holistically mediated indirect effects, and not as a local interaction, is an unexpected result ecologically. It might be argued that compartment models with their requirement for conservation of exchanged medium are inappropriate vehicles for the representation of interactions that are nonconservative (informational) upper scale expressions of lower level direct energy–matter interchange. On the other hand, direct interactions in nature always require in their mechanisms underlying fluxes of physical quantities, such as food, light, sound, chemical compounds, etc.; even competition for space involves the allocation of places in which to live, which could be modeled as conservative 'flows' in a compartment model. Physical flows are always implicit in non-flow oriented interactions. In the first two examples of this section, implicit competitions were observed to arise from convergent and divergent flows; in Example 3 an implicit mutualism developed from sequential flows.

The accepted convention of ignoring flow-graph aspects of effects graph relationships, as for example in Lotka–Volterra modeling of competition which does not involve conservative flows between the competing populations, should not obscure the fact that at the reductionist level of mechanisms, flows of physical quantities always undergird the more subjective behavioral relationships expressed as observables at higher levels of organization. Miller (1978) calls these physical conveyances the energy–matter '*markers*' of information exchange: 'The term *marker* was used by von Neumann to refer to those observable bundles, units or changes of matter–energy whose patterning bears or conveys the informational symbols from an ensemble or repertoire.... *Communication* of almost every sort requires that the marker move in space, from the transmitting system to the receiving system, and this movement follows the same physical laws as the movement of any other sort of matter–energy' (Miller op. cit., p. 12).

Given all of this, and the generality of compartment models for depicting flow–storage processes like those in the five examples of this section, a not unreasonable working conclusion might be that competition, mutualism, commensalism, amensalism, growth and dissipation are all in fact interaction types that arise strictly as consequences of network indirect effects. These interactions are always the outcomes of complex interrelationships between more than two entities, and never the result of simple direct energy–matter interchange between only two things. The shift from lower-level, conservative micro processes to higher-level, distributive (non-conservative) macro ones is a scale phenomenon of hierarchical organization. The transitive closure mathematics of holistic determination can yield all nine of the definable qualitative interaction types as indirect consequences of network organization; these arise from the following pairs of relationships between globally expressed utilities u_{ij} and u_{ji} (or equivalently, their dimensional counterparts v_{ij} and v_{ji}) in $s_{ij}^{v} = (su_{ij}, su_{ji}) = (sv_{ij}, sv_{ji})$:

$$u_{ij} > 0, u_{ji} < 0 \Rightarrow \text{indirect nihilism}$$

$$u_{ij} < 0, u_{ji} > 0 \Rightarrow \text{indirect altruism}$$

$$u_{ij} < 0, u_{ji} < 0 \Rightarrow \text{indirect competition}$$

$$u_{ij} > 0, u_{ji} > 0 \Rightarrow \text{indirect mutualism}$$

$$u_{ij} = 0, u_{ji} > 0 \Rightarrow \text{indirect commensalism}$$

$$u_{ij} = 0, u_{ji} < 0 \Rightarrow \text{indirect amensalism}$$

$$u_{ij} > 0, u_{ji} = 0 \Rightarrow \text{indirect growth}$$

$$u_{ij} < 0, u_{ji} = 0 \Rightarrow \text{indirect dissipation}$$

$$u_{ij} = 0, u_{ji} = 0 \Rightarrow \text{indirect neutralism.}$$

The table on p. 332 summarizing numerical results from Table C.2 shows no neutralisms (or other zero-containing interactions except growth) globally expressed in the Okefenokee model. All 205 local neutralisms are converted to 62 mutualisms, 51 competitions, 45 nihilisms, 24 anabolisms, and 23 altruisms – all indirect. Of the 51 initial predator–prey relationships, 43 remain as global nihilisms, five become mutualisms and three competitions. Of the 44 local altruisms, 26 are also expressed holistically as altruisms, ten become competitions, six mutualisms and two predations. The total of 300 global interactions is distributed among 97 mutualisms, 90 nihilisms, 64 competitions and 49 altruisms. As in the oyster model of

Example 4, there is a distinct weighting in favor of universal positivism also in this model, especially in the transformations from direct neutralisms to direct plus indirect anabolisms and mutualisms.

In holistic analysis, as observed in the list of equality and inequality

Table C.3. *Matrix of differences in local (direct) vs. global (direct plus indirect) interaction types in the Okefenokee Swamp marsh food web model, based on sign pairs* $s_{ij}^{\delta} = (sd_{ij}, sd_{ji}) = (s\delta_{ij}, s\delta_{ji})$ *in the difference matrices* sD_5 *and* $s\Delta_5$*, and* $s_{ij}^{v} = (su_{ij}, su_{ji}) = (sv_{ij}, sv_{ji})$ *in the utility sign matrices* $sU_5 = s\Upsilon_5$*, where* u_{ij} *is set equal to zero producing* $su_{ij} = 0$ *when* $|u_{ij}| < 0.01|u_{ji}|$*, and* u_{ji} *is set to zero giving* $su_{ji} = 0$ *whenever* $|u_{ji}| < 0.01|u_{ij}|$*.* $K = (-, -)$*, competition;* $A = (0, -)$*, amensalism;* $D = (-, 0)$*, dissipation;* $P = (+, -)$*, predation;* $N = (0, 0)$*, neutralism;* $L = (-, +)$*, altruism;* $C = (0, +)$*, commensalism;* $G = (+, 0)$*, growth; and* $M = (+, +)$*, mutualism. The left member of each entry represents direct interaction types, and the right member direct plus indirect types integrated over time passage in the interconnection network.*

	1	2	3	4	5	6	7	8	9	10	11	12
1	NG											
2	NG	NG										
3	PG	PA	NG									
4	PG	LC	PG	NG								
5	NM	PG	PG	NA	NG							
6	NA	LA	LC	PM	PG	NG						
7	LC	NA	NM	NC	NG	LC	NG					
8	NM	LC	PM	NC	NM	LC	LC	NG				
9	NG	LM	PG	NA	NC	LA	NA	NG	NG			
10	NA	LC	NA	NC	NM	PG	NM	LA	PG	NG		
11	NM	LM	NM	PG	PG	LG	NG	LG	NC	NA	NG	
12	NG	LC	NG	NM	NM	LM	NG	LA	NM	PG	PG	NG
13	NA	LC	NM	NC	NM	LA	NG	NA	PG	NA	NM	NG
14	NG	LC	NG	NA	NM	LA	NA	PG	NC	NA	NA	NA
15	NG	PG	NM	PG	PG	LC	NA	NG	NG	NM	NA	NG
16	NG	LA	NG	NM	NG	LM	NG	NC	NM	PG	PG	PG
17	NG	LG	NM	NA	NM	LC	NA	NG	NM	NC	NC	NC
18	NA	LC	NM	NC	NG	LC	NG	NA	PG	NA	NM	NG
19	LA	LC	NG	NA	NG	LC	PA	PG	NC	NA	NC	NA
20	NG	PG	NM	NC	PG	LC	NA	NA	NG	NM	NA	NM
21	NG	LA	NG	NM	NG	LC	NG	NA	NM	PG	PG	PG
22	NG	LC	NM	NG	NG	LC	NA	NA	NG	NM	NM	NM
23	NG	LC	NG	NG	NG	LC	NA	NA	NG	NM	NG	NM
24	NM	LC	NG	NG	NG	LA	NM	NC	NM	NM	NG	NM

pairs above, all pairings of u_{ij} and u_{ji} are possible. However, because the convergence of $\sum_{m=0}^{\infty} D^m$ tends to fill all the cells of the convergent matrix U with nonzero values (in a process we term '*network homogenization*'; Patten *et al.*, 1989, 1990), interactions based on $su_{ij} = 0$ or $su_{ji} = 0$ or both tend to be rare. Thus, as with direct effects in which commensalism, amensalism, growth and dissipation are precluded, these interactions, and in addition neutralism, tend not to be realized either as global types. Indeed, none of them occurs in Table C.2 for the Okefenokee food web model. This is in accordance with strict definitions by the equality and inequality pairs (above) of values of u_{ij} and u_{ji}. However, if the strictness is relaxed to reflect fuzziness or uncertainty, these interaction types can become expressed when near-zero values of u_{ij} or u_{ji} relative to the other are taken as zero. For example, Table C.3 and the summary table below indicate transitions from local to global interaction types in the Okefenokee model assuming that convergent values of u_{ij} and u_{ji} are

Table C.3. (*cont.*)

	13	14	15	16	17	18	19	20	21	22	23	24
1												
2												
3												
4												
5												
6												
7												
8												
9												
10												
11												
12												
13	NG											
14	NA	NG										
15	NM	NG	NG									
16	NC	NA	NC	NG								
17	PG	PG	PG	NA	NG							
18	NA	NA	NG	NG	NA	NG						
19	NA	NA	NC	NC	NC	NA	NG					
20	NG	NG	NA	NA	NA	NA	NA	NG				
21	NC	NM	NG	NA	NM	NC	NA	NM	NG			
22	PG	PG	PG	PG	PG	PG	NA	PG	LC	NG		
23	NM	NM	NG	NM	NM	PA	PG	PG	PG	PG	NG	
24	NG	NG	NG	NG	LC	PG	NM	PG	PM	LM	PG	NG

not significantly different from zero if they are less than one percent of the value of the other member of the pair in $s_{ij}^{v} = (su_{ij}, su_{ji})$; that is,

$$|u_{ij}| < 0.01|u_{ji}| \Rightarrow u_{ij} \equiv 0 \Rightarrow su_{ij} = 0$$

$$|u_{ji}| < 0.01|u_{ij}| \Rightarrow u_{ji} \equiv 0 \Rightarrow su_{ji} = 0.$$

Other percentages could of course be used, just as five and one percent levels of significance are commonly employed in statistical analysis. The summary table below omits precluded direct interactions, which would yield columns of zeros; letter designators of the additional interaction types generated by nonstrict analysis are A = amensalism, D = dissipation, and C = commensalism. The table is:

↓	P	N	L	Direct + indirect
K	0	0	0	0
A	3	51	10	64
D	0	0	0	0
P	0	0	0	0
N	0	0	0	0
L	0	0	0	0
C	0	23	26	49
G	45	82	3	130
M	3	49	5	57
Direct	51	205	44	300

The three underlined entries indicate that none of the 300 original local interactions (51 nihilisms, 44 altruisms and 205 neutralisms) remains unchanged in the global perspective. That is, all the direct interaction types are transformed to other types through the transitive closure process (i.e. convergence of $\sum_{m=0}^{\infty} D^m = U$, which accounts for the direct and indirect propagation of utility over all paths of all lengths in the network while the transferred substance remains within the confines of the defined system). There are no globally expressed dissipations (D). The only negative whole system interaction represented is amensalism (A), and the 64 of these amount to only 21 percent of the 300 total relationships. Thus, the indirectly expressed interaction types are predominantly positive in character. Most telling, no competitive (K) or predative (P) relationships are manifested globally. Of the 51 direct predator–prey nihilisms (P), all disappear at the global level; 48 are changed to positive interactions [growth (G) and mutualism (M)], and only three to amensalism. Of the 205 local neutralisms (N), 154 are transformed to positive relationships [commensalism (C), and growth and mutualism]. Of the 44 local altruisms, which have a negative component although positively directed, only ten amensalisms are produced vs. 34 positive interactions represented by 26 commensalisms, three anabolisms (G) and five mutualisms. A count of individual signs in these interactions yields 293 (48.8 percent) positive, 243 (40.5 percent) zero and 64 (10.7 percent) negative; the positive/negative ratio for this system when the smallest element of each (u_{ij}, u_{ji}) pair is equated to zero if its magnitude is less than one percent of the other is therefore $293/64 = 4.58$. By comparison, of the 600 individual direct effects involved in the 300 local interactions, only 95 (15.8 percent) are positive, 95 are negative, and 410 (68.3 percent) are null.

In terms of utilities, the sum of all positive entries in D_5 of nondimensional direct effects is $\sum_i \sum_j d_{ij}^+ = +16.294$ utiles, while the sum of negative entries is $\sum_i \sum_j d_{ij}^- = -13.601$; the positive/negative ratio of absolute values is 1.24. Corresponding sums of dimensional direct utilities derived from Δ_5 are $\sum_i \sum_j \delta_{ij}^+ = \sum_i \sum_j \delta_{ij}^- = \pm 4913.7$ g m^{-2} y^{-1}-utiles. Globally, direct plus indirect nondimensional utilities and disutilities determined from U_5 are $\sum_i \sum_j u_{ij}^+ = +37.711$ and $\sum_i \sum_j u_{ij}^- = -14.531$ utiles, giving a positive/negative magnitude ratio of 2.60. Corresponding dimensional values from Υ_5 are $\sum_i \sum_j \upsilon_{ij}^+ = +15720.62$ and $\sum_i \sum_j \upsilon_{ij}^- = -3788.883$ g m^{-2} y^{-1}-utiles, for which the positive/negative ratio of absolute values is 4.15. Thus, the transitive closure mechanism of the systemwide network produces substantially positive ultimate utilities throughout this food web, both relatively (proportion 2.60:1, increased from 1.24:1 in the direct case) and as actually experienced by the member compartments (proportion 4.15:1, increased from 1:1 in local interactions).

With these results, Legovic & Patten's 'network serendipity' (Section C.3.2) can now be broadened. This referred to the transformation of not especially positive local interactions, of the kind which historically have given rise to paradigm concepts such as 'the struggle for existence', 'nature red in tooth and claw', 'competition and predation as principal determinants of community structure', etc., into predominantly positive indirect effects. In the present methodology direct interaction types are limited to nihilism, altruism and neutralism; competition can only arise indirectly and did so to a limited extent in the examples of this section. A ravenous, discordant nature in perpetual conflict does not match the '*oikos*' of ecology – home, place of comfort, security, nurturance, serenity. The present analysis supports the view of a nature dominated by *holistic positivism*, and organized in consequence to provide the greatest good for the greatest number of individual lives playing out their collective existences together. By a totally different methodology than that employed by Legovic & Patten, a preponderance of positive qualitative interaction types and positive quantitative utilities to constituents, has been demonstrated to arise as an emergent property of holism in the food web organization of the Little Cooter Prairie ecosystem in Okefenokee Swamp. To acknowledge the good fortune implied for life under network rule, I suggest that the still to be understood mechanism of this emergence might be referred to as a kind of '*hedonic property*' of ecological networks.

C.6 Conclusion: future prospects

The viewpoints and results I have tried to sketch in this chapter are not the products of orthodox twentieth-century ecology. Like the contributions of many of the other authors to this book, they are new ecology. They represent perspectives and results from a systems ecology that seeks to place the excised organism and other living categories back into nature, linked to other entities in myriads of mutual dependency relations. The consequences point to a mind boggling, difficult quest with many unfathomable elements, so why, one might ask, should science go that way? The answer is immediate – because that is the way the world is.

It is not diced up into isolated fragments faintly linked by a few single factor causes, as we in science have described it, or tried to. It is robustly interconnected, and therefore irreducible if its true properties are ever to be discerned. The information for this is straightforward, obvious to those who will only notice and acknowledge. Put radioactive or toxic material, or a pathogen, into the environment and it will trace out tortuous extended paths through organisms and nonliving compartments until it either asymptotically dissipates or becomes transformed into something else unrecognizable. Try to establish a gnotobiotic laboratory, curtail the spread of AIDS, or limit the consequences of an industrial accident like an oil spill or nuclear meltdown. Try to stop change! Perform any of a thousand ritual experiments *in situ* and indirect effects will with certainty be expressed in the data if the interval of observation is long enough. This is the touchstone; there is no end to interconnection and its consequences in real nature. Causal agents may change and be altered as to type, or slowed by system discontinuities or other interruptive or delaying phenomena, but the causal impulse inevitably must travel on interminably in the invisible networks of nature, whose last state of history is the initial state of the future. The propagation of cause in ecosystems is limited only by the dissipative mechanisms that eventually bring it to ground.

I cannot say whether this new ecology, thc legacy of the intellectual ancestors of the Odums and others who marked time for them in this century of gradual awakening to systems and systemness, will be very tractable. I can, however, safely predict that if science does not mount a serious initiative in holism to enable understanding and management of worldly complexity at all levels of organization, then humanity will sooner rather than later have to submit to the classical natural solution for failed or outmoded species. My young colleagues who fostered this book have spoken eloquently in the introductory chapter of the need to understand the ecosystem as both a complex physical entity and a paradigm for science. They have said that the network perspective is not a theory but a means to see nature through the particular formalism of network models that allows rigor in the approach to complexity. No science can continue to indulge in handwaving stasis for more than a few generations and maintain viability; inquiry, like the state of the world, must move on. There are many fascinating areas to move to.

I, for example, do not understand hierarchy and the issues of scale they bring to the surface; I understand less why little attempt has been made to instill rigor into this subject when the likely rewards of doing so are so promising. I do not understand the evolutionary process. I know with reasonable certainty that it is a process unfolding under the impetus of dominant indirect effects within whole ecosystems – environments specified – and thus that the parameters of natural selection are multiple and diffusely distributed over the complexities of hierarchical webs. But I cannot at present figure out these mechanisms. I am quite certain as an article of faith that the whole planet is a coevolved and coevolving entity – Gaia, channeled into unending change but directed by what is present, what has gone before, and what will come in the future. I do not know the mechanisms of such self design, if it can be called so, or the principles of constraint or control that guide it. I am virtually certain, however, that these mechanisms inhere in obscure properties of interconnection.

Although many questions remain about the true significance of indirect effects in real ecological systems, what can almost certainly be generalized at this point is that a shift in the scale of observation from local to global entails a shift also in the kind of determination that appears to operate. We have observed that the logic of networks calls for changes in the qualitative interaction types that interrelate network components. We have also observed that indirect/direct effects ratios tend to increase quantitatively with increasing numbers of components, connections between these components, and storage, cycling and feedback of substance, and in Higashi's parabolic rule, strength of the direct interactions themselves. In the qualitative theory, we have cast some doubts on the significance of competition and predation as organizing forces of communities, at least as these interactions are traditionally interpreted in terms of local expression. In flow models, we have observed that competition cannot even be a direct interaction; some question remains, though, about the universality of flow models.

All the examples of this chapter have had the purpose of demonstrating that a unified qualitative–quantitative methodology applicable to the study of complex ecological interactions is feasible. Close study of these examples reveals that many questions about both details of method and interpretation of results remain to be clarified. The perspectives and methods outlined are not definitive, and present results are only preliminary and suggestive. Even at its present stage of development, however, the theory of network indirect effects, in both its qualitative and quantitative proportions, shows promise as a cornerstone for the future development of a whole new generation of ecological questions concerned with understanding the deep nature of the organism–environment relationship in a complex unreduced world of inseparable holism.

In considering all of this to draw a general conclusion, it would seem that local, pairwise interactions between entities within systems serve little in the direct causal determination of one entity by another, although most of science is based on such an assumption or worldview. Interactive coupling seems instead to have its main significance in *system formation* through the establishment of an interconnecting, cause propagating network. Once this many level reticulum is in place and operating, as it is now and has been over geologic time in the extant ecosystems of the globe, its contribution to determination appears overriding. The control of change is not local, but shifts to the whole system level. General determination is not based on specific, isolated, causal events, but upon the sum of all of them arriving at each point of stimulatory focus at receptor sites of living and nonliving entities, having passed with modification and damping over endless afferent histories within the input environs defined at each point of excitation in the long train of processes begetting processes within evolving ecosystems.

To be clearer, in the causal connective $B \rightarrow C$ there is no ambiguity, only the direct effect of B on C. But in the causal chain $A \rightarrow B(A) \rightarrow C$, B still directly affects C but now A has an influence also, through B, and C has no knowledge of A. Extend this to where A is a system, say $\{A\}$, with a full network of interactions, and we have the potential for holistic determination: $\{A\} \rightarrow B(\{A\}) \rightarrow C$. Embed C within $\{A\}$ and this potential becomes realized, and furthermore self determination is also possible: $\{A(C)\} \rightarrow B(\{A(C)\}) \rightarrow C$. Finally, if B is within A then indirect effects propagated from B to C can happen, $\{A(B)\} \rightarrow B(\{A(B)\}) \rightarrow C$. Now B both affects C directly and influences it indirectly, and because the latter is the resultant of a time integration process, the magnitude of the indirect component tends to exceed that of the direct, and the sign also may change. Whereas without the system B's immediate effect on C may be negative, $B \rightarrow^{-} C$ and $A \rightarrow B(A) \rightarrow^{-} C$, with the system and propagated indirect effects ultimate utility can be positive: $\{A(B)\} \rightarrow B(\{A(B)\}) \rightarrow^{+} C$. This is the tendency the results of this chapter demonstrate.

The human belief that nature in the whole, and man–nature relations in particular, are somehow in harmony is very old. The religions, mythologies and ethical systems of many cultures, ancient and modern, reflect such belief. Intuition outstrips factual knowledge in matters of holism, and it is time for science to catch up. Ecology *is* science, and ecology *is* networks. Let it change.

Acknowledgements

I am grateful to the following colleagues for the use of their unpublished results. Masahiko Higashi for his algebra of network properties contributing to dominant indirect effects. Tarzan Legovic for his contribution to our original theory of qualitative indirect effects in which the present approach, though very different, is anchored. Khokichi Kawasaki for his proof of the convergence theorem that is basic to the unified qualitative–quantitative theory, and for calculation of some examples. Robert E. Ulanowicz for fortuitously paving the way to the unified theory by an approach he took to single-signed digraphs in a manuscript he chanced to send me.

Two students made explicit and implicit contributions. Stuart J. Whipple did all the hard work of extensive software development; he 'cut and tried' the ideas involved in shaping the utility theory and demonstrating the local to global transformation of interaction types. Thomas P. Burns has been a prime discussant, collaborator and sounding board for several years; his imprint is nonspecifically everywhere in the paper.

So is that of my wife, Marie A. Patten, who provided the usual ample doses of support, encouragement and unflinching criticism.

Appendix

AC.1 Higashi's parabolic rule

The most general possible network relationship between a source node j and a terminal one i is that shown in Fig. C.4. For subsequent purposes, the loops σ_j and σ_i in this diagram will be coalesced into the respective cycles ω_j and ω_i. The total influence propagated from j to i over the depicted network is then:

$$\begin{aligned} \mathrm{I}+\mathrm{D} &= (1+\omega_j^{\,2}+\ldots)(g_{ij}+\alpha)(1+\omega_i^{\,2}+\ldots)[1+(g_{ij}+\alpha)\beta \\ &\qquad +\{(g_{ij}+\alpha)\beta\}^2+\ldots] \\ &= (g_{ij}+\alpha)/[(1-\omega_j)(1-\omega_i)\{1-(g_{ij}+\alpha)\beta\}], \end{aligned}$$

making the indirect/direct interaction ratio:

$$\mathrm{I/D} = (g_{ij}+\alpha)/[(1-\omega_j)(1-\omega_i)\{1-(g_{ij}+\alpha)\beta\}\,g_{ij}]-1 \equiv f(g_{ij}),$$

where

$$\beta = \frac{\phi}{(1-\omega_j)(1-\omega_i)}$$

The derivative of $f(g_{ij})$ with respect to g_{ij} is

$$\begin{aligned} f'(g_{ij}) &= \frac{(1-\omega_j)(1-\omega_i)[\beta g_{ij}^{\,2}+2\alpha\beta g_{ij}-\alpha(1-\alpha\beta)]}{\{(1-\omega_j)(1-\omega_i)[1-(g_{ij}+\alpha)\beta]\,g_{ij}\}^2} \\ &= \frac{(1-\omega_j)(1-\omega_i)[(g_{ij}+\alpha)^2\beta-\alpha]}{\{(1-\omega_j)(1-\omega_i)[1-(g_{ij}+\alpha)\beta]\,g_{ij}\}^2}, \end{aligned}$$

Fig. C.4. Depiction of all possible path types directed from node j to i in a general network.

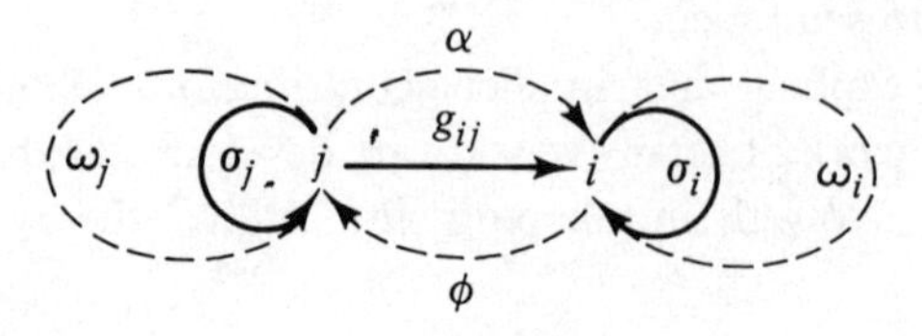

Fig. C.5. (*a*) Rate of change, $h(g_{ij})$, of generalized indirect/direct effects ratio, $f(g_{ij})$. (*b*) Plot of $\beta(\alpha)$, showing small feedbacks (β) required relative to acyclic, forward transition intensities (α), for increasing I/D ratios, $f(g_{ij})$, in the domain $g_{ij}^* < g_{ij} \leqslant 1$.

(*a*)

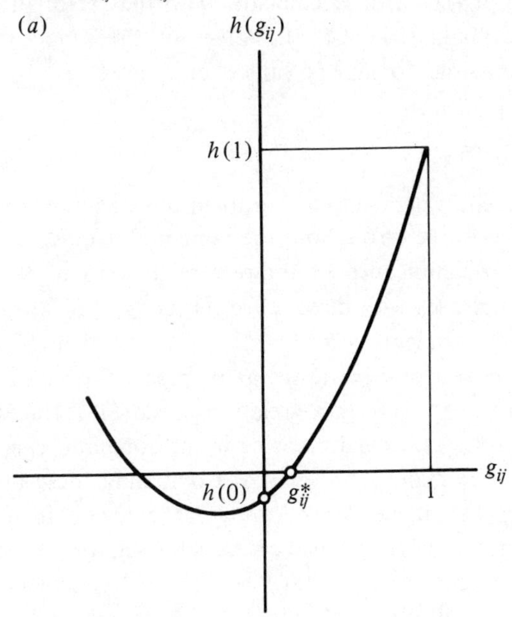

(*b*)

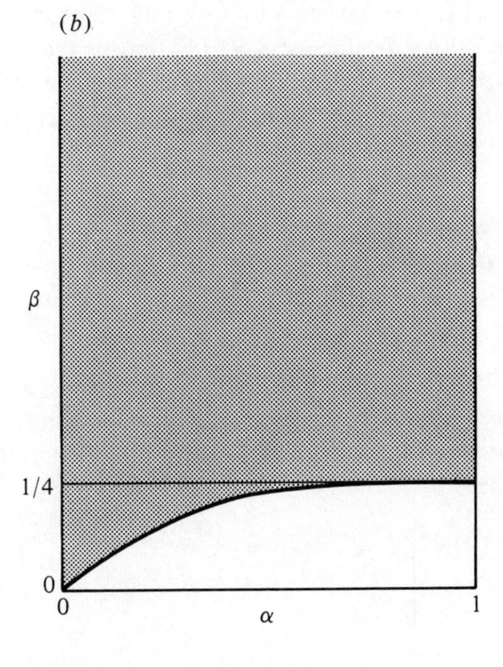

where

$$[(g_{ij}+\alpha)^2\beta-\alpha] \equiv h(g_{ij}).$$

The sign of change in I/D with D coincides with that of this function, which defines a parabola (Fig. C.5*a*). Values of the function at the extremes of the admissible domain of values for g_{ij} are:

$$h(0) = -\alpha(1-\alpha\beta)$$
$$h(1) = (1+\alpha)^2\beta-\alpha.$$

If $h(0) \geqslant 0$, which is equivalent to $\beta \geqslant 1/\alpha$, then obviously $h(1) > 0$, and $h(g_{ij})$ and $f'(g_{ij})$ are positive throughout the domain $0 \leqslant g_{ij} \leqslant 1$, showing that the quantitative significance of indirect relative to direct effects, expressed by I/D, increases with direct effect D (i.e. g_{ij}). If $h(0) < 0$ and $h(1) > 0$, which are equivalent to $\alpha/(1+\alpha)^2 < \beta < 1/\alpha$, then $h(g_{ij})$ must pass from a negative to a positive range as g_{ij} passes through a critical point $g_{ij} = g_{ij}{}^*$ where $h(g_{ij}{}^*) = 0$. These relationships are illustrated in Fig. C.5(*a*). The function $f(g_{ij})$ thus is decreasing in the domain $0 \leqslant g_{ij} \leqslant g_{ij}{}^*$, and increasing in the domain $g_{ij}{}^* < g_{ij} \leqslant 1$. Combining these two cases results in the case $h(1) > 0$, i.e. $\beta > \alpha/(1+\alpha)^2$. The region in the (α, β) space that corresponds to this combined case, i.e. the set, $\{(\alpha, \beta)\}$, of paired (forward, feedback) intensities such that $\beta > \alpha/(1+\alpha)^2$, is depicted as the shaded zone of Fig. C.5(*b*). When the pair (α, β) falls in this shaded region, in the interval $g_{ij}{}^* < g_{ij} \leqslant 1$ ($g_{ij}{}^* = 0$ when $h(0) \geqslant 0$), I/D ratios increase with g_{ij}. This is guaranteed for all values of $\beta > 1/4$ after the maximum value of $\alpha = 1$ has been attained, but in general for lesser intensities of the acyclic process, the feedback required for an increasing indirect/direct effects ratio is small.

AC.2 Kawasaki's convergence theorem

Theorem. Powers of $D = (d_{ij})$, $-1 \leqslant d_{ij} \leqslant +1$, converge to a transitive closure matrix $U = (u_{ij})$, $-\infty \leqslant u_{ij} \leqslant +\infty$, where $U = (I-D)^{-1}$,

$$I+D+D^2+\ldots+D^m+\ldots = (I-D)^{-1} = U,$$

if and only if all absolute values of the eigenvalues of D are less than one.

Proof. Let $\lambda_1, \ldots, \lambda_n$ be the eigenvalues of D. From matrix theory, there exists a matrix S such that

$$SDS^{-1} = \begin{bmatrix} \lambda_1 & & 0 \\ & \lambda_2 & \\ & & \ddots \\ 0 & & \lambda_n \end{bmatrix}.$$

A power series of this matrix can be written:

$$\sum_{m=0}^{\infty}\begin{bmatrix}\lambda_1 & & & 0\\ & \lambda_2 & & \\ & & \ddots & \\ 0 & & & \lambda_n\end{bmatrix}^m = \sum_{m=0}^{\infty}\begin{bmatrix}\lambda_1{}^m & & & 0\\ & \lambda_2{}^m & & \\ & & \ddots & \\ 0 & & & \lambda_n{}^m\end{bmatrix} = \begin{bmatrix}\sum_{m=0}^{\infty}\lambda_1{}^m & & & 0\\ & \sum_{m=0}^{\infty}\lambda_2{}^m & & \\ & & \ddots & \\ 0 & & & \sum_{m=0}^{\infty}\lambda_n{}^m\end{bmatrix},$$

or alternatively, in the form:

$$\begin{aligned}\sum_{m=0}^{\infty}\begin{bmatrix}\lambda_1 & & & 0\\ & \lambda_2 & & \\ & & \ddots & \\ 0 & & & \lambda_n\end{bmatrix}^m &= \sum_{m=0}^{\infty}[SDS^{-1}]^m\\ &= \sum_{m=0}^{\infty} SDS^{-1}\cdot SDS^{-1}\cdot\ldots(m \text{ terms})\ldots\cdot SDS^{-1}\\ &= \sum_{m=0}^{\infty} SD\cdot D\cdot\ldots(m \text{ terms})\ldots\cdot DS^{-1}\\ &= \sum_{m=0}^{\infty} SD^m S^{-1}\\ &= S\left[\sum_{m=0}^{\infty} D^m\right]S^{-1}.\end{aligned}$$

From this and the previous expression,

$$\begin{bmatrix}\sum_{m=0}^{\infty}\lambda_1{}^m & & & 0\\ & \sum_{m=0}^{\infty}\lambda_2{}^m & & \\ & & \ddots & \\ 0 & & & \sum_{m=0}^{\infty}\lambda_n{}^m\end{bmatrix} = S\left[\sum_{m=0}^{\infty} D^m\right]S^{-1},$$

and thus the convergence of $\sum_{m=0}^{\infty} D^m$ is equivalent to that of $\sum_{m=0}^{\infty} \lambda_i{}^m$, $i = 1, 2, \ldots, n$. From $\sum_{m=0}^{\infty} x^m = (1-x)^{-1}$ if and only if $|x| < 1$, $\sum_{m=0}^{\infty} \lambda_i{}^m$ diverges if $|\lambda_i| \geqslant 1$ and converges otherwise. Therefore, if and only if $|\lambda_i| < 1$, $i = 1, 2, \ldots, n$, then and only then does $\sum_{m=0}^{\infty} D^m$ converge, and if it converges it does so to $(I-D)^{-1} = U$. The latter is shown by letting $S_m = I + D + D^2 + \ldots + D^m$, and $DS_m = D + D^2 + \ldots + D^{m+1}$. Cancelling corresponding terms on the right hand sides, $S_m - DS_m = I - D^{m+1}$ or $(I-D)\,S_m = I - D^{m+1}$, making $S_m = (I-D)^{-1}(I-D^{m+1})$. Since $-1 \leqslant d_{ij} \leqslant +1$, $D^{m+1} \to 0$ as $m \to \infty$, and therefore $\lim_{m\to\infty} S_m = (I-D)^{-1} \equiv U$.

References

Archibald, C. P. (1975). Experimental observations on the effects of predation on goldfish (*Carassius auratus*) on the zooplankton of a small saline lake. *J. Fish. Res. Bd. Canada*, **32**, 1589–94.

Bartell, S. M. (1981). Potential impact of size-selective planktivory on phosphorus release by zooplankton. *Hydrobiologia*, **80**, 139–44.

Bohm, D. (1980). *Wholeness and the Implicate Order*. London: Rutledge and Kegan Paul.

Bohnert, H. G. (1954). The logical structure of the utility concept. In *Decision Processes*, ed. R. M. Thrall, C. H. Coombs & R. L. Davis, pp. 221–30. New York: Wiley.

Brabrand, A., Faajeng, B., Kallqvist, T. & Petter Nilssen, J. *et al.* (1984). Can iron defecation from fish influence plankton productivity and biomass in eutrophic lakes? *Limnol. Oceanogr.*, **29**, 1330–4.

Brooks, J. L. (1968). The effects of prey size selection by lake planktivores. *Syst. Zool.*, **17**, 273–91.

Brooks, J. L. & Dodson, S. I. (1965). Predation, body size, and composition of plankton. *Science*, **150**, 28–35.

Brown, J. H. & Munger, J. C. (1985). Experimental manipulation of a desert rodent community: food addition and species removal. *Ecology*, **66**, 1545–63.

Brown, J. H., Davidson, D. W., Munger, J. C. & Inouye, R. S. (1985). Experimental community ecology: the desert granivore system. In *Community Ecology*, ed. T. J. Case & J. M. Diamond, pp. 41–61. New York: Harper and Row.

Bunge, M. (1959). *Causality*. Cambridge, MA: Harvard University Press.

Burns, T. P. (1989). *Toward a Theory of Evolution in and of Ecosystems*. Ph.D. Dissertation, University of Georgia, Athens, Georgia.

Dame, R. F. & Patten, B. C. (1981). Analysis of energy flows in an intertidal oyster reef. *Mar. Ecol. Progr. Ser.*, **5**, 115–24.

Davidson, D. W., Inouye, R. S. & Brown, J. H. (1984). Granivory in a desert ecosystem: experimental evidence for direct facilitation of ants by rodents. *Ecology*, **65**, 1780–6.

Davis, R. L. (1954). Introduction to 'decision processes'. In *Decision Processes*, ed. R. M. Thrall, C. H. Coombs & R. L. Davis, pp. 1–18. New York: Wiley.

Elton, C. (1927). *Animal Ecology*. London: Sidgwick and Jackson.

Fishburn, P. C. (1970). *Utility Theory for Decision Making*. New York: Wiley.

Gibson, J. J. (1977). *The Ecological Approach to Visual Perception*. Boston: Houghton Mifflin.

Goldsmith, E. (1988). Gaia: some implications for theoretical ecology. *The Ecologist*, **18**, 64–74.

Grinnell, J. (1917). The niche-relationships of the California thrasher. *Auk*, **34**, 427–33.

Hannon, B. (1973). The structure of ecosystems. *J. Theor. Biol.*, **41**, 534–46.

Higashi, M. & Patten, B. C. (1986). Further aspects of the analysis of indirect effects in ecosystems. *Ecol. Mod.*, **31**, 69–77.

Higashi, M. & Patten, B. C. (1989). Dominance of indirect causality in ecosystems. *Am. Nat.*, **133**, 288–302.

Higashi, M., Patten, B. C. & Burns, T. P. (1990). Network trophic dynamics: an emerging paradigm in ecosystem ecology. In *Theoretical Studies of Ecosystems. The Network Perspective*, ed. M. Higashi & T. P. Burns. Cambridge: Cambridge University Press. This volume.

Holt, R. D. (1977). Predation, apparent competition and the structure of prey communities. *Theor. Pop. Biol.*, **12**, 197–229.

Hurlbert, S. H. (1975). Secondary effects of pesticides on aquatic ecosystems. *Residue Rev.*, **57**, 81–148.

Inouye, R. S. (1981). Interactions among unrelated species: granivorous rodents, a parasitic fungus, and a shared prey species. *Oecologia*, **49**, 425–7.

Kerfoot, W. C. & Sih, A. (1987). *Predation, Direct and Indirect Impacts on Aquatic Communities*. Hanover: University Press of New England.

Kozlovsky, D. G. (1968). A critical evaluation of the trophic level concept. I. ecological efficiencies. *Ecology*, **49**, 48–60.

Leontief, W. W. (1966). *Input–Output Economics*. Oxford: Oxford University Press.

Levine, S. H. (1977). Exploitation interactions and the structure of ecosystems. *J. Theor. Biol.*, **69**, 345–55.

Levins, R. (1974). The qualitative analysis of partially specified systems. *Ann. New York Acad. Sci.*, **231**, 123–38.

Lindeman, R. L. (1942). The trophic-dynamic aspect of ecology. *Ecology*, **23**, 399–418.

Lovelock, J. E. (1979). *Gaia, A New Look at Natural History*. Oxford: Oxford University Press.

Lovelock, J. E. (1982). *Gaia, A New Look at Life on Earth*. Oxford: Oxford University Press.

Lubchenko, J. (1978). Plant species diversity in a marine rocky intertidal community: importance of herbivore food preference and algal competitive abilities. *Am. Nat.*, **112**, 23–39.

Luce, R. D. & Raiffa, H. (1957). *Games and Decisions, Introduction and Critical Survey*. New York: Wiley.

Maruyama, M. (1963). The second cybernetics: deviation-amplifying mutual causal processes. *Am. Scient.*, **51**, 164–79.

Mattson, W. J. & Addy, N. D. (1975). Phytophagous insects as regulators of forest primary productivity. *Science*, **190**, 515–21.

Miller, J. G. (1978). *Living Systems*. New York: McGraw-Hill.

Miller, T. E. & Kerfoot, W. C. (1987). Redefining indirect effects. In *Predation, Direct and Indirect Impacts on Aquatic Communities*, ed. W. C. Kerfoot & A. Sih, pp. 33–7. Hanover: University Press of New England.

Montague, C. L. (1980). *The Net Influence of the Mud Fiddler Crab*, Uca pugnax, *on Carbon Flow through a Georgia Salt Marsh: The Importance of Work by Macroorganisms to the Metabolism of Ecosystems*. Ph.D. Dissertation, University of Georgia, Athens, Georgia.

Odum, H. T. (1983). *Systems Ecology: An Introduction*. New York: Wiley.

Odum, H. T. (1988). Self-organization, transformity, and information. *Science*, **242**, 1132–9.

Paine, R. T. (1966). Food web complexity and species diversity. *Am. Nat.*, **100**, 65–75.

Patten, B. C. (1978). Systems approach to the concept of environment. *Ohio J. Sci.*, **78**, 206–22.

Patten, B. C. (1981). Environs: the superniches of ecosystems. *Am. Zool.*, **21**, 845–52.

Patten, B. C. (1982*a*). Environs: relativistic elementary particles for ecology. *Am. Nat.*, **119**, 179–219.

Patten, B. C. (1982*b*). On the quantitative dominance of indirect effects in ecosystems. In *Analysis of Ecological Systems: State-of-the-Art in Ecological Modelling*, ed. W. K. Lauenroth, G. V. Skogerboe & M. Flug, pp. 27–37. Amsterdam: Elsevier.

Patten, B. C. (1982*c*). Indirect causality in ecosystems: its significance for environmental protection. In *Research on Fish and Wildlife Habitat*, ed. W. T. Mason & S. Iker, pp. 92–107. Commemorative monograph honoring the first decade of the US Environmental Protection Agency. Office of Research and Development, US Env. Prot. Agency, EPA-600/8-82-022. Washington, D.C.

Patten, B. C. (1984). Toward a theory of the quantitative dominance of indirect effects in ecosystems. *Verh. Ges. f. Œcologie*, **13**, 271–84.

Patten, B. C. (1985). Energy cycling in the ecosystem. *Ecol. Mod.*, **28**, 1–71.

Patten, B. C. & Auble, G. T. (1980). Systems approach to the concept of niche. *Synthese*, **43**, 155–81.

Patten, B. C. & Auble, G. T. (1981). System theory of the ecological niche. *Am. Nat.*, **117**, 893–922.

Patten, B. C., Bosserman, R. W., Finn, J. T. & Cale, W. G. (1976). Propagation of cause in ecosystems. In *Systems Analysis and Simulation in Ecology*, ed. B. C. Patten, vol. 4, pp. 457–579. New York: Academic Press.

Patten, B. C., Higashi, M. & Burns, T. P. (1989). Network trophic dynamics: the food web of an Okefenokee Swamp aquatic bed marsh. In *Freshwater Wetlands and Wildlife*. Proc. Savannah River Ecol. Lab. Conf. on Wetland Ecology, March 24–27, 1986, ed. R. R. Sharitz & J. W. Gibbons. US Dept. of Energy, Aiken, South Carolina.

Patten, B. C., Higashi, M. & Burns, T. P. (1990). Trophic dynamics in ecosystem networks: significance of cycles and storage. *Ecol. Mod.*, **51**, 1–28.

Patten, B. C., Richardson, T. H. & Barber, M. C. (1982). Path analysis of a reservoir ecosystem model. *Can. Water Res. J.*, **7**, 252–82.

Patten, B. C. & Witkamp, M. (1967). Systems analysis of ^{134}cesium kinetics in terrestrial microcosms. *Ecology*, **48**, 813–24.

Puccia, C. J. & Levins, R. (1985). *Qualitative Modelling of Complex Systems, an Introduction to Loop Analysis and Time Averaging*. Cambridge, MA: Harvard University Press.

Roberts, F. S. (1976). *Discrete Mathematical Models, with Applications to Social, Biological and Environmental Problems*. Englewood Cliffs: Prentice-Hall.

Seale, D. B. (1980). Influence of amphibian larvae production on primary production, nutrient flux, and competition in a pond ecosystem. *Ecology*, **61**, 1531–50.

Simberloff, D. (1980). A succession of paradigms in ecology. *Synthese*, **43**, 3–39.

Slobodkin, L. B. (1960). Ecological energy relationships at the population level. *Am. Nat.*, **94**, 213–36.

Smiley, J. T., Horn, J. M. & Rank, N. E. (1985). Ecological effects of salicin at three trophic levels: new problems from old adaptations. *Science*, **229**, 649–51.

Sterner, R. W. (1986). Herbivores' direct and indirect effects on algal populations. *Science*, **231**, 605–7.

Thorpe, W. H. (1965). *Science, Man and Morals.* London: Methuen.

Turvey, M. T. & Shaw, R. (1978). The primacy of perceiving: an ecological reformulation of perception for understanding memory. In *Perspectives on Memory Research: Essays in Honor of Uppsala University's 500th Anniversary*, ed. L.-G. Nilsson, pp. 167–222. Hillsdale: Erlbaum.

Ulanowicz, R. E. (1986). *Growth and Development, Ecosystems Phenomenology.* New York: Springer-Verlag.

Von Neumann, J. D. & Morgenstern, O. (1944, 1947). *Theory of Games and Economic Behavior*, first edition, second edition. Princeton: Princeton University Press.

Werner, E. E., Gilliam, J. F., Hall, D. J. & Mittelbach, G. G. (1983). An experimental test of the effects of predation risk on habitat use in fish. *Ecology*, **64**, 1540–8.

Wiegert, R. G. & Kozlowski, J. (1984). Indirect causality in ecosystems. *Am. Nat.*, **124**, 293–8.

Zadeh, L. A. & Desoer, C. A. (1963). *Linear System Theory, The State Space Approach.* New York: McGraw-Hill.

Zaret, T. & Paine, R. T. (1973). Species introduction in a tropical lake. *Science*, **182**, 449–55.

Author index

Subject index